STUDENT STUDY GUIDE

LINEAR ALGEBRA AND ITS APPLICATIONS
SIXTH EDITION

David C. Lay
University of Maryland–College Park

Steven R. Lay
Lee University

Judi J. McDonald
Washington State University

Pearson

I0037956

The author and publisher of this book have used their best efforts in preparing this book. These efforts include the development, research, and testing of the theories and programs to determine their effectiveness. The author and publisher make no warranty of any kind, expressed or implied, with regard to these programs or the documentation contained in this book. The author and publisher shall not be liable in any event for incidental or consequential damages in connection with, or arising out of, the furnishing, performance, or use of these programs.

Reproduced by Pearson from electronic files supplied by the author.

Copyright © 2021 by Pearson Education, Inc., 221 River Street, Hoboken, NJ 07030. All rights reserved.

No part of this publication may be reproduced, stored in a retrieval system, or transmitted, in any form or by any means, electronic, mechanical, photocopying, recording, or otherwise, without the prior written permission of the publisher.

Pearson

ISBN-13: 978-0-13-585123-4
ISBN-10: 0-13-585123-8

ScoutAutomatedPrintCode

Contents

Copyright © 2021 Pearson Education, Inc.

Copyright © 2021 Pearson Education, Inc.

Introduction

I fell in love with linear algebra when I was an undergraduate student and it has remained a central part of my life since that time. It is an interesting and beautiful subject, with a broad range of applications. In recent years, I consistently hear from industry partners about how much the high-tech industry appreciates individuals having a strong foundation in both technical and theoretical aspects of linear algebra. I hope you will enjoy teaching this course as much as I do. You are also welcome to email me at LLinearAlgebra@gmail.com any time you have comments and suggestions, or just want to talk about linear algebra.

There are many ways in which modern technology can support (or hinder) your student's learning. The interactive figures from the electronic textbook can be used in classroom demonstrations to bring linear algebraic concepts alive and demonstrate numerous examples with the push of a button. Take time to explore with the interactive figures and show your students how to use technology to find key definitions and theorems quickly in the electronic textbook.

If your course uses MyLab for homework, there are several things to be aware of. First, for some exercises, your students will enter only a final answer. To get to that answer, they may have half a page or more of calculations. Encourage them to keep a notebook with the exercise statement, worked solutions, and summary notes about what they learned while solving an exercise. As an instructor, you can choose many settings in the program. You can set how many tries students are allowed to solve each question. Most exercises let the student have three tries before MyLab either records an incorrect answer or offers the student a similar question. In my experience, persistence pays off – if students are allowed to continue to work similar exercises, mastery of the skill will result. I have also found that the "View an Example" and "Help Me Solve It" tab help get students going again when they are stuck.

At the end of each chapter, we have highlighted some of the projects that are available online, but moved away from updating the toolbox and the computer manuals. I find that when I am trying to code almost anything, I go to the help features for the program or open a search engine and enter some key words. There are still tips in the *Student Study Guide* about appropriate MATLAB code for various parts of the course.

Technology also provides students with easy access to a wealth of videos on linear algebra and solutions for some of the exercises. Please refrain from posting portions of this *Instructor's Solution Manual* online, as by doing so you are giving other instructors' students solutions to the exercises. Some of the open-access online videos are amazing. Others contain errors or introduce the material in a different order from how it is covered in this text, leading to confusion. I try to talk to my students about using technology to learn effectively without it becoming a crutch that leaves them with perfect homework and failed exams.

The *Instructor's Solution Manual* contains detailed solutions for all the exercises, as well as advice on the exercises themselves. I am interested to hear from you at LLinearAlgebra@gmail.com as to what types of material you would like to use in your course, additional topics you would like to see covered, any typos you find, or just to talk about my favorite subject – linear algebra.

—Judi J. McDonald

Copyright © 2021 Pearson Education, Inc.

1 Linear Equations in Linear Algebra

As you work through this chapter and the next, your experience may resemble several walks through a village at different seasons of the year. The surroundings will be familiar, but the landscape will change. You will examine various mathematical concepts from several points of view, and one of your goals will be to learn all the new terminology and the many connections between the concepts. In Chapter 4, you will see these ideas in a more abstract setting. Diligent work now will make the trip through Chapter 4 just another walk through the same village.

If you are using MyLab for homework solutions, it is important to keep a notebook with your complete solutions to assigned exercises. It is also a good strategy to include observations and study notes along with your worked exercises.

1.1 - Systems of Linear Equations

The fundamental concepts presented in this section and the next must be mastered for they will be used throughout the course.

STUDY NOTES

Please read **How to Study Linear Algebra**, on the preceding two pages, before you continue.

The text uses boldface type to identify important terms the first time they appear. You need to learn them; some students write selected terms on 3×5 cards, for review. At the end of each chapter in this *Study Guide*, a glossary checklist may help you learn definitions.

The text defines the **size** of a matrix. Don't use the term *dimension*, even though that appears in some computer programming languages, because in linear algebra, *dimension* refers to another concept (in Section 4.5).

The first few examples are so simple that they could be solved by a variety of techniques. But it is important to learn the systematic method presented here, because it easily handles more complicated linear systems, and it works in all cases.

The calculations in this section are based on the following important fact:

> *When elementary row operations are applied to a linear system, the new system has exactly the same solution set.*

(See the text.) The steps in the summary below will be modified slightly in Section 1.2.

Copyright © 2021 Pearson Education, Inc.

Summary of the Elimination Method (for This Section)

1. The first equation must contain an x_1. Interchange equations, if necessary. This will create a nonzero entry in the first row, first column, of the augmented matrix.

2. Eliminate x_1 terms in the other equations. That is, use replacement operations to create zeros in the first column of the matrix below the first row.

3. Obtain an x_2 term in the second equation. (Interchange the second equation with one below, if needed, but don't touch the first equation.) You may scale the second equation, if desired, to create a 1 in the second column and second row of the matrix.

4. Eliminate x_2 terms in equations below the second equation, using replacement operations.

5. Continue with x_3 in the third equation, x_4 in the fourth equation, etc., eliminating these variables in the equations below. This will produce a "triangular" system (at least for systems in this section).

6. Check if the system in triangular form is consistent. If it is, a solution is found by starting with the last nonzero equation and working back up to the first equation. Each variable on the "diagonal" is used to eliminate the terms in that variable above it. The solution to the system becomes apparent when the system is finally transformed into "diagonal" form.

7. Check any solutions you find by substituting them into the original system.

The *solution set* of a system of linear equations either is empty, or contains one solution, or contains infinitely many solutions. When asked to "solve" a system, you may write "inconsistent" if the system has no solution.

As you will see later, determining the number of solutions in the solution set is sometimes more important than actually computing the solution or solutions. For that reason, pay close attention to the subsection on existence and uniqueness questions. Key Exercises: 23–26 and 35.

SOLUTIONS TO EXERCISES

Get into the habit now of working the Practice Problems before you start the exercises. Probably, you should attempt all the Practice Problems before checking the solutions at the end of the exercise set, because once you start reading the first solution, you might tend to read on through the other solutions and spoil your chance to benefit from those problems.

For brevity, the symbols R1, R2, . . ., stand for row 1 (or equation 1), row 2 (or equation 2), and so on.

Copyright © 2021 Pearson Education, Inc.

1. $\begin{array}{l} x_1 + 5x_2 = 7 \\ -2x_1 - 7x_2 = -5 \end{array}$ $\quad \begin{bmatrix} 1 & 5 & 7 \\ -2 & -7 & -5 \end{bmatrix}$.

Replace R2 by R2 + (2)R1: $\quad \begin{array}{l} x_1 + 5x_2 = 7 \\ \phantom{x_1 + {}} 3x_2 = 9 \end{array}$ $\quad \begin{bmatrix} 1 & 5 & 7 \\ 0 & 3 & 9 \end{bmatrix}$

Scale R2 by 1/3: $\quad \begin{array}{l} x_1 + 5x_2 = 7 \\ x_2 = 3 \end{array}$ $\quad \begin{bmatrix} 1 & 5 & 7 \\ 0 & 1 & 3 \end{bmatrix}$

Replace R1 by R1 + (−5)R2: $\quad \begin{array}{l} x_1 = -8 \\ x_2 = 3 \end{array}$ $\quad \begin{bmatrix} 1 & 0 & -8 \\ 0 & 1 & 3 \end{bmatrix}$

The solution is $(x_1, x_2) = (-8, 3)$, or simply $(-8, 3)$. Check: $\quad \begin{array}{l} (-8) + 5(3) = -8 + 15 = 7 \\ -2(-8) - 7(3) = 16 - 21 = -5 \end{array}$

7. $\begin{bmatrix} 1 & 7 & 3 & -4 \\ 0 & 1 & -1 & 3 \\ 0 & 0 & 0 & 1 \\ 0 & 0 & 1 & -2 \end{bmatrix}$. Ordinarily, the next step would be to interchange R3 and R4, to put a 1

in the third row and third column. But in this case, the third row of the augmented matrix corresponds to the equation $0x_1 + 0x_2 + 0x_3 = 1$, or simply, $0 = 1$. A system containing this condition has no solution. Further row operations are unnecessary once an equation such as $0 = 1$ is evident.

The solution set is empty.

Study Tip: When writing a coefficient matrix or augmented matrix for a system of linear equations, be sure that the variables appear *in the same order* in each equation. Arrange the variables in columns, as in the text, placing zeros in the matrix whenever a variable is missing from an equation.

13. $\begin{bmatrix} 1 & 0 & -3 & 8 \\ 2 & 2 & 9 & 7 \\ 0 & 1 & 5 & -2 \end{bmatrix} \sim \begin{bmatrix} 1 & 0 & -3 & 8 \\ 0 & 2 & 15 & -9 \\ 0 & 1 & 5 & -2 \end{bmatrix} \sim \begin{bmatrix} 1 & 0 & -3 & 8 \\ 0 & 1 & 5 & -2 \\ 0 & 2 & 15 & -9 \end{bmatrix} \sim \begin{bmatrix} 1 & 0 & -3 & 8 \\ 0 & 1 & 5 & -2 \\ 0 & 0 & 5 & -5 \end{bmatrix}$

$\sim \begin{bmatrix} 1 & 0 & -3 & 8 \\ 0 & 1 & 5 & -2 \\ 0 & 0 & 1 & -1 \end{bmatrix} \sim \begin{bmatrix} 1 & 0 & 0 & 5 \\ 0 & 1 & 0 & 3 \\ 0 & 0 & 1 & -1 \end{bmatrix}$. The solution is $(5, 3, -1)$.

One of the most important skills, in our quantitative driven society, is the ability to judge if your solutions are accurate or reasonable. With a system of equations, you can determine if your solution is correct by substituting it back into the original system and comparing your answer to the desired right hand side for each equation. You are asked to do this in Exercises 15–18.

Copyright © 2021 Pearson Education, Inc.

17. The solution found in Exercise 13 is (5, 3, −1). Substituting into the equations

$$x_1 \qquad -3x_3 = 8 \qquad\qquad 5 \qquad -3(-1) = 8$$

$$2x_1 + 2x_2 + 9x_3 = 7 \quad \text{results in} \quad 2(5) + 2(3) + 9(-1) = 7 \text{ , which establishes that the answer}$$

$$x_2 + 5x_3 = -2 \qquad\qquad (3) + 5(-1) = -2$$

found in Exercise 13 is indeed correct.

Study Tip: Pay attention to how a problem is worded. If you only need to determine the existence or uniqueness of a solution, you can stop row operations when you reach a "triangular" form. Exercises 19–22 do not require you to solve the systems of equation.

25. $\begin{bmatrix} 1 & 3 & -2 \\ -4 & h & 8 \end{bmatrix} \sim \begin{bmatrix} 1 & 3 & -2 \\ 0 & h+12 & 0 \end{bmatrix}$. Think of $h + 12$ as a constant, c. When c is zero, that is, when $h = -12$, the system has infinitely many solutions. Otherwise, when c is nonzero, that is, when $h \neq 12$ the system has exactly one solution.

David Lay's students always recommended that he never give the complete answers to the true/false questions. We have maintained this tradition in this edition. His students felt that the temptation to read the answers was too great. After working both with and without answers, they realized how much they benefited from doing the true/false work by themselves. So, all you will see here are the places where you can find the answers.

27. See the remarks following the box titled "Elementary Row Operations."

29. The size of a matrix is defined just before the subsection titled "Solving a Linear System."

31. The solution set of a linear system is the set of all solutions of the system.

33. See the box before Example 2.

35. $\begin{bmatrix} 1 & -4 & 7 & g \\ 0 & 3 & -5 & h \\ -2 & 5 & -9 & k \end{bmatrix} \sim \begin{bmatrix} 1 & -4 & 7 & g \\ 0 & 3 & -5 & h \\ 0 & -3 & 5 & k+2g \end{bmatrix} \sim \begin{bmatrix} 1 & -4 & 7 & g \\ 0 & 3 & -5 & h \\ 0 & 0 & 0 & k+2g+h \end{bmatrix}$

Let b denote the number $k + 2g + h$. Then the third equation represented by the augmented matrix above is $0x_3 = b$. If b is nonzero, this equation has no solution, so the system is inconsistent. The system is consistent if b is zero, that is, if $k + 2g + h = 0$, then the system

$$x_1 - 4x_2 + 7x_3 = g$$
$$3x_2 - 5x_3 = h$$
$$0 = 0$$

has a solution no matter what the values of g and h. The text will explore this situation more in Section 1.2. Briefly, here is why this system, and hence the original system, is consistent. In this case, the third equation can be ignored, and the second equation, $3x_2 - 5x_3 = h$ has many solutions. Imagine choosing any values for x_2 and x_3 that satisfy the second equation,

Copyright © 2021 Pearson Education, Inc.

and substituting those values for x_2 and x_3 in the first equation. The resulting first equation can be solved for x_1. These values for x_1, x_2, and x_3 will satisfy all three equations.

41. Look at the first column. The next row operation should replace the 4 in the third row by a 0. To do this, replace R3 by R3 + (–4)R1. To reverse the operation, replace R3 by R3 + (4)R1.

A Mathematical Note: "If . . . , then . . . ."

Many important facts and theorems in the text are written as implication statements, in the form "If P, then Q," where P and Q represent complete sentences. For instance, the statement in the box on page 7 has the form

$$\text{If } \begin{Bmatrix} \text{the augmented matrices} \\ \text{of two linear systems} \\ \text{are row equivalent} \end{Bmatrix}, \text{then} \begin{Bmatrix} \text{the two systems} \\ \text{have the same} \\ \text{solution set} \end{Bmatrix} \qquad (1)$$

An implication statement "If P, then Q" is itself true provided that statement Q is true *whenever* statement P is true. In mathematical terminology, we say that "P implies Q," and we write $P \Rightarrow Q$.

Be careful to distinguish between an implication statement "P implies Q" and the **converse** or "opposite" implication, "Q implies P." The converse may or may not be true when the original implication is true. For instance, the converse of (1) above is not true, because there exist two linear systems with the same solution set but whose augmented matrices are not row equivalent. For example:

$$\begin{array}{rcl} x_1 + x_2 &=& 1 \\ 2x_1 + x_2 &=& 2 \end{array} \qquad \begin{array}{rcl} x_1 + x_2 &=& 1 \\ 2x_1 + x_2 &=& 2 \\ 3x_1 + 3x_2 &=& 3 \end{array}$$

MATLAB   Row Operations
Most of the programming discussed in this *Study Guide* will work in Octave, a free software package that is very similar to MATLAB. Specific commands for MATLAB will be introduced as needed at the end of some sections. Corresponding commands for other matrix programs can be found in the appendices. If you copy and paste information into MATLAB, sometimes special characters, such as the apostrophe, will need to be retyped.

To enter a matrix A, you type it a lot like you would write it in pencil, except that the brackets are not as large. Try entering

```
A=[1 2 3
   4 5 6
   7 8 9]
```

Copyright © 2021 Pearson Education, Inc.

The entry in row r and column c of a matrix A is denoted in MATLAB by **A(r, c)**. If the number stored in **A(r, c)** is displayed with a decimal point, then the displayed value may be accurate to only about five digits.

Row r of A is denoted by **A(r,:)**. To multiply row r by a constant c type **A(r,:)=c*A(r,:)**.

If you want to pick out specific rows, such as rows r and s, type the list in the order you want inside square brackets. For example, to swap rows r and s for the matrix A, type **A([r,s],:)=A([s,r],:)**.

If you wish to add c times row s to row r, type **A(r,:)=A(r,:)+c*A(s,:)**.

Throughout the text, exercises we recommend using technology for are marked with a symbol ⬛. The first such exercise is Exercise 46 in Section 1.2.

There are many excellent websites with additional commands and advice on using MATLAB (or Octave). A search engine will help you find them.

Warning: Using a matrix program such as MATLAB is fun and will save you time, but make sure you can perform row operations rapidly and accurately with pencil and paper. Probably, you should work all the exercises in Section 1.1 by hand and use your matrix program only to check your work.

1.2 - Row Reduction and Echelon Forms

Our interest in the row reduction algorithm lies mostly in the echelon forms that are created by the algorithm. For practical work, a computer should perform the calculations. However, you need to understand the algorithm so you can learn how to use it for various tasks. Also, unless you take your exams at a computer or with a matrix programmable calculator, you must be able to perform row reduction quickly and accurately by hand.

STUDY NOTES

The row reduction algorithm applies to any matrix, not just an augmented matrix for a linear system. In many cases, all you need is an echelon form. The reduced echelon form is mainly used when it comes from an augmented matrix and you have to find all the solutions of a linear system.

Strategies for faster and more accurate row reduction:

- Avoid subtraction in a row replacement. It leads to mistakes in arithmetic. Instead, add a negative multiple of one row to another.

- Always enclose each matrix with brackets or large parentheses.

- To save time, combine all row replacement operations that use the same pivot position, and write just one new matrix. Never "clean out" more than one column at a time. (You

Copyright © 2021 Pearson Education, Inc.

can combine several scaling operations, or combine several interchanges, if you are careful. But that seldom will be necessary.)

- *Never* combine an interchange with a replacement. In general, don't combine different types of row operations. This will be particularly important when you evaluate determinants, in Chapters 3 and 5.

How to avoid copying errors:

- Practice neat writing, not too small. Develop proper habits in homework so your work on tests will be accurate, complete, and readable. Even if you are using an online homework program, keep a notebook with your hand calculations carefully laid out.

- Write a matrix row by row. Your eye may be less likely to read from the wrong row if you place the new matrix beside the old one. Arrange your sequence of matrices across the page, rather than down the page. (Some students prefer to place the matrices in columns. Use whichever method seems to work best for you.)

- Try not to let your work flow from one side of a paper to the reverse side.

Study Tips: Theorem 2 is a key result for future work. Also, study the procedure in the box following Theorem 2. Failure to write out the system of equations (step 4) is a common source of errors.

_____

SOLUTIONS TO EXERCISES

1. To check whether a matrix is in echelon form, ask the questions:

   (i)   Is every nonzero row above the all-zero rows (if any)?

   The matrices (a)-(d) pass this test, but this is *not* enough to show it is in echelon form.

   (ii) Are the leading entries in a stair-step pattern, with zeros below each leading entry?

   The matrices (a)-(d) all pass tests (i) and (ii), so they are in echelon form.

   To check whether a matrix in echelon form is actually in *reduced* echelon form, ask two more questions:

   (iii) Is there a 1 in every pivot position?

   Matrix (d) fails this test, so it is only in echelon form. Finally, ask:

   (iv) Is each leading 1 the only nonzero entry in its column?

   Matrix (b) fails this test because of the one in the first row and third column, so it is only in echelon form. Matrices (a) and (c) pass all four tests, so they are in reduced echelon form.

Study Tip: Exercises 5 and 6 ask you to "visualize" echelon forms and write out matrices whose entries are just symbols. Example 2 suggests the form of your "answer," but it does not show you *how to find* the answer. Later, other exercises will ask you to construct other types of examples. If you look at answers from the text, or the *Study Guide* (or another student), before

Copyright © 2021 Pearson Education, Inc.

you try to write your own answers, you will lose most of the value of such exercises. The *process* of trying to understand the question and writing an example is important.

7. $\begin{bmatrix} 1 & 3 & 4 & 7 \\ 3 & 9 & 7 & 6 \end{bmatrix} \sim \begin{bmatrix} 1 & 3 & 4 & 7 \\ 0 & 0 & -5 & -15 \end{bmatrix} \sim \begin{bmatrix} 1 & 3 & 4 & 7 \\ 0 & 0 & 1 & 3 \end{bmatrix} \sim \begin{bmatrix} 1 & 3 & 0 & -5 \\ 0 & 0 & 1 & 3 \end{bmatrix}$

Corresponding system of equations: $\begin{aligned} x_1 + 3x_2 \quad &= -5 \\ x_3 &= 3 \end{aligned}$

The basic variables (corresponding to the pivot positions) are x_1 and x_3. The remaining variable x_2 is free. Solve for the basic variables in terms of the free variable. The general solution is

$$\begin{cases} x_1 = -5 - 3x_2 \\ x_2 \text{ is free} \\ x_3 = 3 \end{cases}$$

13. $\begin{bmatrix} 1 & -3 & 0 & -1 & 0 & -2 \\ 0 & 1 & 0 & 0 & -4 & 1 \\ 0 & 0 & 0 & 1 & 9 & -4 \\ 0 & 0 & 0 & 0 & 0 & 0 \end{bmatrix} \sim \begin{bmatrix} 1 & -3 & 0 & 0 & 9 & -6 \\ 0 & 1 & 0 & 0 & -4 & 1 \\ 0 & 0 & 0 & 1 & 9 & -4 \\ 0 & 0 & 0 & 0 & 0 & 0 \end{bmatrix} \sim \begin{bmatrix} 1 & 0 & 0 & 0 & -3 & -3 \\ 0 & 1 & 0 & 0 & -4 & 1 \\ 0 & 0 & 0 & 1 & 9 & -4 \\ 0 & 0 & 0 & 0 & 0 & 0 \end{bmatrix}$

Corresponding system: $\begin{aligned} x_1 \quad\quad -3x_5 &= -3 \\ x_2 \quad -4x_5 &= 1 \\ x_4 + 9x_5 &= -4 \\ 0 &= 0 \end{aligned}$

Basic variables: x_1, x_2, x_4; free variables: x_3, x_5. General solution: $\begin{cases} x_1 = -3 + 3x_5 \\ x_2 = 1 + 4x_5 \\ x_3 \text{ is free} \\ x_4 = -4 - 9x_5 \\ x_5 \text{ is free} \end{cases}$

Note: A common error in this exercise is to assume that x_3 is zero. Another common error is to say *nothing* about x_3 and write only x_1, x_2, x_4, and x_5, as above. To avoid these mistakes, identify the basic variables first. Any remaining variables are *free*. (This type of computation will arise in Chapter 5.) See also Exercise 8.

Study Tip: For Exercises 15–18, you first need to write down the system of equations that corresponds to the original matrix. From there, you can check your solutions the same way you did in Section 1.1.

19. **a**. Examine the location of the pivots. Since there is a pivot in every column except the last column, the system is consistent with a unique solution.

Copyright © 2021 Pearson Education, Inc.

b. Again examine the location of the pivots. A pivot in the last column indicates that there is no solution. See Theorem 2.

Study Tip: Be sure to work Exercises 21–24. The experience will help you later. These exercises make nice quiz questions, too.

23. $\begin{bmatrix} 1 & h & 2 \\ 4 & 8 & k \end{bmatrix} \sim \begin{bmatrix} 1 & h & 2 \\ 0 & 8-4h & k-8 \end{bmatrix}$. Look first at $8 - 4h$. If this number is not zero, then the

system must be consistent. Also, the solution will be unique because there are no free variables. This is case (b), when $h \neq 2$. Now, if $8 - 4h$ is zero, that is, if $h = 2$, there are two possibilities—either k equals 8 or k does not equal 8. If $h = 2$ and $k = 8$, the second equation is $0x_2 = 0$. The system is consistent and has a free variable, so the system has infinitely many solutions. This is case (c). When $h = 2$ and $k \neq 8$, the second equation is $0x_2 = b$, with b nonzero, and the system has no solution. This is case (a).

25. See Theorem 1.

27. See the second paragraph of the section.

29. Basic variables are defined after equation (4).

31. See the beginning of the subsection, "Parametric Descriptions of Solution Sets." Actually, this question does not consider the case of an inconsistent system. A better true/false statement would be: "If a linear system is consistent, then finding a parametric description of the solution set is the same as *solving* the system.

33. The row shown corresponds to the equation $0x_4 = 5$. Can there be a solution to an equation of the form $0x_4 = b$, with b nonzero?

37. A full solution is in the text answer section.

Study Tip: Notice from Exercise 38 that the question of uniqueness of the solution of a linear system is not influenced by the numbers in the rightmost column of the augmented matrix.

43. Yes, a system of linear equations with more equations than unknowns can be consistent. The answer in the text includes an example.

46. The data for this exercise comes from one of David's students who was working part time for a private wind tunnel company near the University of Maryland. You will need a matrix program to solve this problem. The basic instructions for MATLAB were given in the *Study Guide* notes for Section 1.1. For Maple, Mathematica, the TI-calculators, see the respective appendices at the end of this *Study Guide*.

Copyright © 2021 Pearson Education, Inc.

A Mathematical Note: "If and only if"

You need to know what the phrase "if and only if" means. It was used earlier in Exercise 39, and you will see it again in theorems and in boxed facts. The phrase "if and only if" always appears between two complete statements. Look at Theorem 2, for instance:

$$\left\{\begin{array}{c} \text{A specific} \\ \text{linear system} \\ \text{is consistent} \end{array}\right\} \text{ if and only if } \left\{\begin{array}{c} \text{the rightmost column of the} \\ \text{augmented matrix is not} \\ \text{a pivot column.} \end{array}\right\} \qquad (1)$$

The entire sentence means that the two statements in parentheses are either both true or both false.

Sentence (1) has the general form

$$P \text{ if and only if } Q \qquad (2)$$

where P denotes the first statement and Q denotes the second statement. This sentence says two things:

> If statement P is true, then statement Q is also true.
> If statement Q is true, then statement P is also true.

A mathematical shorthand for (2) is "$P \Leftrightarrow Q$."

1.3 - Vector Equations

Do not be deceived by the rather simple beginning of Section 1.3. The important material on Span$\{\mathbf{v}_1, \ldots, \mathbf{v}_p\}$ will take time to digest. Figures 8, 10, and 11 are important, along with Exercises 11–14, 17, 18, 33, and 34. Each of the exercises involves an *existence* question about whether a certain vector equation has a solution. (You don't have to find the solution.) Notice how the same basic question can be asked in several different ways.

STUDY NOTES

Develop the habit of reading the section carefully once or twice before looking at the *Study Guide* and before starting the exercises. (Don't just look at the pictures and examples! Important comments lurk in between.)

In nearly all of the text, a *scalar* is just a real number. By convention, scalars are usually written to the left of vectors, such as $5\mathbf{v}$ or $c\mathbf{v}$, rather than $\mathbf{v}5$ or $\mathbf{v}c$. It is important to pay attention to context to determine if a letter represents a matrix, vector, or variable.

Vectors must be the same size to be added or used in a linear combination. For instance, a vector in $\mathbb{R}^3$ cannot be added to a vector in $\mathbb{R}^2$.

Copyright © 2021 Pearson Education, Inc.

SOLUTIONS TO EXERCISES

1. $\mathbf{u} + \mathbf{v} = \begin{bmatrix} -1 \\ 2 \end{bmatrix} + \begin{bmatrix} -3 \\ 3 \end{bmatrix} = \begin{bmatrix} -1 + (-3) \\ 2 + 3 \end{bmatrix} = \begin{bmatrix} -4 \\ 5 \end{bmatrix}$.

Using the definitions carefully,

$\mathbf{u} - 2\mathbf{v} = \begin{bmatrix} -1 \\ 2 \end{bmatrix} + (-2)\begin{bmatrix} -3 \\ 3 \end{bmatrix} = \begin{bmatrix} -1 \\ 2 \end{bmatrix} + \begin{bmatrix} (-2)(-3) \\ (-2)(3) \end{bmatrix} = \begin{bmatrix} -1 + 6 \\ 2 - 6 \end{bmatrix} = \begin{bmatrix} 5 \\ -4 \end{bmatrix}$, or, more quickly,

$\mathbf{u} - 2\mathbf{v} = \begin{bmatrix} -1 \\ 2 \end{bmatrix} - 2\begin{bmatrix} -3 \\ 3 \end{bmatrix} = \begin{bmatrix} -1 + 6 \\ 2 - 6 \end{bmatrix} = \begin{bmatrix} 5 \\ -4 \end{bmatrix}$. The intermediate step is often not written.

7. See the figure below. Since the grid can be extended in every direction, the figure suggests that every vector in $\mathbb{R}^2$ can be written as a linear combination of $\mathbf{u}$ and $\mathbf{v}$. To write a vector $\mathbf{a}$ as a linear combination of $\mathbf{u}$ and $\mathbf{v}$, imagine walking from the origin to $\mathbf{a}$ along the grid "streets" and keep track of how many "blocks" you travel in the $\mathbf{u}$-direction and how many in the $\mathbf{v}$-direction

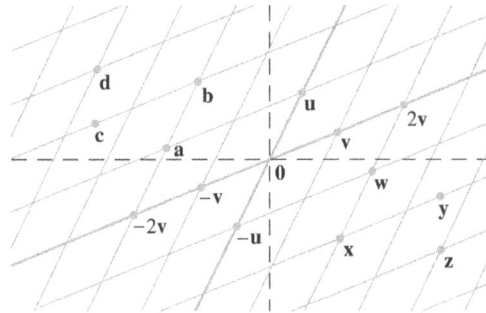

a. To reach $\mathbf{a}$ from the origin, you might travel 1 unit in the $\mathbf{u}$-direction and −2 units in the $\mathbf{v}$-direction (that is, 2 units in the negative $\mathbf{v}$-direction). Hence $\mathbf{a} = \mathbf{u} - 2\mathbf{v}$.

b. To reach $\mathbf{b}$ from the origin, travel 2 units in the $\mathbf{u}$-direction and −2 units in the $\mathbf{v}$-direction. So $\mathbf{b} = 2\mathbf{u} - 2\mathbf{v}$. Or, use the fact that $\mathbf{b}$ is 1 unit in the $\mathbf{u}$-direction from $\mathbf{a}$, so that

$\mathbf{b} = \mathbf{a} + \mathbf{u} = (\mathbf{u} - 2\mathbf{v}) + \mathbf{u} = 2\mathbf{u} - 2\mathbf{v}$

c. The vector $\mathbf{c}$ is −1.5 units from $\mathbf{b}$ in the $\mathbf{v}$-direction, so

$\mathbf{c} = \mathbf{b} - 1.5\mathbf{v} = (2\mathbf{u} - 2\mathbf{v}) - 1.5\mathbf{v} = 2\mathbf{u} - 3.5\mathbf{v}$

d. The "map" suggests that you can reach $\mathbf{d}$ if you travel 3 units in the $\mathbf{u}$-direction and −4 units in the $\mathbf{v}$-direction. If you prefer to stay on the paths displayed on the map, you might travel from the origin to −3$\mathbf{v}$, then move 3 units in the $\mathbf{u}$-direction, and finally move −1 unit in the $\mathbf{v}$-direction. So

$\mathbf{d} = -3\mathbf{v} + 3\mathbf{u} - \mathbf{v} = 3\mathbf{u} - 4\mathbf{v}$

Another solution is

Copyright © 2021 Pearson Education, Inc.

$$\mathbf{d} = \mathbf{b} - 2\mathbf{v} + \mathbf{u} = (2\mathbf{u} - 2\mathbf{v}) - 2\mathbf{v} + \mathbf{u} = 3\mathbf{u} - 4\mathbf{v}$$

9. Here are the intermediate calculations, which usually are not displayed. Check with your instructor whether you need to "show work" on a problem such as this.

$$\begin{aligned} x_2 + 5x_3 &= 0 \\ 4x_1 + 6x_2 - x_3 &= 0, \\ -x_1 + 3x_2 - 8x_3 &= 0 \end{aligned} \qquad \begin{bmatrix} x_2 + 5x_3 \\ 4x_1 + 6x_2 - x_3 \\ -x_1 + 3x_2 - 8x_3 \end{bmatrix} = \begin{bmatrix} 0 \\ 0 \\ 0 \end{bmatrix},$$

$$\begin{bmatrix} 0 \\ 4x_1 \\ -x_1 \end{bmatrix} + \begin{bmatrix} x_2 \\ 6x_2 \\ 3x_2 \end{bmatrix} + \begin{bmatrix} 5x_3 \\ -x_3 \\ -8x_3 \end{bmatrix} = \begin{bmatrix} 0 \\ 0 \\ 0 \end{bmatrix}, \qquad x_1 \begin{bmatrix} 0 \\ 4 \\ -1 \end{bmatrix} + x_2 \begin{bmatrix} 1 \\ 6 \\ 3 \end{bmatrix} + x_3 \begin{bmatrix} 5 \\ -1 \\ -8 \end{bmatrix} = \begin{bmatrix} 0 \\ 0 \\ 0 \end{bmatrix}$$

Helpful Hint: As you work Exercises 11–14, circle the pivots in an echelon form of an appropriate matrix. This will help you visualize the cases when a vector either is or is not a linear combination of other vectors.

13. Denote the columns of A by $\mathbf{a}_1, \mathbf{a}_2, \mathbf{a}_3$. To determine if $\mathbf{b}$ is a linear combination of these columns, use the boxed fact that include equation (5) in the text. Row reduce the augmented matrix until you reach echelon form:

$$\begin{bmatrix} 1 & -4 & 2 & 3 \\ 0 & 3 & 5 & -7 \\ -2 & 8 & -4 & -3 \end{bmatrix} \sim \begin{bmatrix} ① & -4 & 2 & 3 \\ 0 & ③ & 5 & -7 \\ 0 & 0 & 0 & ③ \end{bmatrix}$$

The system for this augmented matrix is inconsistent, so $\mathbf{b}$ is *not* a linear combination of the columns of A.

19. By inspection, $\mathbf{v}_2 = (3/2)\mathbf{v}_1$. Any linear combination of $\mathbf{v}_1$ and $\mathbf{v}_2$ is actually just a multiple of $\mathbf{v}_1$. For instance,

$$a\mathbf{v}_1 + b\mathbf{v}_2 = a\mathbf{v}_1 + b(3/2)\mathbf{v}_1 = (a + 3b/2)\mathbf{v}_1$$

So Span$\{\mathbf{v}_1, \mathbf{v}_2\}$ is the set of points on the line through $\mathbf{v}_1$ and $\mathbf{0}$.

Warning: Although Figures 8 and 11 provide the most common ways to view Span$\{\mathbf{u}, \mathbf{v}\}$, don't forget Exercise 19, which shows that in a special case, Span$\{\mathbf{u}, \mathbf{v}\}$ can be just a line through the origin. In fact, Span$\{\mathbf{u}, \mathbf{v}\}$ can also be just the origin itself. How?

21. Let $\mathbf{y} = \begin{bmatrix} h \\ k \end{bmatrix}$. Then $[\mathbf{u} \quad \mathbf{v} \quad \mathbf{y}] = \begin{bmatrix} 2 & 2 & h \\ -1 & 1 & k \end{bmatrix} \sim \begin{bmatrix} ② & 2 & h \\ 0 & ② & k+h/2 \end{bmatrix}$. This augmented matrix corresponds to a consistent system for all h and k. So $\mathbf{y}$ is in Span$\{\mathbf{u}, \mathbf{v}\}$ for all h and k.

23. The alternative notation for a (column) vector is discussed after Example 1.

25. Plot the points to check the assertion. Or, see the statement preceding Example 3.

27. See the line displayed just before Example 4.

29. See the boxed expression after equation (4).

Copyright © 2021 Pearson Education, Inc.

31. Read the geometric description of Span{**u**, **v**} very carefully.

Study Tip: Students may find it helpful to work by themselves on the true/false questions and then meet together in groups of two or three to compare and discuss their answers.

33. a. There are only three vectors in the set {**a**₁, **a**₂, **a**₃}, and **b** is not one of them.

     b. There are infinitely many vectors in $W = $ Span{**a**₁, **a**₂, **a**₃}. To determine if **b** is in W, use the method of Exercise 13.

$$\begin{bmatrix} 1 & 0 & -4 & 4 \\ 0 & 3 & -2 & 1 \\ -2 & 6 & 3 & -4 \end{bmatrix} \sim \begin{bmatrix} 1 & 0 & -4 & 4 \\ 0 & 3 & -2 & 1 \\ 0 & 6 & -5 & 4 \end{bmatrix} \sim \begin{bmatrix} \boxed{1} & 0 & -4 & 4 \\ 0 & \boxed{3} & -2 & 1 \\ 0 & 0 & \boxed{-1} & 2 \end{bmatrix}$$
$$\begin{matrix} \uparrow & \uparrow & \uparrow & \uparrow \\ \mathbf{a}_1 & \mathbf{a}_2 & \mathbf{a}_3 & \mathbf{b} \end{matrix}$$

     The system for this augmented matrix is consistent, so **b** is in W.

     c. $\mathbf{a}_1 = 1\mathbf{a}_1 + 0\mathbf{a}_2 + 0\mathbf{a}_3$. See the discussion following the definition of Span{$\mathbf{v}_1, \ldots, \mathbf{v}_p$}.

39. a. The center of mass is $\dfrac{1}{3}\left(1 \cdot \begin{bmatrix} 0 \\ 1 \end{bmatrix} + 1 \cdot \begin{bmatrix} 8 \\ 1 \end{bmatrix} + 1 \cdot \begin{bmatrix} 2 \\ 4 \end{bmatrix}\right) = \begin{bmatrix} 10/3 \\ 2 \end{bmatrix}$.

     b. The total mass of the new system is 9 grams. The three masses added, w_1, w_2, and w_3, satisfy the equation

$$\frac{1}{9}\left((w_1+1) \cdot \begin{bmatrix} 0 \\ 1 \end{bmatrix} + (w_2+1) \cdot \begin{bmatrix} 8 \\ 1 \end{bmatrix} + (w_3+1) \cdot \begin{bmatrix} 2 \\ 4 \end{bmatrix}\right) = \begin{bmatrix} 2 \\ 2 \end{bmatrix}$$

     which can be rearranged to

$$(w_1+1) \cdot \begin{bmatrix} 0 \\ 1 \end{bmatrix} + (w_2+1) \cdot \begin{bmatrix} 8 \\ 1 \end{bmatrix} + (w_3+1) \cdot \begin{bmatrix} 2 \\ 4 \end{bmatrix} = \begin{bmatrix} 18 \\ 18 \end{bmatrix}$$

     and

$$w_1 \cdot \begin{bmatrix} 0 \\ 1 \end{bmatrix} + w_2 \cdot \begin{bmatrix} 8 \\ 1 \end{bmatrix} + w_3 \cdot \begin{bmatrix} 2 \\ 4 \end{bmatrix} = \begin{bmatrix} 8 \\ 12 \end{bmatrix}$$

     The condition $w_1 + w_2 + w_3 = 6$ and the vector equation above combine to produce a system of three equations whose augmented matrix is shown below, along with a sequence of row operations:

$$\begin{bmatrix} 1 & 1 & 1 & 6 \\ 0 & 8 & 2 & 8 \\ 1 & 1 & 4 & 12 \end{bmatrix} \sim \begin{bmatrix} 1 & 1 & 1 & 6 \\ 0 & 8 & 2 & 8 \\ 0 & 0 & 3 & 6 \end{bmatrix} \sim \begin{bmatrix} 1 & 1 & 1 & 6 \\ 0 & 8 & 2 & 8 \\ 0 & 0 & 1 & 2 \end{bmatrix}$$

Copyright © 2021 Pearson Education, Inc.

$$\sim \begin{bmatrix} 1 & 1 & 0 & 4 \\ 0 & 8 & 0 & 4 \\ 0 & 0 & 1 & 2 \end{bmatrix} \sim \begin{bmatrix} 1 & 0 & 0 & 3.5 \\ 0 & 8 & 0 & 4 \\ 0 & 0 & 1 & 2 \end{bmatrix} \sim \begin{bmatrix} \text{①} & 0 & 0 & 3.5 \\ 0 & \text{①} & 0 & .5 \\ 0 & 0 & \text{①} & 2 \end{bmatrix}$$

Answer: Add 3.5 g at (0, 1), add .5 g at (8, 1), and add 2 g at (2, 4).

41. a. For $j = 1, \ldots, n$, the jth entry of $(\mathbf{u} + \mathbf{v}) + \mathbf{w}$ is $(u_j + v_j) + w_j$. By associativity of addition in $\mathbb{R}$, this entry equals $u_j + (v_j + w_j)$, which is the jth entry of $\mathbf{u} + (\mathbf{v} + \mathbf{w})$. By definition of equality of vectors, $(\mathbf{u} + \mathbf{v}) + \mathbf{w} = \mathbf{u} + (\mathbf{v} + \mathbf{w})$.

  b. For any scalar c, the jth entry of $c(\mathbf{u} + \mathbf{v})$ is $c(u_j + v_j)$, and the jth entry of $c\mathbf{u} + c\mathbf{v}$ is $cu_j + cv_j$ (by definition of scalar multiplication and vector addition). These entries are equal, by a distributive law in $\mathbb{R}$. So $c(\mathbf{u} + \mathbf{v}) = c\mathbf{u} + c\mathbf{v}$.

MATLAB Constructing a Matrix

Review the MATLAB section in Section 1.1 to recall how to enter a matrix and use MATLAB (or Octave) to perform row reduction operations. The ▮ symbol, like the one next to Exercises 35(c) and 36(c) indicates that using a matrix computation program would be helpful to solve the given exercises.

1.4 - The Matrix Equation $A\mathbf{x} = \mathbf{b}$

The ideas, boxed statements, and theorems in this section are absolutely fundamental for the rest of the text, so you should read the section extremely carefully.

KEY IDEAS

The definition of $A\mathbf{x}$ as a linear combination of the columns of A will be used often. You should learn the definition in *words* as well as symbols. *Note*: It is not wrong to write a scalar on the *right* side of a vector and write $A\mathbf{x}$ as $\mathbf{a}_1 x_1 + \cdots + \mathbf{a}_n x_n$, but the text follows the usual practice of writing a scalar on the *left* side of a vector.

You need to understand *why* Theorem 4 is true. That may take some time and effort. Example 3 should help, along with the proof. Theorem 4(d) can be restated as "The reduced echelon form of A has no row of zeros."

The phrase *logically equivalent* is explained in the statement of Theorem 4. This phrase is used with several statements in the same way that *if and only if* (or the symbol ⇔) is used between two statements. (See the Mathematical Note at the end of Section 1.2 in this *Guide*.)

Saying that statements (a), (b), (c), and (d) are logically equivalent means the same thing as saying that (a) ⇔ (b), (b) ⇔ (c), and (c) ⇔ (d).

Key exercises are 1–20, 37, 38, 41 and 42. Think about 41 and 42, even if they are not assigned, because they introduce ideas you will need soon. (Don't check the solution of Exercise 41 until you have written your own answer.)

Copyright © 2021 Pearson Education, Inc.

Checkpoint 1: True or False? If an augmented matrix [*A* **b**] has a pivot position in every row, then the equation *A***x** = **b** is consistent.

Note: You should work a checkpoint problem when you first see it, provided that you have already read the text at least once. Always *write* your answer before comparing it with the one I have written. The checkpoint answer will be at the end of the solutions for this section.

SOLUTIONS TO EXERCISES

1. The text has the solution. Exercises 1–12 are designed to help you learn Theorem 3 and the definition of *A***x**. If a problem involves vectors—say, **v**₁, **v**₂, **v**₃ — you can place the vectors into a matrix [**v**₁ **v**₂ **v**₃], if that is helpful. If a problem involves a matrix *A*, you can give names to the columns of *A*—say, **a**₁, **a**₂, **a**₃—and reformulate a matrix equation as a vector equation. If a problem leads to a system of linear equations, you may regard it as either a vector equation or a matrix equation, whichever is most useful.

7. The left side of the equation is a linear combination of three vectors. Write the matrix *A* whose columns are those three vectors, and create a variable vector **x** with three entries:

$$A = \begin{bmatrix} \begin{bmatrix} 4 \\ -1 \\ 7 \\ -4 \end{bmatrix} & \begin{bmatrix} -5 \\ 3 \\ -5 \\ 1 \end{bmatrix} & \begin{bmatrix} 7 \\ -8 \\ 0 \\ 2 \end{bmatrix} \end{bmatrix} = \begin{bmatrix} 4 & -5 & 7 \\ -1 & 3 & -8 \\ 7 & -5 & 0 \\ -4 & 1 & 2 \end{bmatrix} \text{ and } \mathbf{x} = \begin{bmatrix} x_1 \\ x_2 \\ x_3 \end{bmatrix}.$$

Thus the equation *A***x** = **b** is

$$\begin{bmatrix} 4 & -5 & 7 \\ -1 & 3 & -8 \\ 7 & -5 & 0 \\ -4 & 1 & 2 \end{bmatrix} \begin{bmatrix} x_1 \\ x_2 \\ x_3 \end{bmatrix} = \begin{bmatrix} 6 \\ -8 \\ 0 \\ -7 \end{bmatrix}$$

Warning: Be careful to distinguish between the *matrix equation* *A***x** = **b** and the *augmented matrix* [**a**₁ ⋯ **a**ₚ **b**], which is used in Theorem 3 to refer to a system of linear equations having this augmented matrix. Thus, the answer to Exercise 7 is *not* the augmented matrix at the right:

$$\begin{bmatrix} 4 & -5 & 7 & 6 \\ -1 & 3 & -8 & -8 \\ 7 & -5 & 0 & 0 \\ -4 & 1 & 2 & -7 \end{bmatrix}$$

13. The vector **u** is in the plane spanned by the columns of *A* if and only if **u** is a linear combination of the columns of *A*. This happens if and only if the equation *A***x** = **u** has a solution. (See the box preceding Example 3 in Section 1.4.) To study this equation, reduce the augmented matrix [*A* **u**]:

$$\begin{bmatrix} 3 & -5 & 0 \\ -2 & 6 & 4 \\ 1 & 1 & 4 \end{bmatrix} \sim \begin{bmatrix} 1 & 1 & 4 \\ -2 & 6 & 4 \\ 3 & -5 & 0 \end{bmatrix} \sim \begin{bmatrix} 1 & 1 & 4 \\ 0 & 8 & 12 \\ 0 & -8 & -12 \end{bmatrix} \sim \begin{bmatrix} ① & 1 & 4 \\ 0 & ⑧ & 12 \\ 0 & 0 & 0 \end{bmatrix}$$

Copyright © 2021 Pearson Education, Inc.

The equation $A\mathbf{x} = \mathbf{u}$ has a solution, so $\mathbf{u}$ is in the plane spanned by the columns of A.

Study Tip: Exercises 17–20 require written explanations as well as calculations. For instance, your calculation for Exercise 17 might show the row reduction

$$A = \begin{bmatrix} 1 & 3 & 0 & 3 \\ -1 & -1 & -1 & 1 \\ 0 & -4 & 2 & -8 \\ 2 & 0 & 3 & -1 \end{bmatrix} \sim \begin{bmatrix} 1 & 3 & 0 & 3 \\ 0 & 2 & -1 & 4 \\ 0 & -4 & 2 & -8 \\ 0 & -6 & 3 & -7 \end{bmatrix} \sim \begin{bmatrix} 1 & 3 & 0 & 3 \\ 0 & 2 & -1 & 4 \\ 0 & 0 & 0 & 0 \\ 0 & 0 & 0 & 5 \end{bmatrix} \sim \begin{bmatrix} \boxed{1} & 3 & 0 & 3 \\ 0 & \textcircled{2} & -1 & 4 \\ 0 & 0 & 0 & \textcircled{5} \\ 0 & 0 & 0 & 0 \end{bmatrix}$$

After this, it is not enough to write "No, by Theorem 4." Instead, you should show that you know *why* Theorem 4 is relevant. For instance, you might write:

The matrix A does *not* have a pivot in every row. By Theorem 4, the equation $A\mathbf{x} = \mathbf{b}$ does *not* have a solution for each $\mathbf{b}$ in $\mathbb{R}^4$.

On a test, you probably would not have to know the theorem number. It might be enough to say "By a theorem," instead of "By Theorem 4." (Check with your instructor.)

19. The work in Exercise 17 shows that the equation $A\mathbf{x} = \mathbf{b}$ does not have a solution for each $\mathbf{b}$. That is, statement (d) in Theorem 4 is false. So all four statements in Theorem 4 are false. Since statement (b) is false, not all vectors in $\mathbb{R}^4$ can be written as a linear combination of the columns of A. Since statement (c) is false, the columns of A do *not* span $\mathbb{R}^4$.

Checkpoint 2: Given $\mathbf{v}_1$, $\mathbf{v}_2$, $\mathbf{v}_3$ as in Exercise 21, find a specific vector in $\mathbb{R}^4$ that is not in Span$\{\mathbf{v}_1, \mathbf{v}_2, \mathbf{v}_3\}$. (If necessary, reread Example 3.)

23. See the paragraph following equation (3).

25. See the box before Example 1. Saying that $\mathbf{b}$ is not in the set spanned by the columns of A is the same as saying that $\mathbf{b}$ is not a linear combination of the columns of A.

27. See the warning following Theorem 4.

29. See Example 4.

31. See Theorem 4.

33. See Theorem 4.

35. By definition, the matrix-vector product on the left is a linear combination of the columns of the matrix, in this case using weights -3, -1, and 2. So $c_1 = -3$, $c_2 = -1$, and $c_3 = 2$.

39. Start with any 3×3 matrix B in echelon form that has three pivot positions. Perform a row operation (a row interchange or a row replacement) that creates a matrix A that is *not* in echelon form. Then A has the desired property. The justification is given by row reducing A

Copyright © 2021 Pearson Education, Inc.

to B, in order to display the pivot positions. Since A has a pivot position in every row, the columns of A span $\mathbb{R}^3$, by Theorem 4.

41. A 3×2 matrix has three rows and two columns. With only two columns, A can have at most two pivot columns, and so A has at most two pivot positions, which is not enough to fill all three rows. By Theorem 4, the equation $A\mathbf{x} = \mathbf{b}$ cannot be consistent for all $\mathbf{b}$ in $\mathbb{R}^3$. If A is any $m \times n$ matrix with $m > n$, the equation $A\mathbf{x} = \mathbf{b}$ cannot be consistent for all $\mathbf{b}$ in $\mathbb{R}^m$.

43. If the equation $A\mathbf{x} = \mathbf{b}$ has a unique solution, then the associated system of equations does not have any free variables. If every variable is a basic variable, then each column of A is a

pivot column. So the reduced echelon form of A must be $\begin{bmatrix} 1 & 0 & 0 \\ 0 & 1 & 0 \\ 0 & 0 & 1 \\ 0 & 0 & 0 \end{bmatrix}$.

47. The original matrix has no pivot in the fourth row, so its columns do not span $\mathbb{R}^4$, by Theorem 4.

Helpful Hint: For Exercises 47 through 50, use a matrix program to obtain an echelon form of the matrix. Try covering various columns of this matrix, one at a time, and ask yourself if the columns of the resulting matrix span $\mathbb{R}^4$. If you can delete one column, can you delete a second column? Why or why not?

The analysis here depends on the following idea, which is fairly obvious but is not explicitly mentioned in the text. When a row operation is performed on a matrix A, the calculations for each new entry depend only on the other entries in the *same column*. If a column of A is removed, forming a new matrix, the absence of this column has no affect on any row-operation calculations for entries in the other columns of A. (The absence of a column might affect the particular *choice* of row operations performed for some purpose, but that is not relevant.)

Answers to Checkpoints:

1. False. See the Warning after Theorem 4. If you missed this, you are not studying the text properly. You should read the text thoroughly *before* you look at the *Study Guide* and before you work on the exercises.

2. Let $A = [\mathbf{v}_1 \quad \mathbf{v}_2 \quad \mathbf{v}_3] = \begin{bmatrix} 1 & 0 & 1 \\ 0 & -1 & 0 \\ -1 & 0 & 0 \\ 0 & 1 & -1 \end{bmatrix}$ and $\mathbf{b} = \begin{bmatrix} b_1 \\ b_2 \\ b_3 \\ b_4 \end{bmatrix}$. Row reduce the augmented matrix for $A\mathbf{x} = \mathbf{b}$ to determine values of $b_1, \ldots, b_4$ that make the equation *in*consistent.

Copyright © 2021 Pearson Education, Inc.

$$
\begin{bmatrix} 1 & 0 & 1 & b_1 \\ 0 & -1 & 0 & b_2 \\ -1 & 0 & 0 & b_3 \\ 0 & 1 & -1 & b_4 \end{bmatrix} \sim \begin{bmatrix} 1 & 0 & 1 & b_1 \\ 0 & 1 & 0 & -b_2 \\ 0 & 0 & 1 & b_3+b_1 \\ 0 & 1 & -1 & b_4 \end{bmatrix} \sim \begin{bmatrix} 1 & 0 & 1 & b_1 \\ 0 & 1 & 0 & -b_2 \\ 0 & 0 & 1 & b_3+b_1 \\ 0 & 0 & -1 & b_4+b_2 \end{bmatrix}
$$

$$
\sim \begin{bmatrix} 1 & 0 & 1 & b_1 \\ 0 & 1 & 0 & -b_2 \\ 0 & 0 & 1 & b_3+b_1 \\ 0 & 0 & 0 & b_4+b_2+b_3+b_1 \end{bmatrix}
$$

Take $\mathbf{b} = (1, 1, 0, 0)$, for example, or any other choice of $b_1, \dots, b_4$ whose sum is *not* zero.

Mastering Linear Algebra Concepts: Span

Please begin by reviewing "How to Study Linear Algebra," at the beginning of this *Study Guide*.

To really understand a key concept, you need to form an image in your mind that consists of the basic definition(s) together with many related ideas. Your goal at this point is to collect various ideas associated with the set Span$\{\mathbf{v}_1, \dots, \mathbf{v}_p\}$ and the concept of a set that "spans" R^n. Here are specific things to do now as you prepare a sheet (or sheets) for review and reference.

- Write the **definition** of Span$\{\mathbf{v}_1, \dots, \mathbf{v}_p\}$. (Learn it word for word.)

- Write the **definition** of the phrase: $\{\mathbf{v}_1, \dots, \mathbf{v}_p\}$ spans R^n. Here *span* is a verb rather than a noun as in Span$\{\mathbf{v}_1, \dots, \mathbf{v}_p\}$.

- Add the **equivalent description** (not definition) of what is meant for a vector **b** to be in Span$\{\mathbf{v}_1, \dots, \mathbf{v}_p\}$.

- Copy **Theorem 4** word for word. (If you try to rephrase or summarize it in your own words, you are likely to change the meaning.)

- Sketch some **geometric interpretations** of Span$\{\mathbf{v}_1, \dots, \mathbf{v}_p\}$. (Select some of Figs. 8, 10, 11, and Exercises 19, 20 in Section 1.3.)

- Identify **special cases**. (Describe Span$\{\mathbf{u}\}$ and Span$\{\mathbf{u}, \mathbf{v}\}$.)

- Summarize **algorithms** or **typical computations** (such as Example 6 and Exercises 11–14, 17, 18, 25, and 26 in Section 1.3, or Example 3 and Exercises 13–22 in Section 1.4.

- Describe connections with other concepts.

Copyright © 2021 Pearson Education, Inc.

Whenever you encounter new examples or situations that help you understand the concept of a spanning set, add them to this review sheet.

> **MATLAB   rref** and matrix multiplication
>
> To solve $A\mathbf{x} = \mathbf{b}$, row reduce the matrix $\mathbf{M} = [\mathbf{A} \quad \mathbf{b}]$. To speed up your calculations, you can use the command **rref(M)** to get the row reduced echelon form of M.
>
> The command $\mathbf{x} = [5;3;-7]$ creates a column vector $\mathbf{x}$ with entries 5, 3, –7. To find $A\mathbf{x}$ in MATLAB or Octave, use $\mathbf{A*x}$.

1.5 - Solution Sets of Linear Systems

Many of the concepts and computations in linear algebra involve sets of vectors which are visualized geometrically as lines and planes. The most important examples of such sets are the solution sets of linear systems.

KEY IDEAS

Visualize the solution set of a homogeneous equation $A\mathbf{x} = \mathbf{0}$ as:

- the single point $\mathbf{0}$, when $A\mathbf{x} = \mathbf{0}$ has only the trivial solution,

- a line through $\mathbf{0}$, when $A\mathbf{x} = \mathbf{0}$ has one free variable,

- a plane through $\mathbf{0}$, when $A\mathbf{x} = \mathbf{0}$ has two free variables.
  (For more than two free variables, you can also visualize it as a plane through $\mathbf{0}$.)

For $\mathbf{b} \neq \mathbf{0}$, visualize the solution set of $A\mathbf{x} = \mathbf{b}$ as:

- empty, if $\mathbf{b}$ is not a linear combination of the columns of A,

- one nonzero point (vector), when $A\mathbf{x} = \mathbf{b}$ has a unique solution,

- a line not through $\mathbf{0}$, when $A\mathbf{x} = \mathbf{b}$ is consistent and has one free variable,

- a plane not through $\mathbf{0}$, when $A\mathbf{x} = \mathbf{b}$ is consistent and has two or more free variables.

The solution set of $A\mathbf{x} = \mathbf{b}$ is said to be described *implicitly*, because the equation is a condition an $\mathbf{x}$ must satisfy in order to be in the set, yet the equation does not show how to find such an $\mathbf{x}$. When the solution set of $A\mathbf{x} = \mathbf{0}$ is written as Span$\{\mathbf{v}_1, \ldots, \mathbf{v}_p\}$, the set is said to be described *explicitly*; each element in the set is produced by forming a linear combination of $\mathbf{v}_1, \ldots, \mathbf{v}_p$.

A common explicit description of a set is an equation in *parametric vector form*. Examples are:

$\mathbf{x} = t\mathbf{v},$ a line through $\mathbf{0}$ in the direction of $\mathbf{v}$,

Copyright © 2021 Pearson Education, Inc.

$$\mathbf{x} = \mathbf{p} + t\mathbf{v}, \qquad\qquad \text{a line through } \mathbf{p} \text{ in the direction of } \mathbf{v},$$

$$\mathbf{x} = x_2\mathbf{u} + x_3\mathbf{v}, \qquad\qquad \text{a plane through } \mathbf{0}, \mathbf{u}, \text{ and } \mathbf{v},$$

$$\mathbf{x} = \mathbf{p} + x_2\mathbf{u} + x_3\mathbf{v}, \qquad\qquad \text{a plane through } \mathbf{p} \text{ parallel to the plane whose equation is } \mathbf{x} = x_2\mathbf{u} + x_3\mathbf{v}.$$

An equation in parametric vector form describes a set explicitly because the equation shows how to produce each **x** in the set.

To *solve* an equation $A\mathbf{x} = \mathbf{b}$ means to find an explicit description of the solution set. If the system is inconsistent, the solution set is empty. Otherwise, the description of all solutions can be written in parametric vector form, in which the parameters are the free variables from the system. *Important*: The number of free variables in $A\mathbf{x} = \mathbf{b}$ depends only on A, not on **b**.

Theorem 6 and the paragraph following it are important. They describe how the solutions of $A\mathbf{x} = \mathbf{0}$ and $A\mathbf{x} = \mathbf{b}$ are related when the solution set of $A\mathbf{x} = \mathbf{b}$ is nonempty. See Figs. 5 and 6. Key exercises: 5–20, 41–44, 49.

SOLUTIONS TO EXERCISES

1. Reduce the augmented matrix to echelon form and circle the pivot positions. If a column of the *coefficient* matrix is not a pivot column, the corresponding variable is free and the system of equations has a nontrivial solution. Otherwise, the system has *only* the trivial solution.

$$\begin{bmatrix} 2 & -5 & 8 & 0 \\ -2 & -7 & 1 & 0 \\ 4 & 2 & 7 & 0 \end{bmatrix} \sim \begin{bmatrix} 2 & -5 & 8 & 0 \\ 0 & -12 & 9 & 0 \\ 0 & 12 & -9 & 0 \end{bmatrix} \sim \begin{bmatrix} \boxed{2} & -5 & 8 & 0 \\ 0 & \boxed{-12} & 9 & 0 \\ 0 & 0 & 0 & 0 \end{bmatrix}$$

The variable x_3 is free, so the system has a nontrivial solution.

7. Always use the *reduced* echelon form of an augmented matrix to find the solutions of a system. See the text's discussion of back substitution.

$$\begin{bmatrix} 1 & 3 & -3 & 7 \\ 0 & 1 & -4 & 5 \end{bmatrix} \sim \begin{bmatrix} \boxed{1} & 0 & 9 & -8 \\ 0 & \boxed{1} & -4 & 5 \end{bmatrix}, \qquad \begin{array}{rcl} \boxed{x_1} \quad + 9x_3 &=& -8 \\ \boxed{x_2} - 4x_3 &=& 5 \end{array}$$

If you wrote something like the system above, then you made a common mistake. The matrix in the text problem is a coefficient matrix, not an augmented matrix. You should row reduce $[A \quad \mathbf{0}]$. The correct system of equations is

$$\begin{array}{rcl} \boxed{x_1} \quad + 9x_3 - 8x_4 &=& 0 \\ \boxed{x_2} - 4x_3 + 5x_4 &=& 0 \end{array}$$

The basic variables are x_1 and x_2, with x_3 and x_4 free. Next, $x_1 = -9x_3 + 8x_4$, and $x_2 = 4x_3 - 5x_4$. The general solution is

$$\mathbf{x} = \begin{bmatrix} x_1 \\ x_2 \\ x_3 \\ x_4 \end{bmatrix} = \begin{bmatrix} -9x_3 + 8x_4 \\ 4x_3 - 5x_4 \\ x_3 \\ x_4 \end{bmatrix} = \begin{bmatrix} -9x_3 \\ 4x_3 \\ x_3 \\ 0 \end{bmatrix} + \begin{bmatrix} 8x_4 \\ -5x_4 \\ 0 \\ x_4 \end{bmatrix} = x_3 \begin{bmatrix} -9 \\ 4 \\ 1 \\ 0 \end{bmatrix} + x_4 \begin{bmatrix} 8 \\ -5 \\ 0 \\ 1 \end{bmatrix}$$

Copyright © 2021 Pearson Education, Inc.

The solution set is the same as Span$\{\mathbf{u}, \mathbf{v}\}$, where $\mathbf{u} = (-9, 4, 1, 0)$ and $\mathbf{v} = (8, -5, 0, 1)$. Originally, the solution set was described implicitly, by a set of equations. Now the solution set is described explicitly, in parametric vector form.

11.

$$\begin{bmatrix} 1 & -4 & -2 & 0 & 3 & -5 & 0 \\ 0 & 0 & 1 & 0 & 0 & -1 & 0 \\ 0 & 0 & 0 & 0 & 1 & -4 & 0 \\ 0 & 0 & 0 & 0 & 0 & 0 & 0 \end{bmatrix} \sim \begin{bmatrix} 1 & -4 & -2 & 0 & 0 & 7 & 0 \\ 0 & 0 & 1 & 0 & 0 & -1 & 0 \\ 0 & 0 & 0 & 0 & 1 & -4 & 0 \\ 0 & 0 & 0 & 0 & 0 & 0 & 0 \end{bmatrix} \sim \begin{bmatrix} ① & -4 & 0 & 0 & 0 & 5 & 0 \\ 0 & 0 & ① & 0 & 0 & -1 & 0 \\ 0 & 0 & 0 & 0 & ① & -4 & 0 \\ 0 & 0 & 0 & 0 & 0 & 0 & 0 \end{bmatrix}$$

$$\begin{aligned}
\textcircled{x_1} - 4x_2 \qquad\qquad + 5x_6 &= 0 \\
\textcircled{x_3} \qquad - x_6 &= 0 \\
\textcircled{x_5} - 4x_6 &= 0 \\
0 &= 0
\end{aligned}$$

Some students are not sure what to do with x_4. Some ignore it; others set it equal to zero. In fact, x_4 is free; there is no constraint on x_4 at all. The basic variables are x_1, x_3, and x_5. The remaining variables are free. So, $x_1 = 4x_2 - 5x_6$, $x_3 = x_6$, and $x_5 = 4x_6$, with x_2, x_4, and x_6 free. In parametric vector form,

$$\mathbf{x} = \begin{bmatrix} x_1 \\ x_2 \\ x_3 \\ x_4 \\ x_5 \\ x_6 \end{bmatrix} = \begin{bmatrix} 4x_2 - 5x_6 \\ x_2 \\ x_6 \\ x_4 \\ 4x_6 \\ x_6 \end{bmatrix} = \begin{bmatrix} 4x_2 \\ x_2 \\ 0 \\ 0 \\ 0 \\ 0 \end{bmatrix} + \begin{bmatrix} 0 \\ 0 \\ 0 \\ x_4 \\ 0 \\ 0 \end{bmatrix} + \begin{bmatrix} -5x_6 \\ 0 \\ x_6 \\ 0 \\ 4x_6 \\ x_6 \end{bmatrix} = x_2 \underset{\mathbf{u}}{\begin{bmatrix} 4 \\ 1 \\ 0 \\ 0 \\ 0 \\ 0 \end{bmatrix}} + x_4 \underset{\mathbf{v}}{\begin{bmatrix} 0 \\ 0 \\ 0 \\ 1 \\ 0 \\ 0 \end{bmatrix}} + x_6 \underset{\mathbf{w}}{\begin{bmatrix} -5 \\ 0 \\ 1 \\ 0 \\ 4 \\ 1 \end{bmatrix}}$$

The solution set is the same as Span$\{\mathbf{u}, \mathbf{v}, \mathbf{w}\}$.

Study Tip: When solving a system, identify (and perhaps circle) the basic variables. All other variables are free.

15. To verify that the answer you found in Exercise 11 is correct, multiply A by the vectors you found:

Copyright © 2021 Pearson Education, Inc.

$$\begin{bmatrix} 1 & -4 & -2 & 0 & 3 & -5 \\ 0 & 0 & 1 & 0 & 0 & -1 \\ 0 & 0 & 0 & 0 & 1 & -4 \\ 0 & 0 & 0 & 0 & 0 & 0 \end{bmatrix} \left(x_2 \begin{bmatrix} 4 \\ 1 \\ 0 \\ 0 \\ 0 \\ 0 \end{bmatrix} + x_4 \begin{bmatrix} 0 \\ 0 \\ 0 \\ 1 \\ 0 \\ 0 \end{bmatrix} + x_6 \begin{bmatrix} -5 \\ 0 \\ 1 \\ 0 \\ 4 \\ 1 \end{bmatrix} \right) = x_2 \begin{bmatrix} 1 & -4 & -2 & 0 & 3 & -5 \\ 0 & 0 & 1 & 0 & 0 & -1 \\ 0 & 0 & 0 & 0 & 1 & -4 \\ 0 & 0 & 0 & 0 & 0 & 0 \end{bmatrix}\begin{bmatrix} 4 \\ 1 \\ 0 \\ 0 \\ 0 \\ 0 \end{bmatrix}$$

$$+ x_4 \begin{bmatrix} 1 & -4 & -2 & 0 & 3 & -5 \\ 0 & 0 & 1 & 0 & 0 & -1 \\ 0 & 0 & 0 & 0 & 1 & -4 \\ 0 & 0 & 0 & 0 & 0 & 0 \end{bmatrix}\begin{bmatrix} 0 \\ 0 \\ 0 \\ 1 \\ 0 \\ 0 \end{bmatrix} + x_6 \begin{bmatrix} 1 & -4 & -2 & 0 & 3 & -5 \\ 0 & 0 & 1 & 0 & 0 & -1 \\ 0 & 0 & 0 & 0 & 1 & -4 \\ 0 & 0 & 0 & 0 & 0 & 0 \end{bmatrix}\begin{bmatrix} -5 \\ 0 \\ 1 \\ 0 \\ 4 \\ 1 \end{bmatrix}$$

$$= x_2 \begin{bmatrix} 0 \\ 0 \\ 0 \\ 0 \end{bmatrix} + x_4 \begin{bmatrix} 0 \\ 0 \\ 0 \\ 0 \end{bmatrix} + x_6 \begin{bmatrix} 0 \\ 0 \\ 0 \\ 0 \end{bmatrix} = \begin{bmatrix} 0 \\ 0 \\ 0 \\ 0 \end{bmatrix}$$

Whenever you are solving a set of homogenous equations, the result of multiplying out $A\mathbf{x}$ should be the zero vector.

17. To write the general solution in parametric vector form, pull out the constant terms that do not involve the free variable:

$$\mathbf{x} = \begin{bmatrix} x_1 \\ x_2 \\ x_3 \end{bmatrix} = \begin{bmatrix} 5 + 4x_3 \\ -2 - 7x_3 \\ x_3 \end{bmatrix} = \begin{bmatrix} 5 \\ -2 \\ 0 \end{bmatrix} + \begin{bmatrix} 4x_3 \\ -7x_3 \\ x_3 \end{bmatrix} = \underset{\underset{\mathbf{p}}{\uparrow}}{\begin{bmatrix} 5 \\ -2 \\ 0 \end{bmatrix}} + x_3 \underset{\underset{\mathbf{q}}{\uparrow}}{\begin{bmatrix} 4 \\ -7 \\ 1 \end{bmatrix}} = \mathbf{p} + x_3\mathbf{q}$$

Geometrically, the solution set is the line through $\begin{bmatrix} 5 \\ -2 \\ 0 \end{bmatrix}$ parallel to $\begin{bmatrix} 4 \\ -7 \\ 1 \end{bmatrix}$.

Checkpoint: Let A be a 2×2 matrix. Answer True or False: If the solution set of $A\mathbf{x} = \mathbf{0}$ is a line through the origin in $\mathbb{R}^2$ and if $\mathbf{b} \neq \mathbf{0}$, then the solution set of $A\mathbf{x} = \mathbf{b}$ is a line not through the origin.

23. The line through $\mathbf{a}$ parallel to $\mathbf{b}$ can be written as $\mathbf{x} = \mathbf{a} + t\mathbf{b}$, where t represents a parameter:

$$\mathbf{x} = \begin{bmatrix} x_1 \\ x_2 \end{bmatrix} = \begin{bmatrix} -2 \\ 0 \end{bmatrix} + t \begin{bmatrix} -5 \\ 3 \end{bmatrix}, \text{ or } \begin{cases} x_1 = -2 - 5t \\ x_2 = 3t \end{cases}$$

27. See the first paragraph of the subsection titled "Homogeneous Linear Systems."

Copyright © 2021 Pearson Education, Inc.

29. See the first two sentences of the subsection titled "Parametric Vector Form."

31. See the box before Example 1.

33. See the paragraph that precedes Fig. 5.

35. See Theorem 6.

37. Suppose $\mathbf{p}$ satisfies $A\mathbf{x} = \mathbf{b}$. Then $A\mathbf{p} = \mathbf{b}$. Theorem 6 says that the solution set of $A\mathbf{x} = \mathbf{b}$ equals the set $S = \{\mathbf{w} : \mathbf{w} = \mathbf{p} + \mathbf{v}_h$ for some $\mathbf{v}_h$ such that $A\mathbf{v}_h = \mathbf{0}\}$. There are two things to prove: (a) every vector in S satisfies $A\mathbf{x} = \mathbf{b}$, (b) every vector that satisfies $A\mathbf{x} = \mathbf{b}$ is in S.

  a. Let $\mathbf{w}$ have the form $\mathbf{w} = \mathbf{p} + \mathbf{v}_h$,   where $A\mathbf{v}_h = \mathbf{0}$. Then

$$A\mathbf{w} = A(\mathbf{p} + \mathbf{v}_h) = A\mathbf{p} + A\mathbf{v}_h \qquad \text{By Theorem 5(a) in Section 1.4}$$
$$= \mathbf{b} + \mathbf{0} = \mathbf{b}$$

  So every vector of the form $\mathbf{p} + \mathbf{v}_h$ satisfies $A\mathbf{x} = \mathbf{b}$.

  b. Now let $\mathbf{w}$ be any solution of $A\mathbf{x} = \mathbf{b}$, and set $\mathbf{v}_h = \mathbf{w} - \mathbf{p}$. Then

$$A\mathbf{v}_h = A(\mathbf{w} - \mathbf{p}) = A\mathbf{w} - A\mathbf{p} = \mathbf{b} - \mathbf{b} = \mathbf{0}$$

  So $\mathbf{v}_h$ satisfies $A\mathbf{x} = \mathbf{0}$. Thus every solution of $A\mathbf{x} = \mathbf{b}$ has the form

$$\mathbf{w} = \mathbf{p} + \mathbf{v}_h.$$

43. A is a 3×2 matrix with two pivot positions.

  a. Since A has a pivot position in each column, each variable in $A\mathbf{x} = \mathbf{0}$ is a basic variable. So the equation $A\mathbf{x} = \mathbf{0}$ has no free variables and hence no nontrivial solution.

  b. With two pivot positions and three rows, A cannot have a pivot in every row. So the equation $A\mathbf{x} = \mathbf{b}$ cannot have a solution for every possible $\mathbf{b}$ (in $\mathbb{R}^3$), by Theorem 4 in Section 1.4.

49. If you worked on the Checkpoint when you first saw it, you should be ready for this exercise. Since the solution set of $A\mathbf{x} = \mathbf{0}$ contains the point $(4, 1)$, the vector $\mathbf{x} = (4, 1)$ satisfies $A\mathbf{x} = \mathbf{0}$. Write this equation as a vector equation, using $\mathbf{a}_1$ and $\mathbf{a}_2$ for the columns of A:

$$4 \cdot \mathbf{a}_1 + 1 \cdot \mathbf{a}_2 = \mathbf{0}$$

Then $\mathbf{a}_2 = -4\mathbf{a}_1$. So choose any nonzero vector for the first column of A and multiply that column by -4 to get the second column of A. For example, set $A = \begin{bmatrix} 1 & -4 \\ 1 & -4 \end{bmatrix}$.

Finally, the only way the solution set of $A\mathbf{x} = \mathbf{b}$ could *not* be parallel to the line through $(4, 1)$ and the origin is for the solution set of $A\mathbf{x} = \mathbf{b}$ to be *empty*. (Theorem 6 applies only to the case when the equation $A\mathbf{x} = \mathbf{b}$ has a nonempty solution set.) For $\mathbf{b}$, take any vector that is *not* a multiple of the columns of A.

Copyright © 2021 Pearson Education, Inc.

Answer to Checkpoint: False. The solution set could be empty. In this case, the solution set of $A\mathbf{x} = \mathbf{b}$ is not produced by translating the (nonempty) solution set of $A\mathbf{x} = \mathbf{0}$. See the Warning after Theorem 6.

MATLAB   Zero Matrices

The command **zeros(m,n)** creates an $m \times n$ matrix of zeros. When solving an equation $A\mathbf{x} = \mathbf{0}$, create an augmented matrix:

  M = [A zeros(m, 1)] m is the number of rows in A.

Then use **rref(M)** to row reduce M completely.

1.6 - Applications of Linear Systems

All of the examples and exercises in this system involve linear systems that have multiple solutions. In each case, make a note of *why* you should expect the system to have many solutions.

STUDY NOTES

The Leontief exchange model concerns the dollar value (called the *price*) of the annual output of each sector of a nation's economy. An equilibrium price vector **p** provides a list of prices, one for each section, such that each sector's expenses and income are in balance. Example 1 shows that there are many equilibrium price vectors; each one is a multiple of a fixed equilibrium price vector. This means that once the prices are all in balance, multiplying all the prices by a fixed constant does not affect the balance. For instance, if all prices are doubled, then each sector's expenses and income are doubled at the same time and hence they remain in balance.

A solution of a chemical equation-balance problem is a list of coefficients that appear on the various terms in the chemical equation. When a chemical equation is balanced, the number of atoms of each type on the left side of the equation matches the number of corresponding atoms on the right side. If the coefficients in the equation are each multiplied by a fixed positive integer, the equation will remain balanced. So, there are many solutions to a chemical equation-balance problem.

The problems here in network flow have multiple solutions for the simple reason that there are more variables than there are constraint equations. The equations for network flow are mostly nonhomogeneous. In contrast, the Leontief model and the chemical equation-balance problem both lead to systems of homogeneous equations.

SOLUTIONS TO EXERCISES

1. Fill in the exchange table one column at a time. The entries in a column describe where a sector's output goes. The decimal fractions in each column sum to 1.

Copyright © 2021 Pearson Education, Inc.

Distribution of
Output From:

	Goods	Services		Purchased by:
output	↓	↓	input	
	.2	.7	→	Goods
	.8	.3	→	Services

Denote the total annual output (in dollars) of the sectors by p_G and p_S. From the first row, the total input to the Goods sector is $.2p_G + .7p_S$. The Goods sector must pay for that. So the equilibrium prices must satisfy

income   expenses

$$p_G \; = \; .2p_G + .7p_S$$

From the second row, the input (that is, the expense) of the Services sector is $.8p_G + .3p_S$. The equilibrium equation for the Services sector is

income   expenses

$$p_S \; = \; .8p_G + .3p_S$$

Move all variables to the left side and combine like terms:

$$.8p_G - .7p_S = 0$$
$$-.8p_G + .7p_S = 0$$

Row reduce the augmented matrix:

$$\begin{bmatrix} .8 & -.7 & 0 \\ -.8 & .7 & 0 \end{bmatrix} \sim \begin{bmatrix} .8 & -.7 & 0 \\ 0 & 0 & 0 \end{bmatrix} \sim \begin{bmatrix} 1 & -.875 & 0 \\ 0 & 0 & 0 \end{bmatrix}$$

The general solution is $p_G = .875p_S$, with p_S free. One equilibrium solution is $p_S = 1000$ and $p_G = 875$. If one uses fractions instead of decimals in the calculations, the general solution would be written $p_G = (7/8)p_S$, and a natural choice of prices might be $p_S = 80$ and $p_G = 70$. Only the *ratio* of the prices is important: $p_G = .875p_S$. The economic equilibrium is unaffected by a proportional change in prices

7. The following vectors list the numbers of atoms of sodium (Na), hydrogen (H), carbon (C), and oxygen (O):

$$\text{NaHCO}_3: \begin{bmatrix} 1 \\ 1 \\ 1 \\ 3 \end{bmatrix}, \; \text{H}_3\text{C}_6\text{H}_5\text{O}_7: \begin{bmatrix} 0 \\ 8 \\ 6 \\ 7 \end{bmatrix}, \; \text{Na}_3\text{C}_6\text{H}_5\text{O}_7: \begin{bmatrix} 3 \\ 5 \\ 6 \\ 7 \end{bmatrix}, \; \text{H}_2\text{O}: \begin{bmatrix} 0 \\ 2 \\ 0 \\ 1 \end{bmatrix}, \; \text{CO}_2: \begin{bmatrix} 0 \\ 0 \\ 1 \\ 2 \end{bmatrix} \begin{matrix} \text{sodium} \\ \text{hydrogen} \\ \text{carbon} \\ \text{oxygen} \end{matrix}$$

The order of the various atoms is not important. The list here was selected by writing the elements in the order in which they first appear in the chemical equation, reading left to right:

Copyright © 2021 Pearson Education, Inc.

$$x_1 \cdot \text{NaHCO}_3 + x_2 \cdot \text{H}_3\text{C}_6\text{H}_5\text{O}_7 \rightarrow x_3 \cdot \text{Na}_3\text{C}_6\text{H}_5\text{O}_7 + x_4 \cdot \text{H}_2\text{O} + x_5 \cdot \text{CO}_2$$

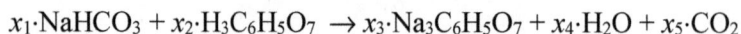

The coefficients $x_1, \ldots, x_5$ satisfy the vector equation

$$x_1 \begin{bmatrix} 1 \\ 1 \\ 1 \\ 3 \end{bmatrix} + x_2 \begin{bmatrix} 0 \\ 8 \\ 6 \\ 7 \end{bmatrix} = x_3 \begin{bmatrix} 3 \\ 5 \\ 6 \\ 7 \end{bmatrix} + x_4 \begin{bmatrix} 0 \\ 2 \\ 0 \\ 1 \end{bmatrix} + x_5 \begin{bmatrix} 0 \\ 0 \\ 1 \\ 2 \end{bmatrix}$$

Move all terms to the left side (changing the sign of each entry in the third, fourth, and fifth vectors) and reduce the augmented matrix:

$$\begin{bmatrix} 1 & 0 & -3 & 0 & 0 & 0 \\ 1 & 8 & -5 & -2 & 0 & 0 \\ 1 & 6 & -6 & 0 & -1 & 0 \\ 3 & 7 & -7 & -1 & -2 & 0 \end{bmatrix} \sim \cdots \sim \begin{bmatrix} 1 & 0 & 0 & 0 & -1 & 0 \\ 0 & 1 & 0 & 0 & -1/3 & 0 \\ 0 & 0 & 1 & 0 & -1/3 & 0 \\ 0 & 0 & 0 & 1 & -1 & 0 \end{bmatrix}$$

The general solution is $x_1 = x_5$, $x_2 = (1/3)x_5$, $x_3 = (1/3)x_5$, $x_4 = x_5$, and x_5 is free. Take $x_5 = 3$. Then $x_1 = x_4 = 3$, and $x_2 = x_3 = 1$. The balanced equation is

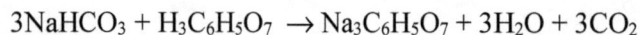

$$3\text{NaHCO}_3 + \text{H}_3\text{C}_6\text{H}_5\text{O}_7 \rightarrow \text{Na}_3\text{C}_6\text{H}_5\text{O}_7 + 3\text{H}_2\text{O} + 3\text{CO}_2$$

13. Write the equations for each intersection (see the diagram for the intersection labels):

Intersection	Flow in		Flow out
A	$x_2 + 30$	=	$x_1 + 80$
B	$x_3 + x_5$	=	$x_2 + x_4$
C	$x_6 + 100$	=	$x_5 + 40$
D	$x_4 + 40$	=	$x_6 + 90$
E	$x_1 + 60$	=	$x_3 + 20$
Total flow:	230	=	230

Rearrange the equations:

$$\begin{aligned} x_1 - x_2 \quad\quad\quad\quad &= -50 \\ x_2 - x_3 + x_4 - x_5 &= 0 \\ x_5 - x_6 &= 60 \\ x_4 \quad - x_6 &= 50 \\ x_1 \quad\quad - x_3 \quad\quad &= -40 \end{aligned}$$

Completely reduce the augmented matrix:

Copyright © 2021 Pearson Education, Inc.

$$\begin{bmatrix} 1 & -1 & 0 & 0 & 0 & 0 & -50 \\ 0 & 1 & -1 & 1 & -1 & 0 & 0 \\ 0 & 0 & 0 & 0 & 1 & -1 & 60 \\ 0 & 0 & 0 & 1 & 0 & -1 & 50 \\ 1 & 0 & -1 & 0 & 0 & 0 & -40 \end{bmatrix} \sim \cdots \sim \begin{bmatrix} 1 & -1 & 0 & 0 & 0 & 0 & -50 \\ 0 & 1 & -1 & 1 & -1 & 0 & 0 \\ 0 & 0 & 0 & 1 & 0 & -1 & 50 \\ 0 & 0 & 0 & 0 & 1 & -1 & 60 \\ 0 & 0 & 0 & 0 & 0 & 0 & 0 \end{bmatrix}$$

$$\sim \cdots \sim \begin{bmatrix} 1 & 0 & -1 & 0 & 0 & 0 & -40 \\ 0 & 1 & -1 & 0 & 0 & 0 & 10 \\ 0 & 0 & 0 & 1 & 0 & -1 & 50 \\ 0 & 0 & 0 & 0 & 1 & -1 & 60 \\ 0 & 0 & 0 & 0 & 0 & 0 & 0 \end{bmatrix}$$

a. The general solution is
$$\begin{cases} x_1 = x_3 - 40 \\ x_2 = x_3 + 10 \\ x_3 \text{ is free} \\ x_4 = x_6 + 50 \\ x_5 = x_6 + 60 \\ x_6 \text{ is free} \end{cases}$$

b. To find minimum flows, note that since x_1 cannot be negative, $x_3 \geq 40$. This implies that $x_2 \geq 50$. Also, since x_6 cannot be negative, $x_4 \geq 50$ and $x_5 \geq 60$. The minimum flows are $x_2 = 50$, $x_3 = 40$, $x_4 = 50$, $x_5 = 60$ (when $x_1 = 0$ and $x_6 = 0$).

MATLAB   Rational Format

Chemical equation-balance problems are studied best using exact or symbolic arithmetic, because the balance variables must be whole numbers (with no round-off allowed). In MATLAB, a simple approach is to execute the command **format rat**, which will make MATLAB display matrix or vector entries as rational numbers. In general, the rational number displayed might be only an approximation for a floating-point number. But since the chemical equations studied here have integer coefficients, **format rat** will make MATLAB display the exact (rational) value of every entry during row reduction. Use **format** or **format short** to return to the standard MATLAB display of numbers.

Once you find a rational solution of a chemical equation-balance problem, you can multiply the entries in the solution vector by a suitable integer to produce a solution that involves only whole numbers.

Copyright © 2021 Pearson Education, Inc.

1.7 - Linear Independence

This section is as important as Section 1.4 and should be studied just as carefully. Full understanding of the concepts will take time, so get started on the section now.

KEY IDEAS

Figures 1 and 2, along with Theorem 7, will help you understand the nature of a linearly dependent set. (Fig. 2 applies only when **u** and **v** are independent.) But you must also learn the *definitions* of linear dependence and linear independence, word for word! Many theoretical problems involving a linearly dependent set are treated by the definition, because it provides an equation (the dependence equation) with which to work. (See the proof of Theorem 7.)

The box before Example 2 contains a very useful fact. Any time you need to study the linear independence of a set of p vectors in $\mathbb{R}^n$, you can always form an $n \times p$ matrix A with those vectors as columns and then study the matrix equation $A\mathbf{x} = \mathbf{0}$. This is not the only method, however. Stay alert for three special situations:

- A set of two vectors. Always check this by inspection; don't waste time on row reduction of $[A \quad \mathbf{0}]$. The set is linearly independent if neither of the vectors is a multiple of the other. (For brevity, I sometimes say that "the vectors are not multiples.") See Example 3.

- A set that contains too many vectors, that is, more vectors than entries in the vectors; the columns of a short, fat matrix. Theorem 8.

- A set that contains the zero vector. Theorem 9.

The most common mistake students make when checking a set of three or more vectors for independence is to think they only have to verify that no vector is a multiple of one of the other vectors. Wrong! Study Example 5 and Figure 4.

Key exercises are 9–20 and 29–34, and 36. Try Exercise 41, even if it is not assigned. Think carefully, and write your answer before checking the answer section.

SOLUTIONS TO EXERCISES

1. Use an augmented matrix to study the solution set of $x_1\mathbf{u} + x_2\mathbf{v} + x_3\mathbf{w} = \mathbf{0}$ (*), where **u**, **v**, and **w** are the three given vectors. Since
$$\begin{bmatrix} 5 & 7 & -2 & 0 \\ 1 & 2 & -1 & 0 \\ 0 & -6 & 6 & 0 \end{bmatrix} \sim \begin{bmatrix} 5 & 7 & 9 & 0 \\ 0 & -3 & 3 & 0 \\ 0 & 0 & 0 & 0 \end{bmatrix},$$
there are free variables. So the homogeneous equation (*) has nontrivial solutions. The vectors are linearly dependent.

Warning: Whenever you study a homogeneous equation, you may be tempted to omit the augmented column of zeros because it never changes under row operations. I urge you to keep the zeros, to avoid possibly misinterpreting your own calculations.

Copyright © 2021 Pearson Education, Inc.

7. Study the equation $A\mathbf{x} = \mathbf{0}$. Some people may start with the method of Example 2:

$$\begin{bmatrix} 1 & 4 & -3 & 0 & 0 \\ -2 & -7 & 5 & 1 & 0 \\ -4 & -5 & 7 & -5 & 0 \end{bmatrix} \sim \begin{bmatrix} 1 & 4 & -3 & 0 & 0 \\ 0 & 1 & -1 & 1 & 0 \\ 0 & 11 & -5 & -5 & 0 \end{bmatrix} \sim \begin{bmatrix} ① & 4 & -3 & 0 & 0 \\ 0 & ① & -1 & 1 & 0 \\ 0 & 0 & ⑥ & -16 & 0 \end{bmatrix}$$

But this is a waste of time. There are only 3 rows, so there are at most three pivot positions. Hence, at least one of the four variables must be free. So the equation $A\mathbf{x} = \mathbf{0}$ has a nontrivial solution and the columns of A are linearly dependent.

Warning: Exercise 9 and Practice Problem 2 emphasize that to check whether a set such as $\{\mathbf{v}_1, \mathbf{v}_2, \mathbf{v}_3\}$ is linearly dependent, it is *not* wise to check instead whether $\mathbf{v}_3$ is a linear combination of $\mathbf{v}_1$ and $\mathbf{v}_2$.

13. To study the linear dependence of three vectors, say $\mathbf{v}_1, \mathbf{v}_2, \mathbf{v}_3$, row reduce the augmented matrix $[\mathbf{v}_1 \quad \mathbf{v}_2 \quad \mathbf{v}_3 \quad \mathbf{0}]$:

$$\begin{bmatrix} 1 & -2 & 3 & 0 \\ 5 & -9 & h & 0 \\ -3 & 6 & -9 & 0 \end{bmatrix} \sim \begin{bmatrix} ① & -2 & 3 & 0 \\ 0 & ① & h-15 & 0 \\ 0 & 0 & 0 & 0 \end{bmatrix}$$

The equation $x_1\mathbf{v}_1 + x_2\mathbf{v}_2 + x_3\mathbf{v}_3 = \mathbf{0}$ has a free variable and hence a nontrivial solution no matter what the value of h. So the vectors are linearly dependent for all values of h.

Checkpoint: What is wrong with the following statement?

The vectors $\begin{bmatrix} 3 \\ -1 \end{bmatrix}, \begin{bmatrix} 2 \\ 8 \end{bmatrix}, \begin{bmatrix} -5 \\ 3 \end{bmatrix}, \begin{bmatrix} 7 \\ -4 \end{bmatrix}$ are linearly dependent "because there is a free variable," or "because there are more variables than equations."

15. The set $\left\{ \begin{bmatrix} 5 \\ 1 \end{bmatrix}, \begin{bmatrix} 2 \\ 8 \end{bmatrix}, \begin{bmatrix} 1 \\ 3 \end{bmatrix}, \begin{bmatrix} -1 \\ 7 \end{bmatrix} \right\}$ is obviously linearly dependent, by Theorem 8, because there are more vectors (4) than entries in the vectors. On a test, you probably will not have to know the theorem number. Check with your instructor.

19. The set is linearly independent because neither vector is a multiple of the other vector. [Two of the entries in the first vector are −4 times the corresponding entry in the second vector. But this multiple does not work for the third entries.]

21. See the box before Example 2.

23. See the warning after Theorem 7.

25. See Fig. 3, after Theorem 8.

27. See the remark following Example 4.

Copyright © 2021 Pearson Education, Inc.

31. $\begin{bmatrix} \blacksquare & * \\ 0 & \blacksquare \\ 0 & 0 \\ 0 & 0 \end{bmatrix}$ and $\begin{bmatrix} 0 & \blacksquare \\ 0 & 0 \\ 0 & 0 \\ 0 & 0 \end{bmatrix}$

37. Think of $A = [\mathbf{a}_1 \quad \mathbf{a}_2 \quad \mathbf{a}_3]$. The text points out that $\mathbf{a}_3 = \mathbf{a}_1 + \mathbf{a}_2$. Rewrite this as $\mathbf{a}_1 + \mathbf{a}_2 - \mathbf{a}_3 = \mathbf{0}$. As a matrix equation, $A\mathbf{x} = \mathbf{0}$ for $\mathbf{x} = (1, 1, -1)$.

39. The text uses Theorem 7 to conclude that $\{\mathbf{v}_1, \ldots, \mathbf{v}_4\}$ is linearly dependent. Another argument is to rewrite the equation $\mathbf{v}_3 = 2\mathbf{v}_1 + \mathbf{v}_2$ as $2\mathbf{v}_1 + 1\mathbf{v}_2 + (-1)\mathbf{v}_3 + 0\mathbf{v}_4 = \mathbf{0}$. This is a linear dependence relation. Some students think of this argument rather than Theorem 7. Did you? (I hope you did not read the answer before trying this problem.)

43. True. The text gives a complete answer.

45. If for all $\mathbf{b}$ the equation $A\mathbf{x} = \mathbf{b}$ has at most one solution, then take $\mathbf{b} = \mathbf{0}$, and conclude that the equation $A\mathbf{x} = \mathbf{0}$ has at most one solution. Then the trivial solution is the only solution, and so the columns of A are linearly independent.

49. Make $\mathbf{v}$ any one of the columns of A that is not in B and row reduce the augmented matrix $[B \quad \mathbf{v}]$. The calculations will show that the equation $B\mathbf{x} = \mathbf{v}$ is consistent, which means that $\mathbf{v}$ is a linear combination of the columns of B. Thus, each column of A that is not a column of B is in the set spanned by the columns of B.

Answer to Checkpoint: The set of four vectors contains only vectors, no variables of any kind, and no equations. It makes no sense to talk about the variables in a set of vectors. Variables appear in an equation. One cannot assume that the writer of the statement has any idea of the appropriate equation. If you want to give an explanation involving variables, then you must specify the equation. One correct answer is: the vectors are linearly dependent because the equation $x_1 \begin{bmatrix} 3 \\ -1 \end{bmatrix} + x_2 \begin{bmatrix} 2 \\ 8 \end{bmatrix} + x_3 \begin{bmatrix} -5 \\ 3 \end{bmatrix} + x_4 \begin{bmatrix} 7 \\ -4 \end{bmatrix} = \begin{bmatrix} 0 \\ 0 \end{bmatrix}$ necessarily has a free variable.

Mastering Linear Algebra Concepts: Linear Independence

In Section 1.4 of this *Guide*, I described how to begin forming a mental image of the concept of a spanning set. The same technique works for linear independence. The goal is to merge all the ideas you find regarding linear independence into a single mental image, with each part immediately available in your mind for use as needed. Start now to organize on paper your understanding of linear independence/dependence, using the following list as a guide. In each case, write information that you think will be helpful. (Definitions and theorems should be copied word-for-word.)

- definitions of linear independence and dependence
- equivalent descriptions   Theorem 7
- geometric interpretations   Figs. 1, 2, 4
- special cases   Theorems 8, 9, Examples 3, 5, 6

Copyright © 2021 Pearson Education, Inc.

- examples and "counterexamples"   Figs. 1, 2, 3, 4, Exercises 9–20, 33–38
- algorithms or typical computations   Examples 1, 2, Exercises 1–8
- connections with other concepts   Examples 2, 4, Exercises 27, 30, 39

As you work on your notes, be careful to use terminology correctly. For instance, the term "linearly independent" may be applied to a set of vectors, but it *never* is applied to a matrix or to an equation. The *columns* of a matrix may be linearly independent, but it is meaningless to refer to a linearly independent matrix. Similarly, *solutions* of a system of linear equations may be linearly independent, but the term "linearly independent equations" has never been defined. Finally, a set of vectors or a matrix cannot have a "nontrivial solution." Only equations have solutions.

> **MATLAB**   Selecting columns from a matrix A
>
> You can create a matrix B using the rows and columns of A by listing the rows you want in square brackets and then columns you want in second set of square brackets. For example, **B=A([1,3,4],[2,5,6])** would form a submatrix of A using the rows 1, 3, and 4 and the columns 2, 5, and 6. If you want to use all m rows, you can use the notation **:** instead of entering a vector. For example, **B=A(:,[1,3,5])** would form a matrix B from columns 1, 3, and 5 of the matrix A.

1.8 - Introduction to Linear Transformations

Linear transformations are important for both the theory and the applications of linear algebra. You will see both uses in a variety of settings throughout the text. The graphical descriptions in this section will be augmented in Section 1.9 and in a later section on computer graphics.

STUDY NOTES

Viewing the correspondence from a vector $\mathbf{x}$ to a vector $A\mathbf{x}$ as a mapping provides a dynamic interpretation of matrix-vector multiplication and a new way to understand the equation $A\mathbf{x} = \mathbf{b}$. Using the language of computer science, we can describe a matrix in two ways—as a data structure (a rectangular array of numbers) and as a program (a prescription for transforming vectors). Strictly speaking, however, the actual linear transformation is the function or mapping $\mathbf{x} \mapsto A\mathbf{x}$ rather than just A itself.

Here is a way to visualize a matrix acting as a linear transformation. The entries in the input vector $\mathbf{x}$ are assigned as weights that multiply the corresponding columns of A, then the resulting weighted columns are added together to produce the output vector $\mathbf{b}$.

Copyright © 2021 Pearson Education, Inc.

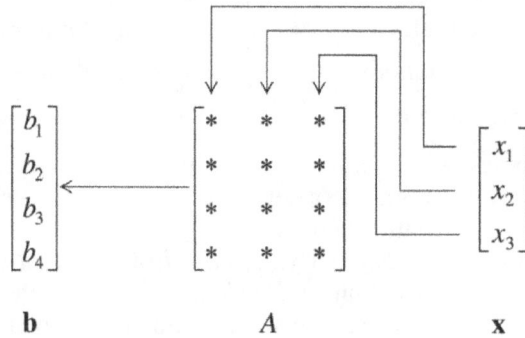

$$\mathbf{b} \qquad\qquad\qquad A \qquad\qquad\qquad \mathbf{x}$$

As you learn the definition of a linear transformation T, don't forget the crucial phrases "for all $\mathbf{u}$ and $\mathbf{v}$ in the domain of T" and "for all $\mathbf{u}$ and all scalars c." The mapping T defined by $T(x_1, x_2) = (|x_2|, |x_1|)$ is *not* a linear mapping, and yet T satisfies the linearity properties for *some* vectors in its domain and *some* scalars.

The key exercises are 17–20, 33 and 39.

SOLUTIONS TO EXERCISES

1. $T(\mathbf{u}) = A\mathbf{u} = \begin{bmatrix} 2 & 0 \\ 0 & 2 \end{bmatrix}\begin{bmatrix} 1 \\ -3 \end{bmatrix} = \begin{bmatrix} 2 \\ -6 \end{bmatrix}$, $T(\mathbf{v}) = \begin{bmatrix} 2 & 0 \\ 0 & 2 \end{bmatrix}\begin{bmatrix} a \\ b \end{bmatrix} = \begin{bmatrix} 2a \\ 2b \end{bmatrix}$

5. $\begin{bmatrix} A & \mathbf{b} \end{bmatrix} = \begin{bmatrix} 1 & -5 & -7 & -2 \\ -3 & 7 & 5 & -2 \end{bmatrix} \sim \begin{bmatrix} 1 & -5 & -7 & -2 \\ 0 & 1 & 2 & 1 \end{bmatrix} \sim \begin{bmatrix} 1 & 0 & 3 & 3 \\ 0 & 1 & 2 & 1 \end{bmatrix}$

Note that a solution is *not* $\begin{bmatrix} 3 \\ 1 \end{bmatrix}$. To avoid this common error, write the equations:

$\begin{aligned} x_1 \quad + 3x_3 &= 3 \\ x_2 + 2x_3 &= 1 \end{aligned}$ and solve for the basic variables: $\begin{cases} x_1 = 3 - 3x_3 \\ x_2 = 1 - 2x_3 \\ x_3 \text{ is free} \end{cases}$

General solution: $\mathbf{x} = \begin{bmatrix} x_1 \\ x_2 \\ x_3 \end{bmatrix} = \begin{bmatrix} 3 - 3x_3 \\ 1 - 2x_3 \\ x_3 \end{bmatrix} = \begin{bmatrix} 3 \\ 1 \\ 0 \end{bmatrix} + x_3 \begin{bmatrix} -3 \\ -2 \\ 1 \end{bmatrix}$. For a particular solution, one might

choose $x_3 = 0$ and $\mathbf{x} = \begin{bmatrix} 3 \\ 1 \\ 0 \end{bmatrix}$.

7. $a = 5$; the domain of T is $\mathbb{R}^5$, because a 6×5 matrix has 5 columns and for $A\mathbf{x}$ to be defined, $\mathbf{x}$ must be in $\mathbb{R}^5$. $b = 6$; the codomain of T is $\mathbb{R}^6$, because $A\mathbf{x}$ is a linear combination of the columns of A, and each column of A is in $\mathbb{R}^6$.

Copyright © 2021 Pearson Education, Inc.

13. The transformation may be described geometrically as a reflection through the origin. Two other correct descriptions are a rotation of π radians about the origin and a rotation of $-\pi$ radians about the origin. See the figure.

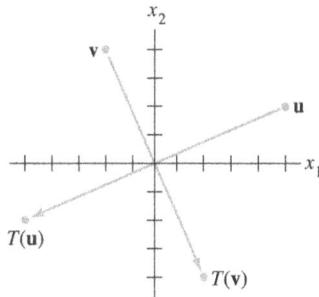

18. *Additional Hint*: Draw a line through **w** parallel to **v**, and draw a line through **w** parallel to **u**. This will help you write **w** as a linear combination of **u** and **v**.

19. All you know are the images of $\mathbf{e}_1$ and $\mathbf{e}_2$ and the fact that T is linear. The key idea is to write

$$\mathbf{x} = \begin{bmatrix} 5 \\ -3 \end{bmatrix} = 5 \begin{bmatrix} 1 \\ 0 \end{bmatrix} - 3 \begin{bmatrix} 0 \\ 1 \end{bmatrix} = 5\mathbf{e}_1 - 3\mathbf{e}_2. \text{ Then, from the linearity of } T, \text{ write}$$

$$T(\mathbf{x}) = T(5\mathbf{e}_1 - 3\mathbf{e}_2) = 5T(\mathbf{e}_1) - 3T(\mathbf{e}_2) = 5\mathbf{y}_1 - 3\mathbf{y}_2 = 5\begin{bmatrix} 2 \\ 5 \end{bmatrix} - 3\begin{bmatrix} -1 \\ 6 \end{bmatrix} = \begin{bmatrix} 13 \\ 7 \end{bmatrix}$$

To find the image of $\begin{bmatrix} x_1 \\ x_2 \end{bmatrix}$, observe that $\mathbf{x} = \begin{bmatrix} x_1 \\ x_2 \end{bmatrix} = x_1 \begin{bmatrix} 1 \\ 0 \end{bmatrix} + x_2 \begin{bmatrix} 0 \\ 1 \end{bmatrix} = x_1 \mathbf{e}_1 + x_2 \mathbf{e}_2.$ Then

$$T(\mathbf{x}) = T(x_1\mathbf{e}_1 + x_2\mathbf{e}_2) = x_1 T(\mathbf{e}_1) + x_2 T(\mathbf{e}_2) = x_1 \begin{bmatrix} 2 \\ 5 \end{bmatrix} + x_2 \begin{bmatrix} -1 \\ 6 \end{bmatrix} = \begin{bmatrix} 2x_1 - x_2 \\ 5x_1 + 6x_2 \end{bmatrix}$$

21. A function is another word for transformation or mapping.

23. See the paragraph before Example 1.

25. See Figure 2. Or, see the paragraph before Example 1.

27. See the paragraph after the definition of a linear transformation.

29. See the paragraph following the box that contains equation (4).

33. Any point **x** on the line through **p** in the direction of **v** satisfies the parametric equation $\mathbf{x} = \mathbf{p} + t\mathbf{v}$ for some value of t. By linearity, the image $T(\mathbf{x})$ satisfies the parametric equation

$$T(\mathbf{x}) = T(\mathbf{p} + t\mathbf{v}) = T(\mathbf{p}) + tT(\mathbf{v}) \tag{*}$$

If $T(\mathbf{v}) = \mathbf{0}$, then $T(\mathbf{x}) = T(\mathbf{p})$ for all values of t, and the image of the original line is just a single point. Otherwise, (*) is the parametric equation of a line through $T(\mathbf{p})$ in the direction of $T(\mathbf{v})$.

Copyright © 2021 Pearson Education, Inc.

Study Tip: Exercise 39 is important, because it will help you to connect the concepts of linear dependence and linear transformation. Be sure to try the exercise first, before looking in the answer section of the text. Don't feel badly if you need to peek at the hint there. Only my best students can do this problem unaided. Once you have seen the hint, try hard to construct the desired explanation without consulting the solution I have written below. Don't give up too soon. Reread the definitions of linear dependence and linear transformation, if necessary.

After you have written your best attempt at an explanation, check it against the *Study Guide* solution. Also, study the strategy there of how I found the solution. Even if your attempt is quite unsatisfactory, the time spent on this problem is worthwhile, because you will learn more from the solution here.

39. To help you use this *Study Guide* properly, I have hidden the solution at the end of the solutions for Section 1.9. *Do not look there until you have followed the instructions above.* (I may not "hide" a solution again, but I wanted this one time to emphasize the importance of working seriously on a problem before checking the solution.)

Mastering Linear Algebra Concepts: Linear Transformation

Start to form a robust mental image of a linear transformation by preparing a review sheet that covers the following categories:

- definition

- equivalent descriptions   Equations (4) and (5)

- geometric interpretations   Figs. 1 or 2

- special cases   Matrix transformation

- examples and "counterexamples"   Superposition, Examples 2-6
                                            Paragraph before Exercise 1, in this *Guide*
                                            Exercises 29, 30, 33

- connections with other concepts   Existence and uniqueness
                                            Linear dependence: Exercise 31

Note: Exercise 31 should enrich your mental image of linear dependence, so add a note about it to your list for "linear independence." If your course does not emphasize the next section, turn now to the end of the *Study Guide* material for Section 1.9 and read the box on *Existence and Uniqueness*.

Copyright © 2021 Pearson Education, Inc.

1.9 – The Matrix of a Linear Transformation

Every matrix transformation is a linear transformation. This section shows that every linear transformation from $\mathbb{R}^n$ to $\mathbb{R}^m$ is a matrix transformation. Chapters 4 and 5 will discuss other examples of linear transformations.

KEY IDEAS

A linear transformation $T : \mathbb{R}^n \to \mathbb{R}^m$ is completely determined by what it does to the columns of the identity matrix I_n. The jth column of the standard matrix for T is $T(\mathbf{e}_j)$, where $\mathbf{e}_j$ is the jth column of I_n.

There are two ways to compute the standard matrix A. Either compute $T(\mathbf{e}_1), \ldots, T(\mathbf{e}_n)$, which is easy to do when T is described geometrically, as in Exercises 1–14, or fill in the entries of A by inspection, which is easy to do when T is described by a formula, as in Exercises 15–22.

Existence and uniqueness questions about the mapping $\mathbf{x} \mapsto A\mathbf{x}$ are determined by properties of A. You should know how this works. The proof of Theorem 11 also applies to linear transformations on the general vector spaces in Chapter 4. Here is another way to understand Theorem 11 [and Theorem 12(b)] using the language of matrix transformations:

Let A be the standard matrix of T. Then T is one-to-one if and only if the equation $A\mathbf{x} = \mathbf{b}$ has at most one solution for each $\mathbf{b}$. This happens if and only if every column of A is a pivot column, which happens if and only if $A\mathbf{x} = \mathbf{0}$ has only the trivial solution.

The "if and only if" phrase in Theorem 11 (and in the proof above) was discussed in *A Mathematical Note*, in Section 1.2 of this *Guide*.

SOLUTIONS TO EXERCISES

1. The columns of the standard matrix A of T are the images of $\mathbf{e}_1$ and $\mathbf{e}_2$. Write these images

    vertically: $T(\mathbf{e}_1) = \begin{bmatrix} 2 \\ 1 \\ 2 \\ 1 \end{bmatrix}$ and $T(\mathbf{e}_2) = \begin{bmatrix} -5 \\ 2 \\ 0 \\ 0 \end{bmatrix}$. Then $A = [T(\mathbf{e}_1) \ T(\mathbf{e}_2)] = \begin{bmatrix} 2 & -5 \\ 1 & 2 \\ 2 & 0 \\ 1 & 0 \end{bmatrix}$.

7. Follow what happens to $\mathbf{e}_1$ and $\mathbf{e}_2$. Since $\mathbf{e}_1$ is on the unit circle in the plane, it rotates through $-3\pi/4$ radians into a point on the unit circle that lies in the third quadrant and on the line $x_2 = x_1$ (that is, $y = x$ in more familiar notation). The point $(-1, -1)$ is on the line $x_2 = x_1$, but its distance from the origin is $\sqrt{2}$. So the rotational image of $\mathbf{e}_1$ is $(-1/\sqrt{2}, -1/\sqrt{2})$. Then this image reflects in the horizontal axis to $(-1/\sqrt{2}, 1/\sqrt{2})$.

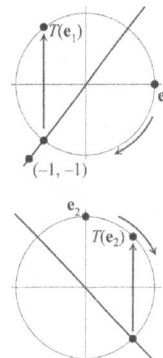

Copyright © 2021 Pearson Education, Inc.

Similarly, e_2 rotates into a point on the unit circle that lies in the second quadrant and on the line $x_2 = -x_1$, namely, $(1/\sqrt{2}, -1/\sqrt{2})$. Then this image reflects in the horizontal axis to $(1/\sqrt{2}, 1/\sqrt{2})$.

When the two calculations described above are written in vertical vector notation, the transformation's standard matrix $[T(e_1)\ T(e_2)]$ is easily seen:

$$e_1 \rightarrow \begin{bmatrix} -1/\sqrt{2} \\ -1/\sqrt{2} \end{bmatrix} \rightarrow \begin{bmatrix} -1/\sqrt{2} \\ 1/\sqrt{2} \end{bmatrix}, \quad e_2 \rightarrow \begin{bmatrix} 1/\sqrt{2} \\ -1/\sqrt{2} \end{bmatrix} \rightarrow \begin{bmatrix} 1/\sqrt{2} \\ 1/\sqrt{2} \end{bmatrix}, \quad A = \begin{bmatrix} -1/\sqrt{2} & 1/\sqrt{2} \\ 1/\sqrt{2} & 1/\sqrt{2} \end{bmatrix}$$

Checkpoint: Use an idea from this section to explain why the linear transformation T that reflects points through the origin, $T(x_1, x_2) = (-x_1, -x_2)$, is the same as the linear transformation R that rotates points about the origin in $\mathbb{R}^2$ through π radians.

13. Since $(2, 1) = 2e_1 + e_2$, the image of $(2, 1)$ under T is $2T(e_1) + T(e_2)$, by linearity of T. On the figure in the exercise, locate $2T(e_1)$ and use it with $T(e_2)$ to form the parallelogram shown in the text's answers.

19. The matrix A that changes (x_1, x_2, x_3) into $(x_1 - 5x_2 + 4x_3, x_2 - 6x_3)$ can be found by inspection when vectors are written in column formation. Write a blank matrix A to the left of the column vector x and fill in the entries of A. Since $T(x)$ has 2 entries, A has 2 rows. Since x has 3 entries, A must have 3 columns.

$$\begin{bmatrix} x_1 - 5x_2 + 4x_3 \\ x_2 - 6x_3 \end{bmatrix} = \begin{bmatrix} & A & \end{bmatrix} \begin{bmatrix} x_1 \\ x_2 \\ x_3 \end{bmatrix} = \begin{bmatrix} 1 & -5 & 4 \\ 0 & 1 & -6 \end{bmatrix} \begin{bmatrix} x_1 \\ x_2 \\ x_3 \end{bmatrix}$$

21. $T(x) = \begin{bmatrix} x_1 + x_2 \\ 4x_1 + 5x_2 \end{bmatrix} = \begin{bmatrix} & A & \end{bmatrix} \begin{bmatrix} x_1 \\ x_2 \end{bmatrix} = \begin{bmatrix} 1 & 1 \\ 4 & 5 \end{bmatrix} \begin{bmatrix} x_1 \\ x_2 \end{bmatrix}$. To solve $T(x) = \begin{bmatrix} 3 \\ 8 \end{bmatrix}$, row reduce the

augmented matrix: $\begin{bmatrix} 1 & 1 & 3 \\ 4 & 5 & 8 \end{bmatrix} \sim \begin{bmatrix} 1 & 1 & 3 \\ 0 & 1 & -4 \end{bmatrix} \sim \begin{bmatrix} 1 & 0 & 7 \\ 0 & 1 & -4 \end{bmatrix}$, $x = \begin{bmatrix} 7 \\ -4 \end{bmatrix}$.

Study Tip: When T is described by a formula, as in Exercises 15–22, you can use the method of Exercise 19 to find an A such that $T(x) = Ax$, *provided* that T is a linear transformation. (Finding A proves that T is linear.) If you can't find the matrix, T is probably *not* a linear transformation. To show that such a T is not linear, you have either to find two vectors u and v such that $T(u + v)$ is not equal to $T(u) + T(v)$ or to find a vector u and scalar c such that $T(cu) \neq cT(u)$.

The text does not give you practice determining whether a transformation is linear because the time needed to develop this skill would have to be taken away from some other topic. If you are expected to have this skill, you will need some exercises (besides Exercises 40 and 41 in Section 1.8). Check with your instructor.

Copyright © 2021 Pearson Education, Inc.

23. See Theorem 10.

25. See Example 3.

27. See the second paragraph in the section Geometric Linear Transformations of $\mathbb{R}^2$.

29. See the definition of *onto*. *Any* function from $\mathbb{R}^n$ to $\mathbb{R}^m$ maps each vector onto another vector.

31. See Example 5.

33. Row reduce the standard matrix A of the transformation T in Exercise 17 to produce

$\begin{bmatrix} ① & 0 & 0 & 0 \\ 0 & ① & 0 & 0 \\ 0 & 0 & 0 & ① \\ 0 & 0 & 0 & 0 \end{bmatrix}$. This matrix shows that A has only three pivot positions, so the equation

$A\mathbf{x} = \mathbf{0}$ has a nontrivial solution. By Theorem 11, the transformation T is *not* one-to-one. Also, since A does not have a pivot in each row, the columns of A do not span $\mathbb{R}^4$. By Theorem 12, T does *not* map $\mathbb{R}^4$ onto $\mathbb{R}^4$.

39. *T is one-to-one if and only if A has n pivot columns.* This statement follows by combining Theorem 12(b) with the statement in Exercise 30 of Section 1.7.

A Mathematical Note: One-to-one

Many students have difficulty with the concept of a one-to-one mapping. Figure 4 should help. The transformation T on the left appears to map three (or even more) points to one image point. In contrast, the transformation T on the right maps six points to six points. You could say that T is six-to-six, but the standard terminology is one-to-one.

---

41. Define $T : \mathbb{R}^n \rightarrow \mathbb{R}^m$ by $T(\mathbf{x}) = B\mathbf{x}$ for some $m \times n$ matrix B, and let A be the standard matrix for T. By definition, $A = [T(\mathbf{e}_1) \cdots T(\mathbf{e}_n)]$, where $\mathbf{e}_j$ is the *j*th column of I_n. However, by matrix-vector multiplication,

$\qquad T(\mathbf{e}_j) = B\mathbf{e}_j = \mathbf{b}_j$, the *j*th column of B. *So* $A = [\mathbf{b}_1 \cdots \mathbf{b}_n] = B$.

43. If $T : \mathbb{R}^n \rightarrow \mathbb{R}^m$ maps $\mathbb{R}^n$ onto $\mathbb{R}^m$, then its standard matrix A has a pivot in each row, by Theorem 12 and by Theorem 4 in Section 1.4. So A must have at least as many columns as rows, so $m \leq n$.

When T is one-to-one, A must have a pivot in each column, by Theorem 12, so $m \geq n$.

45. There is no pivot in the fourth column, so the columns of the matrix are not linearly independent and hence the linear transformation is not one-to-one (Theorem 12). (Or, use the result of Exercise 31.)

Copyright © 2021 Pearson Education, Inc.

47. Row reduction of the matrix shows that columns 1, 2, 3, and 5 contain pivots, but there is no pivot in the fifth row, so the columns of the matrix do not span $\mathbb{R}^5$. By Theorem 12, the linear transformation is not onto.

39. *(This solution is for Section 1.8.) To construct the proof, first write in mathematical terms what is given.*

Since $\{\mathbf{v}_1, \mathbf{v}_2, \mathbf{v}_3\}$ is linearly dependent, there exist scalars c_1, c_2, c_3, not all zero, such that

$$c_1\mathbf{v}_1 + c_2\mathbf{v}_2 + c_3\mathbf{v}_3 = 0 \qquad (*)$$

Next, think about what you must prove. In this problem, to prove that the image points are linearly dependent, you need a dependence relation among $T(\mathbf{v}_1)$, $T(\mathbf{v}_2)$, and $T(\mathbf{v}_3)$. That fact suggests the next step.

Apply T to both sides of (*) and use linearity of T, obtaining

$$T(c_1\mathbf{v}_1 + c_2\mathbf{v}_2 + c_3\mathbf{v}_3) = T(0)$$

and

$$c_1 T(\mathbf{v}_1) + c_2 T(\mathbf{v}_2) + c_3 T(\mathbf{v}_3) = 0$$

Since not all the weights are zero, $\{T(\mathbf{v}_1), T(\mathbf{v}_2), T(\mathbf{v}_3)\}$ is a linearly dependent set. This completes the proof.

Study Tip: Analyze the strategy above for solving Exercise 39 (in Section 1.8). This approach will work later in a variety of situations.

Answer to Checkpoint: The reflection T has the property that $T(\mathbf{e}_1) = -\mathbf{e}_1$ and $T(\mathbf{e}_2) = -\mathbf{e}_2$, while the rotation R has the property that $R(\mathbf{e}_1) = -\mathbf{e}_1$ and $R(\mathbf{e}_2) = -\mathbf{e}_2$. Since a linear transformation is completely determined by what it does to the columns $\mathbf{e}_1$ and $\mathbf{e}_2$ of the identity matrix, T and R must be the same transformation. (You could also explain this by observing that T and R have the same standard matrix, namely, $[-\mathbf{e}_1 \quad -\mathbf{e}_2]$.)

Mastering Linear Algebra Concepts: Existence and Uniqueness

It's time to review and organize what you have learned about existence and uniqueness concepts, if you have not already done so. The review will help to prepare you for an exam on the chapter material.

Search through the chapter and collect all the various ways to express existence and uniqueness statements. Most of them can be found in boxes (and theorems) with an "if and only if" statement. Also, check the exercises. For existence, make two lists—one that concerns the equation $A\mathbf{x} = \mathbf{b}$ for some fixed $\mathbf{b}$ (but not always phrased as a matrix equation), and one that concerns the existence of solutions of $A\mathbf{x} = \mathbf{b}$ for all $\mathbf{b}$.

Copyright © 2021 Pearson Education, Inc.

1.10 - Linear Models in Business, Science, and Engineering

This is the second of twelve sections devoted to uses of linear algebra. The applications in the text were selected to give you an impression of the power of linear algebra. You are likely to encounter some of these topics again—in school or in your career—and the discussions in your text will be valuable references.

The main point of this section is to present several interesting applications in which "linearity" arises naturally.

STUDY NOTES

Nutrition Problem: In some applied problems such as the nutrition problem considered here, the data are already organized naturally in a manner that leads to a vector equation of the type we have discussed. The steps to the solution in this case may be diagrammed as follows:

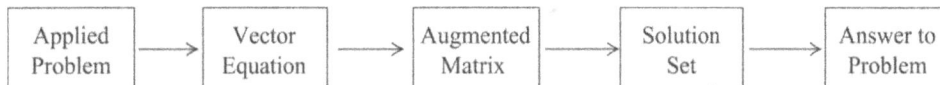

```
Applied    →    Vector    →    Augmented   →   Solution  →   Answer to
Problem         Equation       Matrix           Set           Problem
```

The nutrition model is linear because the nutrients supplied by each foodstuff are *proportional* to the amount of the foodstuff added to the diet mixture, and each nutrient in the mixture is the *sum* of the amounts from each foodstuff. Study equations (1) and (2) in Example 1.

The nutrition problem leads naturally into linear programming, a subject that uses linear algebra and has applications in agriculture, business, engineering, and other areas. In the 1950's and 1960's, one of the most common applications of linear algebra (measured in millions of dollars per year for computer time) was to linear programming problems. Such problems are still of great importance in operations research and management science. The following reference gives an entertaining introduction to linear programming. Matrix notation is used in its appendix.

> Gass, Saul I., *An Illustrated Guide to Linear Programming,* New York: McGraw-Hill, 1970. Republished by Dover Publications, 1990.

Electrical Networks: The linearity of this model, which is evident from the matrix equation $R\mathbf{i} = \mathbf{v}$, comes from the linearity of Ohm's law and Kirchhoff's voltage law. (Kirchhoff's current law, which is also linear, is needed when studying another model that involves branch currents.)

Population Movement: The entries in each *column* of the migration matrix must sum to one because the (decimal) fractions in a column account for the entire population in one region. A certain fraction of the population in a region remains in (or moves within) that region, and other fractions move elsewhere.

Copyright © 2021 Pearson Education, Inc.

SOLUTIONS TO EXERCISES

1. a. If x_1 is the number of servings of Cheerios and x_2 is the number of servings of 100% Natural Cereal, then x_1 and x_2 should satisfy

$$x_1 \begin{bmatrix} \text{nutrients} \\ \text{per serving} \\ \text{of Cheerios} \end{bmatrix} + x_2 \begin{bmatrix} \text{nutrients} \\ \text{per serving of} \\ \text{100\% Natural} \end{bmatrix} = \begin{bmatrix} \text{quantities} \\ \text{of nutrients} \\ \text{required} \end{bmatrix}$$

That is,

$$x_1 \begin{bmatrix} 110 \\ 4 \\ 20 \\ 2 \end{bmatrix} + x_2 \begin{bmatrix} 130 \\ 3 \\ 18 \\ 5 \end{bmatrix} = \begin{bmatrix} 295 \\ 9 \\ 48 \\ 8 \end{bmatrix}$$

b. The equivalent matrix equation is $\begin{bmatrix} 110 & 130 \\ 4 & 3 \\ 20 & 18 \\ 2 & 5 \end{bmatrix} \begin{bmatrix} x_1 \\ x_2 \end{bmatrix} = \begin{bmatrix} 295 \\ 9 \\ 48 \\ 8 \end{bmatrix}$. To solve this, row reduce

the augmented matrix for this equation.

$$\begin{bmatrix} 110 & 130 & 295 \\ 4 & 3 & 9 \\ 20 & 18 & 48 \\ 2 & 5 & 8 \end{bmatrix} \sim \begin{bmatrix} 2 & 5 & 8 \\ 4 & 3 & 9 \\ 20 & 18 & 48 \\ 110 & 130 & 295 \end{bmatrix} \sim \begin{bmatrix} 1 & 2.5 & 4 \\ 4 & 3 & 9 \\ 10 & 9 & 24 \\ 110 & 130 & 295 \end{bmatrix}$$

$$\sim \begin{bmatrix} 1 & 2.5 & 4 \\ 0 & -7 & -7 \\ 0 & -16 & -16 \\ 0 & -145 & -145 \end{bmatrix} \sim \begin{bmatrix} 1 & 2.5 & 4 \\ 0 & 1 & 1 \\ 0 & 0 & 0 \\ 0 & 0 & 0 \end{bmatrix} \sim \begin{bmatrix} 1 & 0 & 1.5 \\ 0 & 1 & 1 \\ 0 & 0 & 0 \\ 0 & 0 & 0 \end{bmatrix}$$

The desired nutrients are provided by 1.5 servings of Cheerios together with 1 serving of 100% Natural Cereal.

Study Tip: Be sure to distinguish between (i) the vector equation, (ii) the matrix equation (which has the form $A\mathbf{x} = \mathbf{b}$), and (iii) the augmented matrix (which has the form $[A \quad \mathbf{b}]$) that represents a system of linear equations.

7. Loop 1: The resistance vector is

Copyright © 2021 Pearson Education, Inc.

$$\mathbf{r}_1 = \begin{bmatrix} 12 \\ -7 \\ 0 \\ -4 \end{bmatrix}$$

Total of three RI voltage drops for current I_1

Voltage drop for I_2 is negative; I_2 flows in opposite direction

Current I_3 does not flow in loop 1

Voltage drop for I_4 is negative; I_4 flows in opposite direction

Loop 2: The resistance vector is

$$\mathbf{r}_2 = \begin{bmatrix} -7 \\ 15 \\ -6 \\ 0 \end{bmatrix}$$

Voltage drop for I_1 is negative; I_1 flows in opposite direction

Total of three RI voltage drops for current I_2

Voltage drop for I_3 is negative; I_3 flows in opposite direction

Current I_4 does not flow in loop 2

Also, $\mathbf{r}_3 = \begin{bmatrix} 0 \\ -6 \\ 14 \\ -5 \end{bmatrix}$, $\mathbf{r}_4 = \begin{bmatrix} -4 \\ 0 \\ -5 \\ 13 \end{bmatrix}$, and $R = [\mathbf{r}_1 \quad \mathbf{r}_2 \quad \mathbf{r}_3 \quad \mathbf{r}_4] = \begin{bmatrix} 12 & -7 & 0 & -4 \\ -7 & 15 & -6 & 0 \\ 0 & -6 & 14 & -5 \\ -4 & 0 & -5 & 13 \end{bmatrix}$.

Note that each off-diagonal entry of R is negative (or zero). This happens because the loop current directions are all chosen in the same direction on the figure. (For each loop j, this choice forces the currents in other loops adjacent to loop j to flow in the direction opposite to current I_j.)

Next, set $\mathbf{v} = \begin{bmatrix} 40 \\ 30 \\ 20 \\ -10 \end{bmatrix}$. Note the negative voltage in loop 4. The current direction chosen in loop 4 is opposed by the orientation of the voltage source in that loop. Thus $R\mathbf{i} = \mathbf{v}$ becomes

$$\begin{bmatrix} 12 & -7 & 0 & -4 \\ -7 & 15 & -6 & 0 \\ 0 & -6 & 14 & -5 \\ -4 & 0 & -5 & 13 \end{bmatrix} \begin{bmatrix} I_1 \\ I_2 \\ I_3 \\ I_4 \end{bmatrix} = \begin{bmatrix} 40 \\ 30 \\ 20 \\ -10 \end{bmatrix}.$$ [M]: The solution is $\mathbf{i} = \begin{bmatrix} I_1 \\ I_2 \\ I_3 \\ I_4 \end{bmatrix} = \begin{bmatrix} 11.43 \\ 10.55 \\ 8.04 \\ 5.84 \end{bmatrix}$.

13. The order of entries in a column of a migration matrix must match the order of the columns. For instance, if the first column concerns the population in the city, then the first entry in *each* column of the matrix must be the fraction of the population that moves to (or remains in) the city. In this case, the data in the exercise leads to

$$M = \begin{bmatrix} .95 & .03 \\ .05 & .97 \end{bmatrix} \text{ and } \mathbf{x}_0 = \begin{bmatrix} 600,000 \\ 400,000 \end{bmatrix}.$$

a. Some of the population vectors are

Copyright © 2021 Pearson Education, Inc.

$$\mathbf{x}_5 = \begin{bmatrix} 523,293 \\ 476,707 \end{bmatrix}, \quad \mathbf{x}_{10} = \begin{bmatrix} 472,737 \\ 527,263 \end{bmatrix}, \quad \mathbf{x}_{15} = \begin{bmatrix} 439,417 \\ 560,583 \end{bmatrix}, \quad \mathbf{x}_{20} = \begin{bmatrix} 417,456 \\ 582,544 \end{bmatrix}$$

The data here shows that the city population is declining and the suburban population is increasing, but the changes in population each year seem to grow smaller.

b. When $\mathbf{x}_0 = \begin{bmatrix} 350,000 \\ 650,000 \end{bmatrix}$, the situation is different. Now

$$\mathbf{x}_5 = \begin{bmatrix} 358,523 \\ 641,477 \end{bmatrix}, \quad \mathbf{x}_{10} = \begin{bmatrix} 364,140 \\ 635,860 \end{bmatrix}, \quad \mathbf{x}_{15} = \begin{bmatrix} 367,843 \\ 632,157 \end{bmatrix}, \quad \mathbf{x}_{20} = \begin{bmatrix} 370,283 \\ 629,717 \end{bmatrix}$$

The city population is increasing slowly and the suburban population is decreasing. No other conclusions are expected. (This example will be analyzed in greater detail later in the text.)

MATLAB   Generating a Sequence

Enter the matrix M and the vector **x = [vector with initial numbers]** to put the initial data into **x**. Then use the command **x = M*x** repeatedly to generate the sequence **xl, x2, ...** You only type the command once. After that, use the up-arrow (↑) key to recall the command, and press <Enter>.

In Exercise 11, you need 6 decimal places to get four significant figures in $M(1, 2)$. Use the command **format long** and then **M** to see more decimal places in **M**. The command **format short** will return MATLAB to the standard four decimal place display. (The display format does not affect MATLAB's accuracy in computations.)

Numbers are entered in MATLAB without commas. The number 600,000 in MATLAB scientific notation is 6e5. A small number such as .00000012 is 1.2e-7.

Chapter 1   Supplementary Exercises

The supplementary exercises at the end of each chapter review material from the chapter, synthesize concepts from several chapters, or supplement the chapter material in some way. The text has solutions for most of the odd-numbered exercises. The *Study Guide* provides solutions for selected odd-numbered exercises that have only an answer or a *Hint* in the regular textbook.

In this chapter, Exercises 1-25 consist of true/false questions, whose level of difficulty varies. Some are similar to the ones that appear in many sections of the text, in which a word or phrase is sometimes missing or slightly misstated. Some follow fairly easily from a theorem: others may need careful reasoning. A few may require an argument that uses several ideas. In each case, think carefully about the statement and attempt to write a solution. The text provides the true/false answer, but you must supply the justification or counterexample. Careful work on these exercises will help you prepare for an exam over the chapter material.

Copyright © 2021 Pearson Education, Inc.

31. a. Set $\mathbf{v}_1 = \begin{bmatrix} 2 \\ -5 \\ 7 \end{bmatrix}$, $\mathbf{v}_2 = \begin{bmatrix} -4 \\ 1 \\ -5 \end{bmatrix}$, $\mathbf{v}_3 = \begin{bmatrix} -2 \\ 1 \\ -3 \end{bmatrix}$ and $\mathbf{b} = \begin{bmatrix} b_1 \\ b_2 \\ b_3 \end{bmatrix}$. "Determine if $\mathbf{v}_1, \mathbf{v}_2, \mathbf{v}_3$ span $\mathbb{R}^3$."

To do this, row reduce $[\mathbf{v}_1 \quad \mathbf{v}_2 \quad \mathbf{v}_3]$:

$$\begin{bmatrix} 2 & -4 & -2 \\ -5 & 1 & 1 \\ 7 & -5 & -3 \end{bmatrix} \sim \begin{bmatrix} 1 & -2 & -1 \\ -5 & 1 & 1 \\ 7 & -5 & -3 \end{bmatrix} \sim \begin{bmatrix} 1 & -2 & -1 \\ 0 & -9 & -4 \\ 0 & 9 & 4 \end{bmatrix} \sim \begin{bmatrix} 1 & -2 & -1 \\ 0 & -9 & -4 \\ 0 & 0 & 0 \end{bmatrix}.$$ The matrix does not have

a pivot in each row, so its columns do not span $\mathbb{R}^3$, by Theorem 4 in Section 1.4.

37. The reduced echelon form of A looks like $E = \begin{bmatrix} 1 & 0 & * \\ 0 & 1 & * \\ 0 & 0 & 0 \end{bmatrix}$. Since E is row equivalent to A,

the equation $E\mathbf{x} = \mathbf{0}$ has the same solutions as $A\mathbf{x} = \mathbf{0}$. Thus $\begin{bmatrix} 1 & 0 & * \\ 0 & 1 & * \\ 0 & 0 & 0 \end{bmatrix}\begin{bmatrix} 3 \\ -2 \\ 1 \end{bmatrix} = \begin{bmatrix} 0 \\ 0 \\ 0 \end{bmatrix}$. By

inspection, the reduced echelon form of A is $E = \begin{bmatrix} 1 & 0 & -3 \\ 0 & 1 & 2 \\ 0 & 0 & 0 \end{bmatrix}$.

41. Here are two arguments. The first is a "direct" proof. The second is called a "proof by contradiction."

    i. Since $\{\mathbf{v}_1, \mathbf{v}_2, \mathbf{v}_3\}$ is a linearly independent set, $\mathbf{v}_1 \neq \mathbf{0}$. Also, Theorem 7 shows that $\mathbf{v}_2$ cannot be a multiple of $\mathbf{v}_1$, and $\mathbf{v}_3$ cannot be a linear combination of $\mathbf{v}_1$ and $\mathbf{v}_2$. By hypothesis, $\mathbf{v}_4$ is not a linear combination of $\mathbf{v}_1, \mathbf{v}_2,$ and $\mathbf{v}_3$. Thus, by Theorem 7, $\{\mathbf{v}_1, \mathbf{v}_2, \mathbf{v}_3, \mathbf{v}_4\}$ cannot be a linearly dependent set and so must be linearly independent.

    ii. Since $\{\mathbf{v}_1, \mathbf{v}_2, \mathbf{v}_3\}$ is a linearly independent set, $\mathbf{v}_1 \neq \mathbf{0}$. Suppose that $\{\mathbf{v}_1, \mathbf{v}_2, \mathbf{v}_3, \mathbf{v}_4\}$ is linearly dependent. Then, by Theorem 7, one of the vectors in the set is a linear combination of the preceding vectors. This vector cannot be $\mathbf{v}_4$ because $\mathbf{v}_4$ is *not* in Span$\{\mathbf{v}_1, \mathbf{v}_2, \mathbf{v}_3\}$. Also, none of the vectors in $\{\mathbf{v}_1, \mathbf{v}_2, \mathbf{v}_3\}$ is a linear combination of the preceding vectors, by Theorem 7. So the linear dependence of $\{\mathbf{v}_1, \mathbf{v}_2, \mathbf{v}_3, \mathbf{v}_4\}$ is impossible, and $\{\mathbf{v}_1, \mathbf{v}_2, \mathbf{v}_3, \mathbf{v}_4\}$ is linearly independent.

Chapter 1 - Glossary Checklist

Check your knowledge by attempting to write definitions of the terms below. Then compare your work with the definitions given in the text's Glossary. Ask your instructor which definitions, if any, might appear on a test.

affine transformation: A mapping $T : \mathbb{R}^n \to \mathbb{R}^m$ of the form $T(\mathbf{x}) = \ldots$.

Copyright © 2021 Pearson Education, Inc.

augmented matrix: A matrix made up of a . . . .

back-substitution (with matrix notation):   The . . . phase of row reduction of an . . . .

basic variable: A variable in a linear system that . . . .

codomain (of $T : \mathbb{R}^n \to \mathbb{R}^m$):   The set . . . that contains . . . .

coefficient matrix: A matrix whose entries are . . . .

consistent linear system: A linear system with . . . .

contraction: A mapping $\mathbf{x} \mapsto$ . . . .

difference equation (or **linear recurrence relation**):   An equation of the form . . . whose solution is
  . . . .

dilation: A mapping $\mathbf{x} \mapsto$ . . . .

domain (of a transformation T):   The set of . . . .

echelon form (or **row echelon form**, of a matrix):   An echelon matrix that . . . .

echelon matrix (or **row echelon matrix**):   A rectangular matrix that has three properties:

elementary row operations: (1) . . . (2) . . . (3) . . . .

equal vectors: Vectors in $\mathbb{R}^n$ whose . . . .

equivalent (linear) systems: Linear systems with the . . . .

existence question: Asks, "Does . . . exist?" or "Is . . .?" Also, "Does . . . exist for . . .?"

floating point arithmetic: Arithmetic with numbers represented as . . . .

flop: One arithmetic operation . . . .

free variable: Any variable in a linear system that . . . .

Gaussian elimination: *See* row reduction algorithm.

general solution (of a linear system):   A . . . description of a solution set that expresses . . . .

homogenous equation: An equation of the . . . .

identity matrix (denoted by I or I_n):   A square matrix . . . .

image (of a vector $\mathbf{x}$ under a transformation T):   The vector . . . . (Use symbols)

inconsistent linear system: A linear system with . . . .

leading entry: The . . . entry in a row of a matrix.

linear combination: A sum of . . . .

linear dependence relation: A . . . equation where . . . .

linear equation (in the variables $x_1, \ldots, x_n$):   An equation that can be written in the form . . . .

linearly dependent (vectors):   An indexed set $\{\mathbf{v}_1, \ldots, \mathbf{v}_p\}$ with the property that . . . .

linearly independent (vectors):   An indexed set $\{\mathbf{v}_1, \ldots, \mathbf{v}_p\}$ with the property . . . .

Copyright © 2021 Pearson Education, Inc.

linear system: A collection of one or more . . . equations involving . . . .

linear transformation: A transformation $T : \mathbb{R}^n \to \mathbb{R}^m$ is linear if (i) . . . , and (ii) . . . .

line through p parallel to v: The set . . . . (Use symbols)

matrix: A rectangular . . . .

matrix equation: An equation that . . . .

matrix transformation: A mapping $\mathbf{x} \mapsto$ . . . .

migration matrix: A matrix that gives the . . . movement between different locations, from . . . .

$m \times n$ matrix: A matrix with . . . .

nontrivial solution: A nonzero solution of . . . .

one-to-one (mapping):   A mapping $T : \mathbb{R}^n \to \mathbb{R}^m$ such that . . . .

onto (mapping):   A mapping $T : \mathbb{R}^n \to \mathbb{R}^m$ such that . . . .

overdetermined system: A system of equations with . . . .

parallelogram rule for addition: A geometric interpretation of . . . .

parametric equation of a line: An equation of the form . . . .

parametric equation of a plane: An equation of the form . . . .

pivot: A . . . number that either is used . . . or is . . . .

pivot column: A column that . . . .

pivot position: A position in a matrix A that corresponds . . . .

plane through u, v, and the origin: A set whose parametric equation is . . . .

product $A\mathbf{x}$:   . . . .

range (of a linear transformation T):   The set of . . . .

reduced echelon form (or **reduced row echelon form,** of a matrix) :   A rectangular matrix in echelon format that has these additional properties . . . .

roundoff error: Error in floating point arithmetic caused when . . . .

row-column rule for computing $A\mathbf{x}$: . . . .

row equivalent (matrices):   Two matrices for which there exists . . . .

row reduction algorithm: A systematic method using . . . .

row replacement: An elementary row operation that . . . .

scalar: . . . .

scalar multiple of u by c: The vector . . . .

set spanned by $\{\mathbf{v}_1, \ldots, \mathbf{v}_p\}$:   . . . .

size (of a matrix):   Two numbers . . . .

Copyright © 2021 Pearson Education, Inc.

solution (of a linear system):   . . . .

solution set: The set of . . . .

Span $\{v_1, \ldots, v_p\}$:   The set . . . .

standard matrix (for a linear transformation T):   The matrix . . . .

system of linear equations (or a linear system):   A collection of . . . .

transformation (or **function** or **mapping**) T **from** $\mathbb{R}^n$ *to* $\mathbb{R}^m$:   A rule that assigns to each vector **x** in $\mathbb{R}^n$ a . . . . Notation: $T : \mathbb{R}^n \rightarrow \mathbb{R}^m$.

translation (by a vector **p**):   The operation of . . . .

trivial solution: The solution . . . of a . . . .

underdetermined system: A system of equations with . . . .

uniqueness question: Asks, "If a solution of a system . . . ?"

vector: . . . .

vector equation: An equation involving . . . .

weights: . . . .

Copyright © 2021 Pearson Education, Inc.

2 Matrix Algebra

2.1 - Matrix Operations

Most of this chapter is an outgrowth of the idea in Section 1.7 that a matrix can transform data. This dynamic role of matrices suggests that we study the *combined effect* of several matrices on data (that is, on a vector or a set of vectors). Sections 2.1 to 2.5 describe this *matrix algebra*.

KEY IDEA

Matrix multiplication corresponds to composition of linear transformations. The definition of *AB*, using the columns of *B*, is critical for the development of both the theory and some of the applications in the text.

STUDY NOTES

Double-subscript notation: The subscripts tell the location of an entry in the matrix—the first subscript identifies the row and the second subscript the column. (Remember: *Row* is shorter than *column*, so *row* goes first.) This convention is opposite to the way a spreadsheet identifies the location of an entry.

In the product *AB*, left-multiplication (that is, multiplication on the left) by *A* acts on the columns of *B*, by definition, while right-multiplication by *B* acts on the rows of *A* (see the box after Example 4). That is,

$$\begin{bmatrix} \text{column } j \\ \text{of } AB \end{bmatrix} = A \begin{bmatrix} \text{column } j \\ \text{of } B \end{bmatrix} \quad \text{and} \quad [\text{row } i \text{ of } AB] = [\text{row } i \text{ of } A]B$$

To compute a specific matrix product by hand, use the Row-Column Rule. If *A* is $m \times n$, then the (i, j)-entry of *AB* is written with sigma notation as

$$(AB)_{ij} = \sum_{k=1}^{n} a_{ik} b_{kj}$$

Remember that if you change the *order* (position) of the factors in a matrix product, the new product may be different, or it may not even be defined. For instance, $(A + C)B$ and $AB + BC$ are probably *not* equal! Also, see the warning box after Example 7.

Copyright © 2021 Pearson Education, Inc.

Notes: Key exercises 13 and 25–30 emphasize the definition of a matrix product. Work at least five of these exercises, for practice.

SOLUTIONS TO EXERCISES

1. $-2A = (-2)\begin{bmatrix} 2 & 0 & -1 \\ 4 & -3 & 2 \end{bmatrix} = \begin{bmatrix} -4 & 0 & 2 \\ -8 & 6 & -4 \end{bmatrix}$. Next, use $B - 2A = B + (-2A)$:

$$B - 2A = \begin{bmatrix} 7 & -5 & 1 \\ 1 & -4 & -3 \end{bmatrix} + \begin{bmatrix} -4 & 0 & 2 \\ -8 & 6 & -4 \end{bmatrix} = \begin{bmatrix} 3 & -5 & 3 \\ -7 & 2 & -7 \end{bmatrix}$$

The product AC is not defined because the number of columns of A does not match the number of rows of C.

$$CD = \begin{bmatrix} 1 & 2 \\ -2 & 1 \end{bmatrix}\begin{bmatrix} 3 & 5 \\ -1 & 4 \end{bmatrix} = \begin{bmatrix} 1 \cdot 3 + 2(-1) & 1 \cdot 5 + 2 \cdot 4 \\ -2 \cdot 3 + 1(-1) & -2 \cdot 5 + 1 \cdot 4 \end{bmatrix} = \begin{bmatrix} 1 & 13 \\ -7 & -6 \end{bmatrix}.$$ For mental computation,

the row-column rule is easier to use than the definition.

7. Since A has 3 columns, B must match with 3 rows. Otherwise, AB is undefined. Since AB has 7 columns, so does B. Thus, B is 3×7.

13. If you had difficulty with this problem, read the definition of AB from *right* to *left*. Here is the definition, written in reverse order:

$$[A\mathbf{b}_1 \cdots A\mathbf{b}_p] = A[\mathbf{b}_1 \cdots \mathbf{b}_p] = AB, \text{ when } B = [\mathbf{b}_1 \cdots \mathbf{b}_p].$$

Thus $[Q\mathbf{r}_1 \cdots Q\mathbf{r}_p] = QR$, when $R = [\mathbf{r}_1 \cdots \mathbf{r}_p]$.

15. See the definition of AB.

17. See the box after Example 3.

19. Read Theorem 2(b) from right to left.

21. The left-to-right order of B and C cannot be changed, in general.

23. See the box after Theorem 3.

27. A solution is in the text. The main point is that the columns of AB are $A\mathbf{b}_1, \ldots, A\mathbf{b}_p$.

Checkpoint: Show that if $\mathbf{y}$ is a linear combination of the columns of AB, then $\mathbf{y}$ is a linear combination of the columns of A.

29. Let $\mathbf{b}_p$ be the last column of B. By hypothesis, the last column of AB is zero. Thus, $A\mathbf{b}_p = \mathbf{0}$. However, $\mathbf{b}_p$ is not the zero vector, because B has no column of zeros. Thus, the equation $A\mathbf{b}_p = \mathbf{0}$ is a linear dependence relation among the columns of A, and so the columns of A are linearly dependent.

Copyright © 2021 Pearson Education, Inc.

31. If $\mathbf{x}$ satisfies $A\mathbf{x} = \mathbf{0}$, then $CA\mathbf{x} = C\mathbf{0} = \mathbf{0}$ and so $I_n\mathbf{x} = \mathbf{0}$ and $\mathbf{x} = \mathbf{0}$. This shows that the equation $A\mathbf{x} = \mathbf{0}$ has no free variables. So every variable is a basic variable and every column of A is a pivot column. (A variation of this argument could be made using linear independence and Exercise 36 in Section 1.7.) Since each pivot is in a different row, A must have at least as many rows as columns.

33. By Exercise 31, the equation $CA = I_n$ implies that (number of rows in A) ≥ (number of columns), that is, $m \geq n$. By Exercise 32, the equation $AD = I_m$ implies that (number of rows in A) ≤ (number of columns), that is, $m \leq n$. Thus $m = n$. To prove the second statement, observe that $CAD = C(AD) = CI_m = C$, and also $CAD = (CA)D = I_nD = D$. Thus $C = D$. A shorter calculation is

$$C = CI_n = C(AD) = (CA)D = I_nD = D$$

Study Tip: In Exercises 35 and 36, *inner* products ($\mathbf{u}^T\mathbf{v}$ and $\mathbf{v}^T\mathbf{u}$) have the transpose symbol in the middle. *Outer* products ($\mathbf{uv}^T$ and $\mathbf{vu}^T$) have the transpose symbol on the outside.

37. The (i, j)-entry of $A(B + C)$ equals the (i, j)-entry of $AB + AC$, because

$$\sum_{k=1}^{n} a_{ik}(b_{kj} + c_{kj}) = \sum_{k=1}^{n} a_{ik}b_{kj} + \sum_{k=1}^{n} a_{ik}c_{kj}$$

The (i, j)-entry of $(B + C)A$ equals the (i, j)-entry of $BA + CA$, because

$$\sum_{k=1}^{n} (b_{ik} + c_{ik})a_{kj} = \sum_{k=1}^{n} b_{ik}a_{kj} + \sum_{k=1}^{n} c_{ik}a_{kj}$$

39. Use the definition of the product I_mA and the fact that $I_m\mathbf{x} = \mathbf{x}$ for $\mathbf{x}$ in $\mathbf{R}^m$.

$$I_mA = I_m[\mathbf{a}_1 \cdots \mathbf{a}_n] = [I_m\mathbf{a}_1 \cdots I_m\mathbf{a}_n] = [\mathbf{a}_1 \cdots \mathbf{a}_n] = A$$

41. The (i, j)-entry of $(AB)^T$ is the (j, i)-entry of AB, which is

$$a_{j1}b_{1i} + \cdots + a_{jn}b_{ni}$$

The entries in row i of B^T are $b_{1i}, \ldots, b_{ni}$, because they come from column i of B. Likewise, the entries in column j of A^T are $a_{j1}, \ldots, a_{jn}$, because they come from row j of A. Thus the (i, j)-entry in B^TA^T is $a_{j1}b_{1i} + \cdots + a_{jn}b_{ni}$, as above.

49. There are eight 4×1 vectors where each entry is either a zero or a one. Since the technique used in pattern recognition is looking for a nonzero vector, there are seven vectors to try.

$$[1 \quad 1 \quad 1 \quad 1]M\begin{bmatrix} 1 \\ 1 \\ 1 \\ 1 \end{bmatrix} = 2; \quad [0 \quad 1 \quad 1 \quad 1]M\begin{bmatrix} 0 \\ 1 \\ 1 \\ 1 \end{bmatrix} = 3; \cdots$$

Copyright © 2021 Pearson Education, Inc.

We are looking for the nonzero vector $\mathbf{w}$ of zeros and ones with the property that $\mathbf{w}^T M \mathbf{w} = 0$.

$$\mathbf{w} = \begin{bmatrix} 1 \\ 0 \\ 1 \\ 0 \end{bmatrix}$$

The correct vector is $\mathbf{w} = \begin{bmatrix} 1 \\ 0 \\ 1 \\ 0 \end{bmatrix}$, corresponding to the colored boxes   s   .

Answer to Checkpoint: If $\mathbf{y}$ is a linear combination of the columns of AB, then there is a vector $\mathbf{x}$ such that $\mathbf{y} = (AB)\mathbf{x}$. By definition of matrix multiplication, $\mathbf{y} = A(B\mathbf{x})$. This expresses $\mathbf{y}$ as a linear combination of the columns of A using the entries in the vector $B\mathbf{x}$ as weights.

MATLAB   Matrix Notation and Operations

To create a matrix, enter the data row-by-row, with a space between entries and a semicolon between rows. For instance, the command

        A=[1 2 3; 4 5 –6]   Use brackets around the data.

creates a 2×3 matrix A. If A is $m \times n$, then **size(A)** is the row vector $[m \quad n]$. The (i, j)-entry in A is **A(i, j)**. If i or j is replaced by a colon, the result is a column or row of A, respectively. Examples:

        A(:,3)   Column 3 of A
        A(2,:)   Row 2 of A

To specify columns 3, 4 and 5 of A, you can use

        A(:,[3 4 5])   or   **A(:,3:5)**

The symbols 3:5 (read "3 to 5") stand for the vector [3 4 5]. Similar notation works for selected rows of A.

    MATLAB uses +, –, and * to denote matrix addition, subtraction, and multiplication, respectively. If A is square and k is a positive integer, **A^k** denotes the kth power of A. The transpose of A is **A′** (with an apostrophe for the prime symbol). Note: when A has complex entries, the (i, j)-entry of **A′** is the complex conjugate of the (j, i)-entry of A.

    Use a single column (or row) matrix for a vector. If $\mathbf{u}$ and $\mathbf{v}$ are column vectors of the same size, then **u′*v** is their inner product, and **u*v′** is an outer product.

    MATLAB has commands that construct many special matrices. For example,

        M=zeros(5,6)   A 5×6 matrix of zeros
        M=ones(3,5)   A 3×5 matrix of ones
        M=eye(6)   The 6×6 identity matrix
        M=diag([3 5 7 2 4])   A 5×5 diagonal matrix
        M=rand(6)   A 6×6 matrix with random entries
        M=randomint(6,4)   A 6×4 matrix with random integer entries

Place **help** in front of any command to learn all the features of the command. The former name for **randomint** was **randint**, but MATLAB now uses **randint** for a slightly different

Copyright © 2021 Pearson Education, Inc.

command in its Communications Toolbox. You can easily find additional MATLAB or Octave commands on the internet using a search engine.

2.2 - The Inverse of a Matrix

Matrix inverses are essential for many discussions in linear algebra. This section and the next describe the main properties of invertible matrices.

STUDY NOTES

The inverse formula for a 2×2 matrix will be used frequently in exercises later in the text. (See Theorem 4.) To invert a 2×2 matrix, interchange the diagonal entries, reverse the signs of the off-diagonal entries, and divide each entry by the determinant (assuming $ad - bc \neq 0$).

Theorem 5 and its proof are important. The phrase "has a unique solution" includes the assertion that a solution exists, so the proof has two parts. The equation $AA^{-1} = I$ is used to prove that a solution exists, and the equation $A^{-1}A = I$ is used to show that the solution is unique.

Except when A is 2×2, Theorem 5 is practically never used to solve $A\mathbf{x} = \mathbf{b}$. Row reduction of $[A \quad \mathbf{b}]$ is faster. Actually, in practical work, you will seldom need to compute A^{-1}. (However, Example 3 illustrates a case in which the entries of A^{-1} could be useful.)

When using an inverse in matrix algebra, remember that matrix multiplication is not commutative. The phrase "left-multiply B by A^{-1}" means to multiply B on its left side by A^{-1}. *Never* write $\dfrac{B}{A}$ (or B/A) because it could stand for $A^{-1}B$ or BA^{-1}.

Elementary matrices are used in this text mainly to link row reduction to matrix multiplication. Each elementary row operation amounts to left-multiplication by an elementary matrix. So, if A can be row reduced to U, then there is a product F of elementary matrices such that $FA = U$.

Theorem 7 includes an *if and only if* statement, which was discussed in the Appendix to Section 1.2 in this *Study Guide*. The proof of this statement in Theorem 7 has two parts: (1) assume that A is invertible and prove that $A \sim I_n$; and (2) assume that $A \sim I_n$ and prove that A is invertible.

SOLUTIONS TO EXERCISES

1. $\begin{bmatrix} 8 & 3 \\ 5 & 2 \end{bmatrix}^{-1} = \dfrac{1}{16-15}\begin{bmatrix} 2 & -3 \\ -5 & 8 \end{bmatrix} = \begin{bmatrix} 2 & -3 \\ -5 & 8 \end{bmatrix}$

7. $\begin{bmatrix} 8 & 3 \\ 5 & 2 \end{bmatrix}^{-1}\begin{bmatrix} 2 \\ -1 \end{bmatrix} = \begin{bmatrix} 2 & -3 \\ -5 & 8 \end{bmatrix}\begin{bmatrix} 2 \\ -1 \end{bmatrix} = \begin{bmatrix} 7 \\ -18 \end{bmatrix}$

Copyright © 2021 Pearson Education, Inc.

9. a. $\begin{bmatrix} 1 & 2 \\ 5 & 12 \end{bmatrix}^{-1} = \dfrac{1}{1\cdot 12 - 2\cdot 5}\begin{bmatrix} 12 & -2 \\ -5 & 1 \end{bmatrix} = \dfrac{1}{2}\begin{bmatrix} 12 & -2 \\ -5 & 1 \end{bmatrix}$ or $\begin{bmatrix} 6 & -1 \\ -2.5 & .5 \end{bmatrix}$

$\mathbf{x} = A^{-1}\mathbf{b}_1 = \dfrac{1}{2}\begin{bmatrix} 12 & -2 \\ -5 & 1 \end{bmatrix}\begin{bmatrix} -1 \\ 3 \end{bmatrix} = \dfrac{1}{2}\begin{bmatrix} -18 \\ 8 \end{bmatrix} = \begin{bmatrix} -9 \\ 4 \end{bmatrix}$. Similar calculations give

$A^{-1}\mathbf{b}_2 = \begin{bmatrix} 11 \\ -5 \end{bmatrix}, A^{-1}\mathbf{b}_3 = \begin{bmatrix} 6 \\ -2 \end{bmatrix}, A^{-1}\mathbf{b}_4 = \begin{bmatrix} 13 \\ -5 \end{bmatrix}$.

b. $[A \quad \mathbf{b}_1 \quad \mathbf{b}_2 \quad \mathbf{b}_3 \quad \mathbf{b}_4] = \begin{bmatrix} 1 & 2 & -1 & 1 & 2 & 3 \\ 5 & 12 & 3 & -5 & 6 & 5 \end{bmatrix}$

$\sim \begin{bmatrix} 1 & 2 & -1 & 1 & 2 & 3 \\ 0 & 2 & 8 & -10 & -4 & -10 \end{bmatrix} \sim \begin{bmatrix} 1 & 2 & -1 & 1 & 2 & 3 \\ 0 & 1 & 4 & -5 & -2 & -5 \end{bmatrix}$

$\sim \begin{bmatrix} 1 & 0 & -9 & 11 & 6 & 13 \\ 0 & 1 & 4 & -5 & -2 & -5 \end{bmatrix}$

The solutions are $\begin{bmatrix} -9 \\ 4 \end{bmatrix}, \begin{bmatrix} 11 \\ -5 \end{bmatrix}, \begin{bmatrix} 6 \\ -2 \end{bmatrix}$, and $\begin{bmatrix} 13 \\ -5 \end{bmatrix}$, the same as in part (a).

Note: This exercise was designed to make the arithmetic simple for both methods, but (a) requires more arithmetic than (b). In fact, (a) requires 22 multiplications or divisions and 9 additions or subtractions. In general, the arithmetic for method (b) can be unpleasant for hand calculation. However, when A is larger than 2×2, method (b) is *much* faster than (a).

Study Tip: Notice in Exercise 9(a) how the 1/2 in the formula for A^{-1} was kept outside the matrix $\begin{bmatrix} 12 & -2 \\ -5 & 1 \end{bmatrix}$ when computing $A^{-1}\mathbf{b}$. This trick sometimes simplifies hand calculations (on exams!) by postponing the arithmetic with fractions (or decimals) until the end.

11. See the definition of *invertible.*

13. See Theorem 6(b).

15. See Theorem 4.

17. See Theorem 5.

19. See the box just before Example 6.

21. (See the proof of Theorem 5.) The $n \times p$ matrix B is given (but is arbitrary). Since A is invertible, the matrix $A^{-1}B$ satisfies $AX = B$, because $A(A^{-1}B) = AA^{-1}B = IB = B$. To show this solution is unique, let X be any solution of $AX = B$. Then, left-multiplication of each side by A^{-1} shows that X must be $A^{-1}B$:

$A^{-1}(AX) = A^{-1}B$, $\quad IX = A^{-1}B$, $\quad$ and $\quad X = A^{-1}B$.

Copyright © 2021 Pearson Education, Inc.

Study Tip: Whenever you are told "A is invertible," you know that A^{-1} exists, and you may use A^{-1} to solve an equation or to make appropriate calculations.

23. Left-multiply each side of the equation $AB = AC$ by A^{-1} to obtain

$$A^{-1}AB = A^{-1}AC, \quad IB = IC, \quad \text{and} \quad B = C.$$

This conclusion does not always follow when A is singular. The matrices in Exercise 10 of Section 2.1 provide a counterexample.

Warning: A common mistake in Exercise 26 is to try to use the formula $(AB)^{-1} = B^{-1}A^{-1}$. But, this formula can be used only when you know, in advance, that both A and B are invertible. In Exercise 26, you must *prove* that A is invertible.

29. Unlike Exercise 27, this exercise asks two things, "Does a solution exist?" and "What is the solution?" First, find what the solution must be, if it exists. That is, suppose X satisfies the equation $C^{-1}(A + X)B^{-1} = I$. Left-multiply each side by C, and then right-multiply each side by B:

$$CC^{-1}(A + X)B^{-1} = CI, \quad I(A + X)B^{-1} = C, \quad (A + X)B^{-1}B = CB, \quad (A + X)I = CB$$

Expand the left side and then subtract A from both sides:

$$AI + XI = CB, \quad A + X = CB, \quad X = CB - A$$

If a solution exists, it must be $CB - A$. To *show* that $CB - A$ really *is* a solution, substitute it for X:

$$C^{-1}[A + (CB - A)]B^{-1} = C^{-1}[CB]B^{-1} = C^{-1}CBB^{-1} = II = I.$$

After this section, your instructor may permit you to include fewer details in your calculations. (Check on this.) For instance, after some practice with algebra, an expression such as $CC^{-1}(A + X)B^{-1}$ could be simplified directly to $(A + X)B^{-1}$ without first replacing CC^{-1} by I.

31. Suppose A is invertible. By Theorem 5, the equation $A\mathbf{x} = \mathbf{0}$ has only one solution, namely, the zero solution. This means that the columns of A are linearly independent, by a remark in Section 1.7.

33. Suppose A is $n \times n$ and the equation $A\mathbf{x} = \mathbf{0}$ has only the trivial solution. Then there are no free variables in this equation, and so A has n pivot columns. Since A is *square* and the n pivot positions must be in different rows, the pivots in an echelon form of A must be on the main diagonal. Hence A is row equivalent to the $n \times n$ identity matrix.

35. Suppose $A = \begin{bmatrix} a & b \\ c & d \end{bmatrix}$ and $ad - bc = 0$. If $a = b = 0$, then examine $\begin{bmatrix} 0 & 0 \\ c & d \end{bmatrix}\begin{bmatrix} x_1 \\ x_2 \end{bmatrix} = \begin{bmatrix} 0 \\ 0 \end{bmatrix}$. This

has the solution $\mathbf{x} = \begin{bmatrix} d \\ -c \end{bmatrix}$. This solution is nonzero, except when $c = d = 0$. In that case, however, A is the zero matrix, and $A\mathbf{x} = \mathbf{0}$ for *every* vector $\mathbf{x}$. Finally, if a and b are not both

Copyright © 2021 Pearson Education, Inc.

zero, set $\mathbf{u} = \begin{bmatrix} -b \\ a \end{bmatrix}$. Then $A\mathbf{u} = \begin{bmatrix} a & b \\ c & d \end{bmatrix} \begin{bmatrix} -b \\ a \end{bmatrix} = \begin{bmatrix} -ab + ba \\ -cb + da \end{bmatrix} = \begin{bmatrix} 0 \\ 0 \end{bmatrix}$, because $-cb + da = 0$.

Thus, $\mathbf{u}$ is a nontrivial solution of $A\mathbf{x} = \mathbf{0}$. So, in all cases, the equation $A\mathbf{x} = \mathbf{0}$ has more than one solution. This is impossible when A is invertible (by Theorem 5), so A is *not* invertible.

37. **a.** Interchange A and B in equation (2) after Example 6 of Section 2.1:

$\text{row}_i(BA) = \text{row}_i(B) \cdot A$. Then replace B by the identity matrix:

$\text{row}_i(A) = \text{row}_i(IA) = \text{row}_i(I) \cdot A$.

b. Using part (a), when rows 1 and 2 of A are interchanged, write the result as

$$\begin{bmatrix} \text{row}_2(A) \\ \text{row}_1(A) \\ \text{row}_3(A) \end{bmatrix} = \begin{bmatrix} \text{row}_2(I) \cdot A \\ \text{row}_1(I) \cdot A \\ \text{row}_3(I) \cdot A \end{bmatrix} = \begin{bmatrix} \text{row}_2(I) \\ \text{row}_1(I) \\ \text{row}_3(I) \end{bmatrix} A = EA \qquad (*)$$

Here, E is obtained by interchanging rows 1 and 2 of I. The second equality in (*) is a consequence of the fact that $\text{row}_i(EA) = \text{row}_i(E) \cdot A$.

c. Using part (a), when row 3 of A is multiplied by 5, write the result as

$$\begin{bmatrix} \text{row}_1(A) \\ \text{row}_2(A) \\ 5 \cdot \text{row}_3(A) \end{bmatrix} = \begin{bmatrix} \text{row}_1(I) \cdot A \\ \text{row}_2(I) \cdot A \\ 5 \cdot \text{row}_3(I) \cdot A \end{bmatrix} = \begin{bmatrix} \text{row}_1(I) \\ \text{row}_2(I) \\ 5 \cdot \text{row}_3(I) \end{bmatrix} A = EA$$

Here, E is obtained by multiplying row 3 of I by 5.

41. $\begin{bmatrix} A & I \end{bmatrix} = \begin{bmatrix} 1 & 0 & -2 & 1 & 0 & 0 \\ -3 & 1 & 4 & 0 & 1 & 0 \\ 2 & -3 & 4 & 0 & 0 & 1 \end{bmatrix} \sim \begin{bmatrix} 1 & 0 & -2 & 1 & 0 & 0 \\ 0 & 1 & -2 & 3 & 1 & 0 \\ 0 & -3 & 8 & -2 & 0 & 1 \end{bmatrix}$

$\sim \begin{bmatrix} 1 & 0 & -2 & 1 & 0 & 0 \\ 0 & 1 & -2 & 3 & 1 & 0 \\ 0 & 0 & 2 & 7 & 3 & 1 \end{bmatrix} \sim \begin{bmatrix} 1 & 0 & 0 & 8 & 3 & 1 \\ 0 & 1 & 0 & 10 & 4 & 1 \\ 0 & 0 & 2 & 7 & 3 & 1 \end{bmatrix}$

$\sim \begin{bmatrix} 1 & 0 & 0 & 8 & 3 & 1 \\ 0 & 1 & 0 & 10 & 4 & 1 \\ 0 & 0 & 1 & 7/2 & 3/2 & 1/2 \end{bmatrix}$. $A^{-1} = \begin{bmatrix} 8 & 3 & 1 \\ 10 & 4 & 1 \\ 7/2 & 3/2 & 1/2 \end{bmatrix}$

Copyright © 2021 Pearson Education, Inc.

43. Let $B = \begin{bmatrix} 1 & 0 & 0 & \cdots & 0 \\ -1 & 1 & 0 & & 0 \\ 0 & -1 & 1 & & \\ \vdots & & \ddots & \ddots & \vdots \\ 0 & 0 & \cdots & -1 & 1 \end{bmatrix}$, and for $j = 1, \ldots, n$, let $\mathbf{a}_j$, $\mathbf{b}_j$, and $\mathbf{e}_j$ denote the jth columns

of A, B, and I, respectively. Note that for $j = 1, \ldots, n - 1$, $\mathbf{a}_j - \mathbf{a}_{j+1} = \mathbf{e}_j$ (because $\mathbf{a}_j$ and $\mathbf{a}_{j+1}$ have the same entries except for the jth row), $\mathbf{b}_j = \mathbf{e}_j - \mathbf{e}_{j+1}$ and $\mathbf{a}_n = \mathbf{b}_n = \mathbf{e}_n$.

To show that $AB = I$, it suffices to show that $A\mathbf{b}_j = \mathbf{e}_j$ for each j. For $j = 1, \ldots, n - 1$,

$$A\mathbf{b}_j = A(\mathbf{e}_j - \mathbf{e}_{j+1}) = A\mathbf{e}_j - A\mathbf{e}_{j+1} = \mathbf{a}_j - \mathbf{a}_{j+1} = \mathbf{e}_j$$

and $A\mathbf{b}_n = A\mathbf{e}_n = \mathbf{a}_n = \mathbf{e}_n$.

47. There are many possibilities for C, but $C = \begin{bmatrix} 1 & 1 & -1 \\ -1 & 1 & 0 \end{bmatrix}$ is the only one whose entries are

$1, -1$, and 0. With only three possibilities for each entry, the construction of C can be done by trial and error. This is probably faster than setting up a system of 4 equations in 6 unknowns. The fact that A cannot be invertible follows from Exercise 33 in Section 2.1, because A is not square.

MATLAB   Constructing A^{-1}

If A is a 5×5 matrix, then the command **M = [A eye(5)]** creates the augmented matrix $[A \; I]$. Use **rref** to reduce $[A \; I]$.

　　MATLAB has other commands that row reduce matrices, invert matrices, and solve equations $A\mathbf{x} = \mathbf{b}$. They will be introduced later, after you have studied the concepts and algorithms in this section.

2.3 - Characterization of Invertible Matrices

In many linear algebra texts, the equivalent of Chapter 4 is extremely difficult for students. But you won't have problems if you prepare well now, because you are already learning basic ideas that will be presented again in Chapter 4. Review the major concepts from the previous sections, and plan for more study time here than you ordinarily spend on one section.

KEY IDEAS

The Invertible Matrix Theorem (IMT) only applies to square matrices. However, some groups of these statements in the IMT are also equivalent for rectangular matrices. The following table will help you remember other important theorems as well as the IMT. (See Theorem 4 in Section 1.4,

Copyright © 2021 Pearson Education, Inc.

Theorems 11 and 12 in Section 1.9, Theorem 5 in Section 2.2, and Theorem 9 in Section 2.3.) All of the statements in the table are equivalent when A is square ($m = n = p$).

STATEMENTS FROM THE INVERTIBLE MATRIX THEOREM

Equivalent statements for an $m \times n$ matrix A.	Equivalent statements for an $n \times n$ square matrix A.	Equivalent statements for an $n \times p$ matrix A.
k. There is a matrix D such that $AD = I$. *. A has a pivot position in every row. h. The columns of A span $\mathbb{R}^m$. g. The equation $A\mathbf{x} = \mathbf{b}$ has at least one solution for each $\mathbf{b}$ in $\mathbb{R}^m$. i. The transformation $\mathbf{x} \mapsto A\mathbf{x}$ maps $\mathbb{R}^n$ onto $\mathbb{R}^m$.	a. A is an invertible matrix. c. A has n pivot positions. b. A is row equivalent to the identity matrix. *. The equation $A\mathbf{x} = \mathbf{b}$ has a unique solution for each $\mathbf{b}$ in $\mathbb{R}^n$. *. The transformation $\mathbf{x} \mapsto A\mathbf{x}$ is invertible. l. A^T is invertible.	j. There is a matrix C such that $CA = I$. *. A has a pivot position in every column. e. The columns of A are linearly independent. d. The equation $A\mathbf{x} = \mathbf{0}$ has only the trivial solution. f. The transformation $\mathbf{x} \mapsto A\mathbf{x}$ is one-to-one.

The four statements denoted by (*) were not listed in the text as part of the IMT, mainly to avoid intimidating you with so many statements in one theorem. Note: the text did not actually prove that for a *rectangular* matrix, statements (j) and (k) are each equivalent to the other statements in their respective columns. (Exercises 31, 32, and 34 in Section 2.1 contain most of the facts needed to prove this.) A matrix C such that $CA = I$ is called a **left-inverse** of A, and a matrix D such that $AD = I$ is called a **right-inverse** of A.

Checkpoint: What can you say about the statements in the first column when A has more rows than columns? (Why?) What about the statements in the third column when A has more columns than rows? (Why?)

SOLUTIONS TO EXERCISES

1. The columns of the matrix $\begin{bmatrix} 5 & 7 \\ -3 & -6 \end{bmatrix}$ are not multiples, so they are linearly independent. By (e) in the IMT, the matrix is invertible. Also, the matrix is invertible by Theorem 4 in Section 2.2 because the determinant is nonzero.

Copyright © 2021 Pearson Education, Inc.

7. $\begin{bmatrix} -1 & -3 & 0 & 1 \\ 3 & 5 & 8 & -3 \\ -2 & -6 & 3 & 2 \\ 0 & -1 & 2 & 1 \end{bmatrix} \sim \begin{bmatrix} -1 & -3 & 0 & 1 \\ 0 & -4 & 8 & 0 \\ 0 & 0 & 3 & 0 \\ 0 & -1 & 2 & 1 \end{bmatrix} \sim \begin{bmatrix} -1 & -3 & 0 & 1 \\ 0 & -4 & 8 & 0 \\ 0 & 0 & 3 & 0 \\ 0 & 0 & 0 & 1 \end{bmatrix}$

The 4×4 matrix has four pivot positions and so is invertible by (c) of the IMT.

Study the Invertible Matrix Theorem. The statements there are true *only for an invertible matrix*. Also, if one of the statements is true about a *square* matrix A, then all statements in the theorem are true; if one of the statements is false, then all are false.

11. See statements (d) and (b) of the IMT.

13. See statements (h) and (e).

15. See statement (g).

17. See statements (d) and (c).

19. See statement (1).

Study Tip: Learn how to recognize when a square matrix is *not* invertible. If A is an $n \times n$ matrix, then each of the following statements is true if and only if A is **not** invertible.

- The matrix A has *fewer* than n pivot positions.
- The equation $A\mathbf{x} = \mathbf{0}$ has a *nontrivial* (nonzero) solution.
- The columns of A are linearly *dependent*.
- The linear transformation $\mathbf{x} \mapsto A\mathbf{x}$ is *not* one-to-one.
- The equation $A\mathbf{x} = \mathbf{b}$ has *no* solution (is *inconsistent*) for *some* $\mathbf{b}$ in $\mathbb{R}^n$.
- The equation $A\mathbf{x} = \mathbf{b}$ has *more than one* solution for some $\mathbf{b}$ in $\mathbb{R}^n$.
- The columns of A *do not* span $\mathbb{R}^n$.
- The linear transformation $\mathbf{x} \mapsto A\mathbf{x}$ *does not* map $\mathbb{R}^n$ onto $\mathbb{R}^n$.

21. If a square upper triangular $n \times n$ matrix has nonzero diagonal entries, then because it is already in echelon form, the matrix is row equivalent to I_n and hence is invertible, by the IMT. Conversely, if the matrix is invertible, it has n pivots on the diagonal and hence the diagonal entries are nonzero.

Study Tip: If you check your answer for odd exercises between 21 and 41, be careful not to read any other answers or hints. You *must* try to write your own solutions first. **27.**   By (e) of the IMT, D is invertible. Thus the equation $D\mathbf{x} = \mathbf{b}$ has a solution for each $\mathbf{b}$ in $\mathbb{R}^7$, by (g) of the IMT. Even better, the equation $D\mathbf{x} = \mathbf{b}$ has a *unique* solution for each $\mathbf{b}$ in $\mathbb{R}^7$, by Theorem 5 in Section 2.2. (See the paragraph following the proof of the IMT.)

33. Suppose that A is square and $AB = I$. Then A is invertible, by the (k) of the IMT. Left-multiplying each side of the equation $AB = I$ by A^{-1}, one has
$$A^{-1}AB = A^{-1}I, \quad IB = A^{-1}, \quad \text{and} \quad B = A^{-1}.$$

Copyright © 2021 Pearson Education, Inc.

By Theorem 6 in Section 2.2, the matrix B (which is A^{-1}) is invertible, and its inverse is $(A^{-1})^{-1} = A$. Note: Exercise 33 makes a good test question.

35. Let W be the inverse of AB. Then $ABW = I$ and $A(BW) = I$. This equation, *by itself*, does not prove that A is invertible. However, since A is *square*, the IMT does apply and by statement (k), A is invertible.

Of course, in this exercise set there is an overall assumption that matrices in this section are square unless otherwise stated. So, with that given, you do not really have to mention here that A is square. However, I put that question "Why not?" in the answer to make you think about this. Look back at Exercise 48 in Section 2.2. There, $AD = I$, which certainly makes AD invertible, yet A is not invertible.

39. Since the equation $A\mathbf{x} = \mathbf{b}$ has a solution for each $\mathbf{b}$, the matrix A has a pivot in each row (Theorem 4 in Section 1.4). Since A is square, A has a pivot in each column, and so there are no free variables in the equation $A\mathbf{x} = \mathbf{b}$, which shows that the solution is unique.

The preceding argument shows that the (square) shape of A plays a crucial role. A less revealing proof is to use the "pivot in each row" and the IMT to conclude that A is invertible. Then Theorem 5 in Section 2.2 shows that the solution of $A\mathbf{x} = \mathbf{b}$ is unique.

41. The standard matrix of T is $A = \begin{bmatrix} -5 & 9 \\ 4 & -7 \end{bmatrix}$, which is invertible because $\det A \neq 0$. By Theorem 9, the transformation T is invertible and the standard matrix of T^{-1} is A^{-1}. From the formula for a 2×2 inverse, $A^{-1} = \begin{bmatrix} 7 & 9 \\ 4 & 5 \end{bmatrix}$. So

$$T^{-1}(x_1, x_2) = \begin{bmatrix} 7 & 9 \\ 4 & 5 \end{bmatrix} \begin{bmatrix} x_1 \\ x_2 \end{bmatrix} = (7x_1 + 9x_2, 4x_1 + 5x_2)$$

43. To show that T is one-to-one, suppose that $T(\mathbf{u}) = T(\mathbf{v})$ for some vectors $\mathbf{u}$ and $\mathbf{v}$ in $\mathbb{R}^n$. Then $S(T(\mathbf{u})) = S(T(\mathbf{v}))$, where S is the inverse of T. By Equation (1), $\mathbf{u} = S(T(\mathbf{u}))$ and $S(T(\mathbf{v})) = \mathbf{v}$, so $\mathbf{u} = \mathbf{v}$. Thus T is one-to-one. To show that T is onto, suppose $\mathbf{y}$ represents an arbitrary vector in $\mathbb{R}^n$ and define $\mathbf{x} = S(\mathbf{y})$. Then, using Equation (2), $T(\mathbf{x}) = T(S(\mathbf{y})) = \mathbf{y}$, which shows that T maps $\mathbb{R}^n$ onto $\mathbb{R}^n$.

Second proof: By Theorem 9, the standard matrix A of T is invertible. By the IMT, the columns of A are linearly independent and span $\mathbb{R}^n$. By Theorem 12 in Section 1.9, T is one-to-one and maps $\mathbb{R}^n$ onto $\mathbb{R}^n$.

45. Let A and B be the standard matrices of T and U, respectively. Then AB is the standard maix of the mapping $\mathbf{x} \mapsto T(U(\mathbf{x}))$, because of the way matrix multiplication is defined (in Section 2.1). By hypothesis, this mapping is the identity mapping, so $AB = I$. Since A and B are square, they are invertible, by the IMT, and $B = A^{-1}$. Thus, $BA = I$. This means that the mapping $\mathbf{x} \mapsto U(T(\mathbf{x}))$ is the identity mapping, i.e., $U(T(\mathbf{x})) = \mathbf{x}$ for all $\mathbf{x}$ in $\mathbb{R}^n$.

Answers to Checkpoint: If A has more rows than columns, then all statements in the first column of the table must be false, because they are equivalent and the statement about a pivot

Copyright © 2021 Pearson Education, Inc.

position in each row cannot be true. If A has more columns than rows, then all statements in the third column of the table must be false, because A cannot have a pivot in each of its columns.

Mastering Linear Algebra: Reviewing and Reflecting

Two important steps to mastery of linear algebra are periodic review of earlier material and reflection on its relation to new material. When you reread the basic conceptual material from Chapter 1, you may be surprised to discover new insights that you missed earlier. Your broader experience now should give you a better framework within which to understand concepts such as spanning and linear independence.

Compare the review you conducted in Section 1.9 (see the *Study Guide* appendix to that section) with the three-part table at the beginning of this *Study Guide* section. (You did carry out that review, didn't you?) The left and right columns of the table should match some of your "existence" and "uniqueness" statements, respectively.

If your review in Section 1.9 was thorough, you probably anticipated some of the content of the Invertible Matrix Theorem. Existence and uniqueness threads run through the fabric of linear algebra, and they intertwine when related to square matrices (the middle column of the table). A good review procedure now is to expand the table to include references to theorems, examples, and counterexamples. This will occupy several pages. The process of constructing this table is what will help you most.

MATLAB   inv, cond, and hilb

Determining whether a matrix is invertible is not always a simple matter. A fast and fairly reliable method is to use the command **inv(A)**, which computes the inverse of A. A warning is given if the matrix is singular (noninvertible) or nearly singular.

For Exercises 49–53, the command **cond(A)** computes the condition number of a matrix A, using what are called the singular values of A (discussed in Section 7.4.) To perform the experiment described in Exercise 50, you can use the following MATLAB instructions

   x=rand(4,1);   b=A*x;   x1=inv(A)*b;   x–x1

Use **format long**. Displaying the value of **x–x1** is the best way to compare **x** and **x1**. Press the up-arrow key (↑) to repeat this instruction line.

For Exercise 53, the commands **format rat; hilb(n)** produce the $n \times n$ Hilbert matrix, with its entries displayed as rational numbers. Enter **format short** to return to the standard display of numbers as decimals.

2.4 - Partitioned Matrices

The ideas in this section are fairly simple. However, mark them for future reference, because you are likely to use this notation after you leave school. Partitioned matrices arise in theoretical discussions in essentially every field that makes use of matrices. Here are two examples.

Copyright © 2021 Pearson Education, Inc.

1. The modern *state space* approach to control systems engineering depends on matrix calculations.[1] The problem of determining whether a system is *controllable* amounts to calculating the number of pivot positions in a *controllability matrix*

$$[B \quad AB \quad A^2B \quad \cdots \quad A^{n-1}B]$$

   where A is $n \times n$, B has n rows, and the matrices come from an equation of the form (8) in the discussion preceding Exercise 21.

2. Discussions of modern algorithms and computer software design for scientific computing naturally use the "language" of partitioned matrices. For instance, common techniques for parallel processing of large matrix calculations, such as *slicing* and *crinkling*, are described with partitioned matrices.[2] Also, the standard computer science reference on matrix - calculations relies heavily on partitioned matrices.[3]

KEY IDEAS

The column-row evaluation of AB is the last of five different "views" of matrix multiplication. All five are special cases of the *block matrix* version of the *row-column* rule for matrix multiplication. Here they are:

(1) The definition of $A\mathbf{x}$ amounts to block multiplication of AB where B has only one column:

$$A\mathbf{x} = [\mathbf{a}_1 \mid \cdots \mid \mathbf{a}_n] \begin{bmatrix} x_1 \\ \vdots \\ x_n \end{bmatrix} = [x_1\mathbf{a}_1 + \cdots + x_n\mathbf{a}_n]$$

(2) Partition A as *one* row and *one* column. Then the definition of the usual product AB is a row-column block product:

$$AB = A[\mathbf{b}_1 \mid \mathbf{b}_2 \mid \cdots \mid \mathbf{b}_p] = [A\mathbf{b}_1 \mid A\mathbf{b}_2 \mid \cdots \mid A\mathbf{b}_p]$$

(3) Likewise, we observed in Section 2.1 that if B is partitioned as one row and one column, then

$$AB = \begin{bmatrix} \text{row}_1(A) \\ \text{row}_2(A) \\ \vdots \\ \text{row}_m(A) \end{bmatrix} B = \begin{bmatrix} \text{row}_1(A)B \\ \text{row}_2(A)B \\ \vdots \\ \text{row}_m(A)B \end{bmatrix}$$

---

[1] An understanding of control systems is important in the design of filtering circuitry, robots, process control systems, and spacecraft. Thus a control systems course is often part of the undergraduate curriculum for electrical, mechanical, chemical, and aerospace engineering. See *Control Systems Engineering*, 3rd ed., by Norman S. Nise, John Wiley & Sons, New York, 2000.

[2] *Parallel Algorithms and Matrix Computations*, by Jagdish J. Modi, Oxford Applied Mathematics and Computing Science Series, Clarendon Press, Oxford, 1988, pp. 73–75.

[3] *Matrix Computations*, 3rd ed., by Gene H. Golub and Charles F. Van Loan, The Johns Hopkins Press, Baltimore, 1996.

Copyright © 2021 Pearson Education, Inc.

(4) The next display can be viewed either as just the standard row-column rule in which each entry of AB is computed as the product of a row of A and a column of B, or as a multiplication of block matrices (with A having only one column of blocks and B having only one row of blocks):

$$AB = \begin{bmatrix} \text{row}_1(A) \\ \text{row}_2(A) \\ \vdots \\ \text{row}_m(A) \end{bmatrix} [\text{col}_1(B) \quad \text{col}_2(B) \quad \cdots \quad \text{col}_p(B)]$$

$$= \begin{bmatrix} \text{row}_1(A)\text{col}_1(B) & \cdots & \text{row}_1(A)\text{col}_j(B) & \cdots & \text{row}_1(A)\text{col}_p(B) \\ \vdots & & \vdots & & \vdots \\ \text{row}_i(A)\text{col}_1(B) & \cdots & \text{row}_i(A)\text{col}_j(B) & \cdots & \text{row}_i(A)\text{col}_p(B) \\ \vdots & & \vdots & & \vdots \\ \text{row}_m(A)\text{col}_1(B) & \cdots & \text{row}_m(A)\text{col}_j(B) & \cdots & \text{row}_m(A)\text{col}_p(B) \end{bmatrix}$$

(5) The final display is the column-row expansion of AB (Theorem 10 in this section). In this view, AB is expressed as a sum of *outer products* of the form $\mathbf{u}\mathbf{v}^T$, with $\mathbf{u}$ a column of A and $\mathbf{v}^T$ a row of B. But the display can also be viewed as the block version of the row-column product in which A has one row (of blocks) and B has one column (of blocks):

$$AB = [\text{col}_1(A) \quad \text{col}_2(A) \quad \cdots \quad \text{col}_n(A)] \begin{bmatrix} \text{row}_1(B) \\ \text{row}_2(B) \\ \vdots \\ \text{row}_n(B) \end{bmatrix}$$

$$= \text{col}_1(A) \cdot \text{row}_1(B) + \cdots + \text{col}_n(A) \cdot \text{row}_n(B)$$

You might say that the row-column rule computes AB as an array of inner products (view 4 above), while the column-row expansion displays AB as a sum of arrays (view 5).

SOLUTIONS TO EXERCISES

1. Apply the row-column rule as if the matrix entries were numbers, but for each product (such as EA below), always write the entry of the left block-matrix on the *left*.

$$\begin{bmatrix} I & 0 \\ E & I \end{bmatrix}\begin{bmatrix} A & B \\ C & D \end{bmatrix} = \begin{bmatrix} IA+0C & IB+0D \\ EA+IC & EB+ID \end{bmatrix} = \begin{bmatrix} A & B \\ EA+C & EB+D \end{bmatrix}$$

This must be EA, not AE.

Copyright © 2021 Pearson Education, Inc.

Checkpoint: Notice in Exercises 1 and 3 that $\begin{bmatrix} I & 0 \\ E & I \end{bmatrix}$ and $\begin{bmatrix} 0 & I \\ I & 0 \end{bmatrix}$ act as block-matrix generalizations of elementary matrices. What sort of 2×2 block matrix is the appropriate generalization of an elementary matrix that acts as a scaling operation? (Answer this carefully.)

7. Compute the left side of the equation:

$$\begin{bmatrix} X & 0 & 0 \\ Y & 0 & I \end{bmatrix} \begin{bmatrix} A & Z \\ 0 & 0 \\ B & I \end{bmatrix} = \begin{bmatrix} XA+0+0B & XZ+0+0 \\ YA+0+IB & YZ+0+I \end{bmatrix}$$

Set this equal to the right side of the equation:

$$\begin{bmatrix} XA & XZ \\ YA+B & YZ+I \end{bmatrix} = \begin{bmatrix} I & 0 \\ 0 & I \end{bmatrix} \text{ so that } \begin{array}{ll} XA = I & XZ = 0 \\ YA+B=0 & YZ+I=I \end{array}$$

Since the (1, 1)-blocks are equal, $XA = I$. Since X and A are square, the IMT implies that A and X are invertible, and hence $X = A^{-1}$. From the (1, 2)-entries, $XZ = 0$. Since X is invertible, Z must be 0. Therefore, the (2, 2)-entries give no new information. Finally, from the (2, 1)-entries, $YA + B = 0$ and $YA = -B$. Right-multiplication by A^{-1} shows that $Y = -BA^{-1}$. The order of the factors for Y is crucial.

Study Tip: Problems such as 5–10 make good exam questions. Remember to mention the IMT when appropriate, and remember that matrix multiplication is generally not commutative.

11. See the subsection Addition and Scalar Multiplication.

13. See the paragraph before Example 3.

15. You are asked to establish an "if and only if" statement. First, suppose that A is invertible, and let $A^{-1} = \begin{bmatrix} D & E \\ F & G \end{bmatrix}$. Then

$$\begin{bmatrix} B & 0 \\ 0 & C \end{bmatrix} \begin{bmatrix} D & E \\ F & G \end{bmatrix} = \begin{bmatrix} BD & BE \\ CF & CG \end{bmatrix} = \begin{bmatrix} I & 0 \\ 0 & I \end{bmatrix}$$

Since B is square, the equation $BD = I$ implies that B is invertible, by the IMT. Similarly, $CG = I$ implies that C is invertible. Also, the equation $BE = 0$ implies that $E = B^{-1}0 = 0$. Similarly $F = 0$. Thus

$$A^{-1} = \begin{bmatrix} B & 0 \\ 0 & C \end{bmatrix}^{-1} = \begin{bmatrix} D & E \\ E & G \end{bmatrix} = \begin{bmatrix} B^{-1} & 0 \\ 0 & C^{-1} \end{bmatrix} \tag{*}$$

This proves that A is invertible *only if* B and C are invertible. For the "*if*" part of the statement, suppose that B and C are invertible. Then (*) provides a likely candidate for A^{-1} which can be used to show that A is invertible. Compute:

Copyright © 2021 Pearson Education, Inc.

$$\begin{bmatrix} B & 0 \\ 0 & C \end{bmatrix} \begin{bmatrix} B^{-1} & 0 \\ 0 & C^{-1} \end{bmatrix} = \begin{bmatrix} BB^{-1} & 0 \\ 0 & CC^{-1} \end{bmatrix} = \begin{bmatrix} I & 0 \\ 0 & I \end{bmatrix}$$

Since A is square, this calculation and the IMT imply that A is invertible. (Don't forget this final sentence. Without it, the argument is incomplete.) Instead of that sentence, you could add the equation:

$$\begin{bmatrix} B^{-1} & 0 \\ 0 & C^{-1} \end{bmatrix} \begin{bmatrix} B & 0 \\ 0 & C \end{bmatrix} = \begin{bmatrix} B^{-1}B & 0 \\ 0 & C^{-1}C \end{bmatrix} = \begin{bmatrix} I & 0 \\ 0 & I \end{bmatrix}$$

21. The matrix equation (8) in the text is equivalent to

$$(A - sI_n)\mathbf{x} + B\mathbf{u} = \mathbf{0} \quad \text{and} \quad C\mathbf{x} + \mathbf{u} = \mathbf{y}$$

Rewrite the first equation as $(A - sI_n)\mathbf{x} = -B\mathbf{u}$. When $A - sI_n$ is invertible,

$$\mathbf{x} = (A - sI_n)^{-1}(-B\mathbf{u}) = -(A - sI_n)^{-1}B\mathbf{u}$$

Substitute this formula for $\mathbf{x}$ into the second equation above:

$$C(-(A - sI_n)^{-1}B\mathbf{u}) + \mathbf{u} = \mathbf{y}, \quad \text{so that} \quad I_m\mathbf{u} - C(A - sI_n)^{-1}B\mathbf{u} = \mathbf{y}$$

Thus $\mathbf{y} = (I_m - C(A - sI_n)^{-1}B)\mathbf{u}$. If $W(s) = I_m - C(A - sI_n)^{-1}B$, then $\mathbf{y} = W(s)\mathbf{u}$. The matrix $W(s)$ is the Schur complement of the matrix $A - sI_n$ in the system matrix in equation (8).

25. To prove a statement by induction, a good first step is to write the statement that depends on n but exclude the phrase "for all n," and label the statement for reference:

The product of two $n \times n$ lower triangular matrices is lower triangular. (*)

Second, verify that the statement is true for $n = 1$. In this particular case, (*) is obviously true, because every 1×1 matrix is lower triangular. The "induction step" is next.

Suppose that (*) is true when n is some positive integer k, and consider any $(k+1) \times (k+1)$ lower-triangular matrices A_1 and B_1. Partition these matrices as

$$A_1 = \begin{bmatrix} a & \mathbf{0}^T \\ \mathbf{v} & A \end{bmatrix}, \quad B_1 = \begin{bmatrix} b & \mathbf{0}^T \\ \mathbf{w} & B \end{bmatrix}$$

where A and B are $k \times k$ matrices, $\mathbf{v}$ and $\mathbf{w}$ are in $\mathbb{R}^k$, and a and b are scalars. Since A_1 and B_1 are lower triangular, so are A and B. Now

$$A_1 B_1 = \begin{bmatrix} a & \mathbf{0}^T \\ \mathbf{v} & A \end{bmatrix} \begin{bmatrix} b & \mathbf{0}^T \\ \mathbf{w} & B \end{bmatrix} = \begin{bmatrix} ab + \mathbf{0}^T\mathbf{w} & a\mathbf{0}^T + \mathbf{0}^T B \\ \mathbf{v}b + A\mathbf{w} & \mathbf{v}\mathbf{0}^T + AB \end{bmatrix} = \begin{bmatrix} ab & \mathbf{0}^T \\ b\mathbf{v} + A\mathbf{w} & AB \end{bmatrix}$$

Assuming (*) is true for $n = k$, AB must be lower triangular. The form of A_1B_1 shows that it, too, is lower triangular. Thus the statement (*) about lower triangular matrices is true for $n = k + 1$ if it is true for $n = k$. By the principle of induction, (*) is true for all $n \geq 1$.

27. First, visualize a partition of A as a 2×2 block-diagonal matrix, as below, and then visualize the (2,2) block-entry A_{22} itself as a block-diagonal matrix. That is,

Copyright © 2021 Pearson Education, Inc.

$$A = \begin{bmatrix} 1 & 2 & 0 & 0 & 0 \\ 3 & 5 & 0 & 0 & 0 \\ 0 & 0 & 2 & 0 & 0 \\ 0 & 0 & 0 & 7 & 8 \\ 0 & 0 & 0 & 5 & 6 \end{bmatrix} = \begin{bmatrix} A_{11} & 0 \\ 0 & A_{22} \end{bmatrix}, \quad \text{where} \quad A_{22} = \begin{bmatrix} 2 & 0 & 0 \\ 0 & 7 & 8 \\ 0 & 5 & 6 \end{bmatrix} = \begin{bmatrix} 2 & 0 \\ 0 & B \end{bmatrix}$$

Observe that B is invertible and $B^{-1} = \dfrac{1}{2}\begin{bmatrix} 6 & -8 \\ -5 & 7 \end{bmatrix} = \begin{bmatrix} 3 & -4 \\ -2.5 & 3.5 \end{bmatrix}$. By Exercise 15, the block

diagonal matrix A_{22} is invertible, and

$$A_{22}^{-1} = \begin{bmatrix} .5 & 0 \\ 0 & \begin{matrix} 3 & -4 \\ -2.5 & 3.5 \end{matrix} \end{bmatrix} = \begin{bmatrix} .5 & 0 & 0 \\ 0 & 3 & -4 \\ 0 & -2.5 & 3.5 \end{bmatrix}$$

Next, observe that A_{11} is also invertible, with inverse $\begin{bmatrix} -5 & 2 \\ 3 & -1 \end{bmatrix}$. By Exercise 15, A itself is

invertible, and its inverse is block diagonal:

$$A^{-1} = \begin{bmatrix} A_{11}^{-1} & 0 \\ 0 & A_{22}^{-1} \end{bmatrix} = \begin{bmatrix} \begin{matrix} -5 & 2 \\ 3 & -1 \end{matrix} & 0 \\ 0 & \begin{matrix} .5 & 0 & 0 \\ 0 & 3 & -4 \\ 0 & -2.5 & 3.5 \end{matrix} \end{bmatrix} = \begin{bmatrix} -5 & 2 & 0 & 0 & 0 \\ 3 & -1 & 0 & 0 & 0 \\ 0 & 0 & .5 & 0 & 0 \\ 0 & 0 & 0 & 3 & -4 \\ 0 & 0 & 0 & -2.5 & 3.5 \end{bmatrix}$$

A somewhat less detailed solution would be to write (without formal proof) that the result of Exercise 15 seems to generalize to any block-diagonal matrix. Such a matrix A is invertible if and only if each of the diagonal blocks is invertible, and the inverse of A is the block-diagonal matrix formed from the inverses of the diagonal blocks. View the 5×5 matrix in this exercise as a 3×3 block matrix:

$$A = \begin{bmatrix} 1 & 2 & 0 & 0 & 0 \\ 3 & 5 & 0 & 0 & 0 \\ 0 & 0 & 2 & 0 & 0 \\ 0 & 0 & 0 & 7 & 8 \\ 0 & 0 & 0 & 5 & 6 \end{bmatrix} = \begin{bmatrix} A_{11} & 0 & 0 \\ 0 & A_{22} & 0 \\ 0 & 0 & A_{33} \end{bmatrix}$$

Finish by inverting each of the diagonal blocks and use the results to assemble A^{-1}, as above.

Answer to Checkpoint: The block diagonal matrices $\begin{bmatrix} E & 0 \\ 0 & I \end{bmatrix}$ and $\begin{bmatrix} I & 0 \\ 0 & E \end{bmatrix}$ are obvious

choices. Less obvious is the requirement that E be invertible, in order to make these block

Copyright © 2021 Pearson Education, Inc.

matrices invertible. (Recall that the invertibility of elementary matrices was essential for the theory in Section 2.2.)

Appendix: The Principle of Induction

Consider a statement "(*)" that depends on a positive integer n, as in Exercise 25. To prove "by induction" that (*) is true for all positive integers, you must prove two things:

(a)   Statement (*) is true for $n = 1$.

(b)   (The induction step) If (*) is true for any positive integer $n = k$, then (*) is also true for the next integer $n = k + 1$.

A property or axiom of the real number system, called the *principle of mathematical induction*, says that if (a) and (b) are true, then (*) is true for all integers $n \geq 1$. This is reasonable, because if (*) is true for $n = 1$, then (b) shows that (*) is true for $n = 2$. Applying (b) again with $n = 2$, we see that (*) is true for $n = 3$. Applying (b) repeatedly, we see that (*) is true for $2, 3, 4, 5, \ldots$.

MATLAB   Partitioned Matrices

MATLAB uses partitioned matrix notation. For example, if A, B, C, D, E, and F are matrices of appropriate sizes, then the command

M = [A B C; D E F]

creates a larger matrix of the form $M = \begin{bmatrix} A & B & C \\ D & E & F \end{bmatrix}$. Once M is formed, there is no record of the partition that was used to create M. For instance, although B was the (1, 2)-block used to form M, the number $M(1, 2)$ is the same as the (1, 2)-entry of A.

2.5 - Matrix Factorizations

In a sense, Section 2.5 is the most up-to-date section in the text, because matrix factorizations lie at the heart of modern uses of matrix algebra. For instance, they are indispensable for the analysis of computational algorithms and research in parallel processing. The text focuses here on triangular factorizations, but the exercises introduce you to other important factorizations that you may encounter later.

Copyright © 2021 Pearson Education, Inc.

KEY IDEAS

When a matrix A is factored as $A = LU$, the data in A are preprocessed in a way that makes the equation $Ax = b$ easier to solve. Write $LUx = b$, or $L(Ux) = b$, and let $y = Ux$. Solve $Ly = b$ for y and then solve $Ux = y$ for x. The two-step process is fast when L and U are triangular.

Finding L and U requires the same number of multiplications and divisions as row reducing A to an echelon form U (about $n^3/3$ operations when A is $n \times n$). After that, L and U are available for solving other equations involving A. The key to finding L is to place entries in L in such a way that the sequence of row operations reducing A to U also reduces L to the identity. In this case, LU must equal A.

The text discusses how to build L when no row interchanges are needed to reduce A to U. In this case, L can be unit lower triangular. An appendix below describes how to build L in permuted unit triangular form when row interchanges are needed (or desired, for numerical reasons).

SOLUTIONS TO EXERCISES

1. $L = \begin{bmatrix} 1 & 0 & 0 \\ -1 & 1 & 0 \\ 2 & -5 & 1 \end{bmatrix}$, $U = \begin{bmatrix} 3 & -7 & -2 \\ 0 & -2 & -1 \\ 0 & 0 & -1 \end{bmatrix}$, $b = \begin{bmatrix} -7 \\ 5 \\ 2 \end{bmatrix}$. First, solve $Ly = b$.

$$[L \quad b] = \begin{bmatrix} 1 & 0 & 0 & -7 \\ -1 & 1 & 0 & 5 \\ 2 & -5 & 1 & 2 \end{bmatrix} \sim \begin{bmatrix} 1 & 0 & 0 & -7 \\ 0 & 1 & 0 & -2 \\ 0 & -5 & 1 & 16 \end{bmatrix} \quad \text{The only arithmetic is in column 4}$$

$$\sim \begin{bmatrix} 1 & 0 & 0 & -7 \\ 0 & 1 & 0 & -2 \\ 0 & 0 & 1 & 6 \end{bmatrix}, \text{ so } y = \begin{bmatrix} -7 \\ -2 \\ 6 \end{bmatrix}.$$

Next, solve $Ux = y$, using back-substitution (with matrix notation).

$$[U \quad y] = \begin{bmatrix} 3 & -7 & -2 & -7 \\ 0 & -2 & -1 & -2 \\ 0 & 0 & -1 & 6 \end{bmatrix} \sim \begin{bmatrix} 3 & -7 & -2 & -7 \\ 0 & -2 & -1 & -2 \\ 0 & 0 & 1 & -6 \end{bmatrix} \sim \begin{bmatrix} 3 & -7 & 0 & -19 \\ 0 & -2 & 0 & -8 \\ 0 & 0 & 1 & -6 \end{bmatrix}$$

$$\sim \begin{bmatrix} 3 & -7 & 0 & -19 \\ 0 & 1 & 0 & 4 \\ 0 & 0 & 1 & -6 \end{bmatrix} \sim \begin{bmatrix} 3 & 0 & 0 & 9 \\ 0 & 1 & 0 & 4 \\ 0 & 0 & 1 & -6 \end{bmatrix} \sim \begin{bmatrix} 1 & 0 & 0 & 3 \\ 0 & 1 & 0 & 4 \\ 0 & 0 & 1 & -6 \end{bmatrix}$$

So $x = (3, 4, -6)$.

Checkpoint: Exercise 22 in Section 2.2 shows how to compute $A^{-1}B$ by row reduction. Describe how you could speed up this calculation if you have an LU factorization of A available (and A is invertible).

Copyright © 2021 Pearson Education, Inc.

7. Place the first pivot column of $\begin{bmatrix} 2 & 5 \\ -3 & -4 \end{bmatrix}$ into L, after dividing the column by 2 (the pivot),

then add 3/2 times row 1 to row 2, yielding U.

$$A = \begin{bmatrix} ② & 5 \\ -3 & -4 \end{bmatrix} \sim \begin{bmatrix} 2 & 5 \\ 0 & ⑺/2 \end{bmatrix} = U$$

$$\begin{bmatrix} ② \\ -3 \end{bmatrix} \quad \boxed{7/2}$$

$$\div 2 \quad \div 7/2$$

$$\begin{bmatrix} 1 & \\ -3/2 & 1 \end{bmatrix}, \quad L = \begin{bmatrix} 1 & 0 \\ -3/2 & 1 \end{bmatrix}$$

13. $\begin{bmatrix} ① & 3 & -5 & -3 \\ -1 & -5 & 8 & 4 \\ 4 & 2 & -5 & -7 \\ -2 & -4 & 7 & 5 \end{bmatrix} \sim \begin{bmatrix} 1 & 3 & -5 & -3 \\ 0 & ⑵ & 3 & 1 \\ 0 & -10 & 15 & 5 \\ 0 & 2 & -3 & -1 \end{bmatrix} \sim \begin{bmatrix} 1 & 3 & -5 & -3 \\ 0 & -2 & 3 & 1 \\ 0 & 0 & 0 & 0 \\ 0 & 0 & 0 & 0 \end{bmatrix} = U$ No more pivots!

$$\begin{bmatrix} ① \\ -1 \\ 4 \\ -2 \end{bmatrix} \quad \begin{bmatrix} ⑵ \\ -10 \\ 2 \end{bmatrix}$$

Use the last two columns of I_4 to make L unit lower triangular.

$$\div 1 \quad \div -2$$

$$\begin{bmatrix} 1 & & & \\ -1 & 1 & & \\ 4 & 5 & 1 & \\ -2 & -1 & 0 & 1 \end{bmatrix}, \quad L = \begin{bmatrix} 1 & 0 & 0 & 0 \\ -1 & 1 & 0 & 0 \\ 4 & 5 & 1 & 0 \\ -2 & -1 & 0 & 1 \end{bmatrix}$$

19. A good answer will require a written paragraph or two. If you have not tried to *write* your answer, do so now, *without reading the solution below*. Explain how you would row reduce [A I], knowing that A is lower triangular. Your answer to this question should contain some of the ideas shown below, although your wording might be quite different.

Let A be a lower-triangular $n \times n$ matrix with nonzero entries on the diagonal, and consider the augmented matrix [A I].

Copyright © 2021 Pearson Education, Inc.

a. The (1, 1)-entry can be scaled to 1 and the entries below it can be changed to 0 by adding multiples of row 1 to the rows below. This affects only the first column of A and the first column of I. So the (2, 2)-entry in the new matrix is still nonzero and now is the only nonzero entry of row 2 in the first n columns (because A was lower triangular).

The (2, 2)-entry can be scaled to 1, and the entries below it can be changed to 0 by adding multiples of row 2 to the rows below. This affects only columns 2 and $n + 2$ of the augmented matrix. Now the (3, 3) entry in A is the only nonzero entry of the third row in the first n columns, so it can be scaled to 1 and then used as a pivot to zero out entries below it. Continuing in this way, A is eventually reduced to I, by scaling each row with a pivot and then using only row operations that add multiples of the pivot row to rows below.

b. The row operations just described only add rows to rows below, so the I on the right in $[A \quad I]$ changes into a lower triangular matrix. By Theorem 7 in Section 2.2, that matrix is A^{-1}.

21. Suppose $A = BC$, with B invertible. Then there exist elementary matrices $E_1, \dots, E_p$ corresponding to row operations that reduce B to I, in the sense that $E_p \cdots E_1 B = I$. Applying the same sequence of row operations to A amounts to left-multiplying A by the product $E_p \cdots E_1$. By associativity of matrix multiplication,

$$E_p \cdots E_1 A = E_p \cdots E_1 BC = IC = C$$

so the same sequence of row operations reduces A to C.

25. $A = UDV^T$. Since U and V^T are square, the equations $U^T U = I$ and $V^T V = I$ imply that U and V^T are invertible, by the IMT, and hence $U^{-1} = U^T$ and $(V^T)^{-1} = V$. Since the diagonal entries $\sigma_1, \dots, \sigma_n$ in D are nonzero, D is invertible, with the inverse of D being the diagonal matrix with $\sigma_1^{-1}, \dots, \sigma_n^{-1}$ on the diagonal. Thus A is a product of invertible matrices. By Theorem 6, A is invertible and $A^{-1} = (UDV^T)^{-1} = (V^T)^{-1} D^{-1} U^{-1} = VD^{-1}U^T$.

Answer to Checkpoint: If A is an invertible $n \times n$ matrix, with an LU factorization $A = LU$, and if B is $n \times p$, then $A^{-1}B$ can be computed by first row reducing $[L \quad B]$ to a matrix $[I \quad Y]$ for some Y and then reducing $[U \quad Y]$ to $[I \quad A^{-1}B]$. One way to see that this algorithm works is to view $A^{-1}B$ as $[A^{-1}\mathbf{b}_1 \ \cdots \ A^{-1}\mathbf{b}_p]$ and use the LU algorithm to solve simultaneously the set of equations $A\mathbf{x} = \mathbf{b}_1, \dots, A\mathbf{x} = \mathbf{b}_p$. MATLAB uses this approach to compute $A^{-1}B$ (after first finding L and U).

Appendix: Permuted LU Factorizations

Any $m \times n$ matrix A admits a factorization $A = LU$, with U in echelon form and L a *permuted unit lower triangular* matrix. That is, L is a matrix such that a permutation (rearrangement) of its rows (using row interchanges) will produce a lower triangular matrix with 1's on the diagonal.

The construction of L and U, illustrated below, depends on first using row replacements to reduce A to a *permuted echelon form* V and then using row interchanges to reduce V to an echelon form U. By watching the reduction of A to V, we can easily construct a permuted unit lower triangular matrix L with the property that the sequence of operations changing A into U

Copyright © 2021 Pearson Education, Inc.

also changes L into I. This property will guarantee that $A = LU$. (See the paragraph before Example 2 in the text.)

The following algorithm reduces any matrix to a permuted echelon form. In the algorithm when a row is covered, we ignore it in later calculations.

1. *Begin with the leftmost nonzero column. Choose any nonzero entry as the pivot. Designate the corresponding row as a pivot row.*

2. *Use row replacements to create zeros above and below the pivot (in all uncovered rows). Then cover that pivot row.*

3. *Repeat steps 1 and 2 on the uncovered submatrix, if any, until all nonzero entries are covered.*

This algorithm forces each pivot to be to the right of the preceding pivots; when the rows are rearranged with the pivots in stair-step fashion, all entries below each pivot will be zero. Thus, the algorithm produces a permuted echelon matrix. Whenever a pivot is selected, the column containing the pivot will be used to construct a column of L, as we shall see.

As an example, choose any entry in the first column of the following matrix as the first pivot, and use the pivot to create zeros in the rest of column 1. We choose the $(3, 1)$-entry.

$$A = \begin{bmatrix} 1 & -1 & 5 & -8 & -7 \\ -2 & -1 & -4 & 9 & 1 \\ \boxed{4} & 8 & -4 & 0 & -8 \\ 2 & 3 & 0 & -5 & 3 \end{bmatrix} \sim \begin{bmatrix} 0 & -3 & 6 & -8 & -5 \\ 0 & 3 & -6 & 9 & -3 \\ 4 & 8 & -4 & 0 & -8 \\ 0 & -1 & 2 & -5 & 7 \end{bmatrix}$$

(call this column a pointing to first column; $\leftarrow$ 1st pivot row for the row $4\ 8\ -4\ 0\ -8$)

Row 3 is the first pivot row. Choose the $(2, 2)$-entry as the second pivot, and create zeros in the rest of column 2, excluding the first pivot row.

$$= \begin{bmatrix} 0 & -3 & 6 & -8 & -5 \\ 0 & \boxed{3} & 6 & 9 & -3 \\ 4 & 8 & -4 & 0 & -8 \\ 0 & -1 & 2 & -5 & 7 \end{bmatrix} \sim \begin{bmatrix} 0 & 0 & 0 & 1 & -8 \\ 0 & 3 & -6 & 9 & -3 \\ 4 & 8 & -4 & 0 & -8 \\ 0 & 0 & 0 & -2 & 6 \end{bmatrix}$$

(call this column b pointing to second column; $\leftarrow$ 2nd pivot row for the row $0\ 3\ -6\ 9\ -3$; $\leftarrow$ 1st pivot row for the row $4\ 8\ -4\ 0\ -8$)

Copyright © 2021 Pearson Education, Inc.

Cover row 2 and choose the (4, 4)-entry as the pivot. (The row index of the pivot is relative to the original matrix.) Create zeros in the other rows (in the pivot column), excluding the first two pivot rows.

$$= \begin{bmatrix} 0 & 0 & 0 & 1 & -8 \\ 0 & 3 & -6 & 9 & -3 \\ 4 & 8 & -4 & 0 & -8 \\ 0 & 0 & 0 & -2 & 6 \end{bmatrix} \sim \begin{bmatrix} 0 & 0 & 0 & 0 & -5 \\ 0 & 3 & -6 & 9 & -3 \\ 4 & 8 & -4 & 0 & -8 \\ 0 & 0 & 0 & -2 & 6 \end{bmatrix} \begin{matrix} \leftarrow \text{4th pivot row} \\ \leftarrow \text{2nd pivot row} \\ \leftarrow \text{1st pivot row} \\ \leftarrow \text{3rd pivot row} \end{matrix}$$

(call this column c; column d)

Let V denote this permuted echelon form, and permute the rows of V to create an echelon form. The first pivot row goes to the top, the second pivot row goes next, and so on. The resulting echelon matrix U is

$$\begin{bmatrix} 4 & 8 & -4 & 0 & -8 \\ 0 & 3 & -6 & 9 & -3 \\ 0 & 0 & 0 & -2 & 6 \\ 0 & 0 & 0 & 0 & -5 \end{bmatrix} = U$$

The last step is to create L. Go back and watch the reduction of A to V. As each pivot is selected, take the pivot column, and divide the pivot into each entry in the column that is not yet in a pivot row. Place the resulting column into L. At the end, fill the holes in L with zeros.

Column: a   b   c   d

$$\begin{bmatrix} 1 \\ -2 \\ 4 \\ 2 \end{bmatrix} \quad \begin{bmatrix} -3 \\ 3 \\ \\ -1 \end{bmatrix} \quad \begin{bmatrix} 1 \\ \\ \\ -2 \end{bmatrix} \quad \begin{bmatrix} -5 \\ \\ \\ \end{bmatrix}$$

$$\begin{matrix} \div 4 & \div 3 & \div -2 & \div -5 \\ \downarrow & \downarrow & \downarrow & \downarrow \end{matrix}$$

$$\begin{bmatrix} 1/4 & -1 & -1/2 & 1 \\ -1/2 & 1 & & \\ 1 & & & \\ 1/2 & -1/3 & 1 & \end{bmatrix}, \quad L = \begin{bmatrix} 1/4 & -1 & -1/2 & 1 \\ -1/2 & 1 & 0 & 0 \\ 1 & 0 & 0 & 0 \\ 1/2 & -1/3 & 1 & 0 \end{bmatrix}$$

Copyright © 2021 Pearson Education, Inc.

You can check that $LU = A$. To see why this is so, observe that L is constructed so the operations that reduce A to V also reduce L to a permuted identity matrix. Since the pivots in L are in exactly the same rows as in V, the sequence of row interchanges that reduces V to U also reduces the permuted identity matrix to I. Thus, the full sequence of operations that reduces A to U also reduces L to I, so that $A = LU$.

The next example illustrates what to do when V has one or more rows of zeros. The matrix is from the Practice Problem for Section 2.5. For the reduction of A to V, pivots were chosen to have the largest possible magnitude (the choice used for "partial pivoting"). Of course, other pivots could have been selected.

$$A = \begin{bmatrix} 2 & -4 & -2 & 3 \\ 6 & -9 & -5 & 8 \\ 2 & -7 & -3 & 9 \\ 4 & -2 & -2 & -1 \\ -6 & 3 & 3 & 4 \end{bmatrix} \sim \begin{bmatrix} 0 & -1 & -1/3 & 1/3 \\ 6 & -9 & -5 & 8 \\ 0 & -4 & -4/3 & 19/3 \\ 0 & 4 & 4/3 & -19/3 \\ 0 & -6 & -2 & 12 \end{bmatrix} \sim \begin{bmatrix} 0 & 0 & 0 & -5/3 \\ 6 & -9 & -5 & 8 \\ 0 & 0 & 0 & -5/3 \\ 0 & 0 & 0 & 5/3 \\ 0 & -6 & -2 & 12 \end{bmatrix}$$

$$\sim V = \begin{bmatrix} 0 & 0 & 0 & 0 \\ 6 & -9 & -5 & 8 \\ 0 & 0 & 0 & 0 \\ 0 & 0 & 0 & 5/3 \\ 0 & -6 & -2 & 12 \end{bmatrix} \begin{matrix} \\ \leftarrow \text{1st pivot row} \\ \\ \leftarrow \text{3rd pivot row} \\ \leftarrow \text{2nd pivot row} \end{matrix} \qquad \sim U = \begin{bmatrix} 6 & -9 & -5 & 8 \\ 0 & -6 & -2 & 12 \\ 0 & 0 & 0 & 5/3 \\ 0 & 0 & 0 & 0 \\ 0 & 0 & 0 & 0 \end{bmatrix}$$

The first three columns of L come from the three pivot columns above.

$$\begin{bmatrix} 2 \\ 6 \\ 2 \\ 4 \\ -6 \end{bmatrix} \quad \begin{bmatrix} -1 \\ \\ -4 \\ 4 \\ -6 \end{bmatrix} \quad \begin{bmatrix} -5/3 \\ \\ -5/3 \\ 5/3 \\ \end{bmatrix}$$
$$\div 6 \qquad \div -6 \qquad \div 5/3$$
$$\downarrow \qquad \downarrow \qquad \downarrow$$

$$\begin{bmatrix} 1/3 & 1/6 & -1 \\ 1 & & \\ 1/3 & 2/3 & -1 \\ 2/3 & -2/3 & 1 \\ -1 & 1 & \end{bmatrix} \begin{matrix} \\ \leftarrow \text{1st pivot row} \\ \\ \leftarrow \text{3rd pivot row} \\ \leftarrow \text{2nd pivot row} \end{matrix}$$

Copyright © 2021 Pearson Education, Inc.

The matrix L needs two more columns. Use columns 1 and 3 of the 5×5 identity matrix to place 1's in the "nonpivot" rows 1 and 3. Fill in the remaining holes with zeros.

$$\begin{bmatrix} 1/3 & 1/6 & -1 & 1 & 0 \\ \boxed{1} & 0 & 0 & 0 & 0 \\ 1/3 & 2/3 & -1 & 0 & 1 \\ 2/3 & -2/3 & \boxed{1} & 0 & 0 \\ -1 & \boxed{1} & 0 & 0 & 0 \end{bmatrix} \sim L = \begin{bmatrix} 1 & 0 & 0 & 0 & 0 \\ -1 & 1 & 0 & 0 & 0 \\ 2/3 & -2/3 & 1 & 0 & 0 \\ 1/3 & 1/6 & -1 & 1 & 0 \\ 1/3 & 2/3 & -1 & 0 & 1 \end{bmatrix}$$

Row reduction of L using only row replacements produces a permuted identity matrix. Moving the 1's in the "pivot rows" 2, 5, and 4 into rows 1, 2, and 3 of the identity requires the same row swaps as reducing V to U. If a further row interchange on the permuted identity is required, it will involve the bottom two rows, which came from the "nonpivot" rows 1 and 3. A corresponding interchange of the bottom two rows of U has no effect on U (and the product LU is unaffected). As a result, L is reduced to I by the same operations that reduce A to V and then to U. Check that $A = LU$.

**MATLAB   LU Factorization and the Backslash Operator **

The MATLAB command **[L U]=lu(A)** produces a permuted LU factorization for any square matrix A, but it does not handle the general case.

When A is invertible, the best way to solve $Ax = b$ with MATLAB is to use the backslash command **x=A\b**. MATLAB proceeds to compute a permuted LU factorization of A and then use L and U to compute **x**. The alternative command **x=inv(A)*b** is less efficient and can be less accurate. The command **inv(A)** uses the LU factorization to compute A^{-1} in the form $U^{-1}L^{-1}$.

2.6 -The Leontief Input-Output Model

If you are in economics, you definitely will need the material in this section for later work. Although most of the discussion concerns economics, the formula for the inverse of $I - C$ is used in a variety of applications.

STUDY NOTES

The power of Leontief's model of the economy is that it compresses hundreds of equations in hundreds of variables into the simple matrix equation $(I - C)x = d$. You should know how to construct the consumption matrix C and know the algebra that leads from the matrix equation $x = Cx + d$ to its solution $x = (I - C)^{-1}d$, under the assumption that the column sums of C are less than one.

Copyright © 2021 Pearson Education, Inc.

You may need to know the formula (8) for $(I - C)^{-1}$. (Check with your instructor.) The formula is analogous to the formula for the sum of a geometric series of positive numbers:

$$1 + r + r^2 + r^3 + \cdots = (1 - r)^{-1} \text{ when } |r| < 1.$$

SOLUTIONS TO EXERCISES

1. Fill in C one column at a time, since each column is a unit consumption vector for one sector. Make sure that the order of the sectors is the same for the rows and columns of C. From the way the data are presented, we use the order: manufacturing, agriculture, and services. Read the sentences carefully, to get the data arranged correctly.

Purchased	Unit consumption vectors		
from:	Manuf.	Agric.	Serv.
Manufacturing	.10	.60	.60
Agriculture	.30	.20	.00
Services	.30	.10	.10

The intermediate demands created by a production vector $\mathbf{x}$ are given by $C\mathbf{x}$. If agriculture plans to produce 100 units (and the other sectors plan to produce nothing), then the intermediate demand is

$$C\mathbf{x} = \begin{bmatrix} .10 & .60 & .60 \\ .30 & .20 & .00 \\ .30 & .10 & .10 \end{bmatrix} \begin{bmatrix} 0 \\ 100 \\ 0 \end{bmatrix} = \begin{bmatrix} 60 \\ 20 \\ 10 \end{bmatrix}$$

7. $C = \begin{bmatrix} .0 & .5 \\ .6 & .2 \end{bmatrix}, \mathbf{d} = \begin{bmatrix} 50 \\ 30 \end{bmatrix}$. Let $\mathbf{d}_1 = \begin{bmatrix} 1 \\ 0 \end{bmatrix}$, the demand for 1 unit of output of sector 1.

   a. The production required to satisfy the demand $\mathbf{d}_1$ is the vector $\mathbf{x}_1$ such that $(I - C)\mathbf{x}_1 = \mathbf{d}_1$, namely, $\mathbf{x}_1 = (I - C)^{-1}\mathbf{d}_1$. From Exercise 5,

   $$I - C = \begin{bmatrix} 1 & -.5 \\ -.6 & .8 \end{bmatrix} \quad \text{and} \quad (I - C)^{-1} = \begin{bmatrix} 1.6 & 1 \\ 1.2 & 2 \end{bmatrix}$$

   so

   $$\mathbf{x}_1 = \begin{bmatrix} 1.6 & 1 \\ 1.2 & 2 \end{bmatrix} \begin{bmatrix} 1 \\ 0 \end{bmatrix} = \begin{bmatrix} 1.6 \\ 1.2 \end{bmatrix}$$

   b. For the final demand $\mathbf{d}_2 = \begin{bmatrix} 51 \\ 30 \end{bmatrix}$, the corresponding production $\mathbf{x}_2$ is given by

Copyright © 2021 Pearson Education, Inc.

$$\mathbf{x}_2 = (I - C)^{-1}\mathbf{d}_2 = \begin{bmatrix} 1.6 & 1 \\ 1.2 & 2 \end{bmatrix}\begin{bmatrix} 51 \\ 30 \end{bmatrix} = \begin{bmatrix} 111.6 \\ 121.2 \end{bmatrix}$$

c. From Exercise 5, the production $\mathbf{x}$ corresponding to the demand $\mathbf{d}$ is given by $\mathbf{x} = \begin{bmatrix} 110 \\ 120 \end{bmatrix}$.

Observe from (a) and (b) that $\mathbf{x}_2 = \mathbf{x} + \mathbf{x}_1$. Also, as pointed out in the text, $\mathbf{d}_2 = \mathbf{d} + \mathbf{d}_1$. The sum of the production vectors $\mathbf{x}$ and $\mathbf{x}_1$ gives the production needed to satisfy the sum of the demands $\mathbf{d}$ and $\mathbf{d}_1$. This is expressing the *linearity* between final demand and production. This relation is true in general, because

$$\mathbf{x}_2 = (I - C)^{-1}\mathbf{d}_2 = (I - C)^{-1}(\mathbf{d} + \mathbf{d}_1)$$

$$= (I - C)^{-1}\mathbf{d} + (I - C)^{-1}\mathbf{d}_1$$

$$= \mathbf{x} + \mathbf{x}_1$$

Warning: In Exercise 9, don't multiply the consumption matrix C by 10, to get rid of the decimals. That changes the equation $C\mathbf{x} = \mathbf{x} + \mathbf{d}$ into $10C\mathbf{x} = \mathbf{x} + \mathbf{d}$, whose solution is different. However, you *may* multiply the *augmented matrix* $[(I - C) \quad \mathbf{0}]$ by 10, because the solution of an equation is not affected when both sides are multiplied by a nonzero number.

11. Following the hint in the text, you should obtain $\mathbf{p}^T\mathbf{x} = \mathbf{p}^T C\mathbf{x} + \mathbf{v}^T\mathbf{x}$ (from the price equation). Then, from the production equation, $\mathbf{p}^T\mathbf{x} = \mathbf{p}^T(C\mathbf{x} + \mathbf{d}) = \mathbf{p}^T C\mathbf{x} + \mathbf{p}^T\mathbf{d}$. Equate the two expressions for $\mathbf{p}^T\mathbf{x}$ to yield $\mathbf{p}^T\mathbf{d} = \mathbf{v}^T\mathbf{x}$.

Another solution: Take transposes in the price equation,

$$\mathbf{p}^T = (C^T\mathbf{p})^T + \mathbf{v}^T = \mathbf{p}^T C + \mathbf{v}^T, \text{ so } \mathbf{v}^T = \mathbf{p}^T - \mathbf{p}^T C$$

and right-multiply by $\mathbf{x}$ to obtain

$$\mathbf{v}^T\mathbf{x} = \mathbf{p}^T\mathbf{x} - \mathbf{p}^T C\mathbf{x} = \mathbf{p}^T(I - C)\mathbf{x} = \mathbf{p}^T\mathbf{d} \quad \text{From the production equation}$$

13. To solve the equation $(I - C)\mathbf{x} = \mathbf{d}$, row reduce the augmented matrix $[(I - C) \quad \mathbf{d}]$ rather than compute $(I - C)^{-1}$. (Another reasonable solution method is suggested in Exercise 15.) The numerical solution is given in the text.

2.7 - Applications to Computer Graphics

According to my students over the past few years, this section is one of the most interesting application sections in the text, because it shows how matrix calculations, performed millions of times per second, can create the illusion of 3D-motion on a computer screen or in a movie theater. Of course, one short section cannot begin to indicate the vast scope of computer graphics. I encourage you to look at the book by Foley et al., referenced in your text. Chapters 5, 6, and 11 are filled with matrices! The rest of the 1100 pages in the book contains lots of interesting mathematics, detailed discussions of computer algorithms, and scores of spectacular (in some cases, almost unbelievable) color plates.

Copyright © 2021 Pearson Education, Inc.

STUDY NOTES

When a graphical object is represented by a set of polygons, each vertex can be stored as one column of a data matrix D. When a linear transformation T acts on the graphical object, the transformed object is determined by the images of the vertices, because each line segment between vertices is transformed into a line segment between the image vertices. If A is the matrix of the transformation T, then the image vertices are the columns of the matrix AD, by definition of the product AD.

Homogeneous coordinates are needed to make translation act as a linear transformation. Translation by a vector $\mathbf{p}$ is illustrated by the computation $\begin{bmatrix} I & \mathbf{p} \\ \mathbf{0}^T & 1 \end{bmatrix}\begin{bmatrix} \mathbf{x} \\ 1 \end{bmatrix} = \begin{bmatrix} \mathbf{x}+\mathbf{p} \\ 1 \end{bmatrix}$. When several transformations are composed, the order of matrix products is important. See Example 6.

For 3D-graphics, homogeneous coordinates are used to compute perspective projections. The text only considers a perspective projection whose center of projection is at $(0, 0, d)$. The matrix for this is displayed just before Example 8. Check with your instructor whether you should memorize this matrix.

SOLUTIONS TO EXERCISES

1. From Example 5, the matrix $\begin{bmatrix} 1 & .25 & 0 \\ 0 & 1 & 0 \\ 0 & 0 & 1 \end{bmatrix}$ has the same effect on homogeneous coordinates

for $\mathbb{R}^2$ that the matrix $\begin{bmatrix} 1 & .25 \\ 0 & 1 \end{bmatrix}$ of Example 2 has on ordinary vectors in $\mathbb{R}^2$. Partitioned

matrix notation explains why this is true. Let A be a 2×2 matrix. The following diagram

shows that the action of $\begin{bmatrix} A & 0 \\ 0^T & 1 \end{bmatrix}$ on $\begin{bmatrix} \mathbf{x} \\ 1 \end{bmatrix}$ corresponds to the action of A on $\mathbf{x}$.

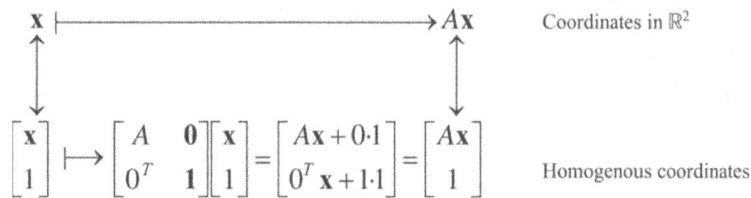

$$\mathbf{x} \longmapsto A\mathbf{x} \quad \text{Coordinates in } \mathbb{R}^2$$

$$\begin{bmatrix} \mathbf{x} \\ 1 \end{bmatrix} \longmapsto \begin{bmatrix} A & \mathbf{0} \\ 0^T & 1 \end{bmatrix}\begin{bmatrix} \mathbf{x} \\ 1 \end{bmatrix} = \begin{bmatrix} A\mathbf{x}+0\cdot 1 \\ 0^T\mathbf{x}+1\cdot 1 \end{bmatrix} = \begin{bmatrix} A\mathbf{x} \\ 1 \end{bmatrix} \quad \text{Homogenous coordinates}$$

7. A 60% rotation about the origin in $\mathbb{R}^2$ is given by

$\begin{bmatrix} \cos 60° & -\sin 60° \\ \sin 60° & \cos 60° \end{bmatrix} = \begin{bmatrix} 1/2 & -\sqrt{3}/2 \\ \sqrt{3}/2 & 1/2 \end{bmatrix}$, so the 3×3 matrix for rotation about $\begin{bmatrix} 6 \\ 8 \end{bmatrix}$ is

Copyright © 2021 Pearson Education, Inc.

$$\begin{bmatrix} 1 & 0 & 6 \\ 0 & 1 & 8 \\ 0 & 0 & 1 \end{bmatrix} \begin{bmatrix} 1/2 & -\sqrt{3}/2 & 0 \\ \sqrt{3}/2 & 1/2 & 0 \\ 0 & 0 & 1 \end{bmatrix} \begin{bmatrix} 1 & 0 & -6 \\ 0 & 1 & -8 \\ 0 & 0 & 1 \end{bmatrix}$$

Finally, translate back	Then, rotate about the origin	First, translate by $-\mathbf{p}$

$$= \begin{bmatrix} 1 & 0 & 6 \\ 0 & 1 & 8 \\ 0 & 0 & 1 \end{bmatrix} \begin{bmatrix} 1/2 & -\sqrt{3}/2 & -3+4\sqrt{3} \\ \sqrt{3}/2 & 1/2 & -4-3\sqrt{3} \\ 0 & 0 & 1 \end{bmatrix} = \begin{bmatrix} 1/2 & -\sqrt{3}/2 & 3+4\sqrt{3} \\ \sqrt{3}/2 & 1/2 & 4-3\sqrt{3} \\ 0 & 0 & 1 \end{bmatrix}$$

13. The answer is given in the text. Notice that the order of the transformations is important. If the translation is done first (that is, if the matrix for the translation is on the right), then

$$\begin{bmatrix} A & \mathbf{0} \\ \mathbf{0}^T & 1 \end{bmatrix} \begin{bmatrix} I & \mathbf{p} \\ \mathbf{0}^T & 1 \end{bmatrix} = \begin{bmatrix} AI + \mathbf{0}\mathbf{0}^T & A\mathbf{p} + \mathbf{0}\cdot 1 \\ \mathbf{0}^T I + 1\mathbf{0}^T & \mathbf{0}^T\mathbf{p} + 1\cdot 1 \end{bmatrix} = \begin{bmatrix} A & A\mathbf{p} \\ \mathbf{0}^T & 1 \end{bmatrix} \neq \begin{bmatrix} A & \mathbf{p} \\ \mathbf{0}^T & 1 \end{bmatrix}$$

Here, $\mathbf{0}^T$ is a zero row vector, and so the outer product $\mathbf{0}\mathbf{0}^T$ is a zero matrix.

19. The matrix P for the perspective transformation with center of projection at (0, 0, 10) and the data matrix D using homogeneous coordinates are shown below. The data matrix for the image of the triangle is PD:

$$PD = \begin{bmatrix} 1 & 0 & 0 & 0 \\ 0 & 1 & 0 & 0 \\ 0 & 0 & 0 & 0 \\ 0 & 0 & -.1 & 1 \end{bmatrix} \begin{bmatrix} 4.2 & 6 & 2 \\ 1.2 & 4 & 2 \\ 4 & 2 & 6 \\ 1 & 1 & 1 \end{bmatrix} = \begin{bmatrix} 4.2 & 6 & 2 \\ 1.2 & 4 & 2 \\ 0 & 0 & 0 \\ .6 & .8 & .4 \end{bmatrix}$$

The $\mathbb{R}^3$ coordinates of the image points come from the top three entries in each column, divided by the corresponding entries in the fourth row.

$$\begin{bmatrix} 4.2/.6 & 6/.8 & 2/.4 \\ 1.2/.6 & 4/.8 & 2/.4 \\ 0 & 0 & 0 \end{bmatrix} = \begin{bmatrix} 7 & 7.5 & 5 \\ 2 & 5 & 5 \\ 0 & 0 & 0 \end{bmatrix}$$

2.8 - Subspaces of $\mathbb{R}^n$

This section presents the basic ideas from Sections 4.1–4.3 that are needed for Chapters 5 to 7. You should study this section and the next only if your course will omit most or all of Chapter 4. If you reviewed carefully after Sections 1.9 and 2.3, you should be well prepared for the new material here.

Copyright © 2021 Pearson Education, Inc.

KEY IDEAS

There are four fundamental concepts in this section: subspace, column space, null space, and basis. The best mental image for a subspace is a plane in $\mathbb{R}^3$ through the origin. (See Fig. 1.) The distinguishing feature of such a plane is that the sum of any two vectors in the plane is another vector in the same plane (by the parallelogram rule), and any scalar multiple of any vector in the plane is also in the plane. Other subspaces of $\mathbb{R}^3$ are lines through the origin, the zero subspace, and $\mathbb{R}^3$ itself.

The main examples of subspaces of $\mathbb{R}^3$ are column spaces (defined explicitly) and null spaces (defined implicitly). Example 6 is probably the most important example in the section. It illustrates a type of computation needed frequently in Chapters 5 and 7.

Actually, there really is not so much to learn here, because you have already been using these concepts for several weeks, without the terminology. (For instance, the notion of a basis is a combination of the ideas of linear independence and spanning.) But you do need to know the precise definitions of these four terms, and you must move beyond mechanical computations. See "Mastering Linear Algebra Concepts" at the end of this section for help.

SOLUTIONS TO EXERCISES

1. The set is closed under sums but not under multiplication by a negative scalar. A counterexample to the subspace condition is shown at the right. You may also give an algebraic example, such as $\mathbf{x} = (2, 1)$ and $c = -1$. Then $\mathbf{x}$ is in H and $c\mathbf{x} = (-2, -1)$ is not in H. Ask your instructor what type of counterexample would be acceptable if this question were on a test.

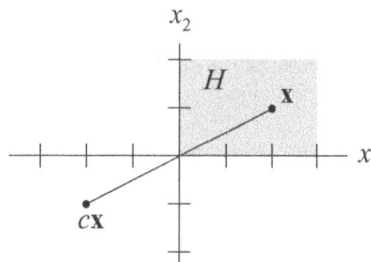

7. **a.** There are three vectors, $\mathbf{v}_1$, $\mathbf{v}_2$, and $\mathbf{v}_3$, in the set $\{\mathbf{v}_1, \mathbf{v}_2, \mathbf{v}_3\}$.

   b. There are infinitely many vectors in Span$\{\mathbf{v}_1, \mathbf{v}_2, \mathbf{v}_3\}$ = Col A.

   c. Deciding whether $\mathbf{p}$ is in Col A requires calculation:

$$[A \quad \mathbf{p}] \sim \begin{bmatrix} 2 & -3 & -4 & 6 \\ -8 & 8 & 6 & -10 \\ 6 & -7 & -7 & 11 \end{bmatrix} \sim \begin{bmatrix} 2 & -3 & -4 & 6 \\ 0 & -4 & -10 & 14 \\ 0 & 2 & 5 & -7 \end{bmatrix} \sim \begin{bmatrix} 2 & -3 & -4 & 6 \\ 0 & -4 & -10 & 14 \\ 0 & 0 & 0 & 0 \end{bmatrix}$$

   The equation $A\mathbf{x} = \mathbf{p}$ has a solution, so $\mathbf{p}$ is in Col A.

13. To produce a vector in Col A, select any column of A. For Nul A, solve the equation $A\mathbf{x} = \mathbf{0}$. (Include an augmented column of zeros, to avoid errors.)

$$\begin{bmatrix} 3 & 2 & 1 & -5 & 0 \\ -9 & -4 & 1 & 7 & 0 \\ 9 & 2 & -5 & 1 & 0 \end{bmatrix} \sim \begin{bmatrix} 3 & 2 & 1 & -5 & 0 \\ 0 & 2 & 4 & -8 & 0 \\ 0 & -4 & -8 & 16 & 0 \end{bmatrix} \sim \begin{bmatrix} 3 & 2 & 1 & -5 & 0 \\ 0 & 2 & 4 & -8 & 0 \\ 0 & 0 & 0 & 0 & 0 \end{bmatrix}$$

Copyright © 2021 Pearson Education, Inc.

$$\sim \begin{bmatrix} 3 & 2 & 1 & -5 & 0 \\ 0 & 1 & 2 & -4 & 0 \\ 0 & 0 & 0 & 0 & 0 \end{bmatrix} \sim \begin{bmatrix} 1 & 0 & -1 & 1 & 0 \\ 0 & 1 & 2 & -4 & 0 \\ 0 & 0 & 0 & 0 & 0 \end{bmatrix}, \qquad \begin{aligned} x_1 \quad - \quad x_3 \, + \, x_4 &= 0 \\ x_2 + 2x_3 - 4x_4 &= 0 \\ 0 &= 0 \end{aligned}$$

The general solution is $x_1 = x_3 - x_4$, and $x_2 = -2x_3 + 4x_4$, with x_3 and x_4 free. The general solution in parametric vector form is not needed. All that is required here is one nonzero vector. So choose any values for x_3 and x_4 (not both zero). For instance, set $x_3 = 1$ and $x_4 = 0$ to obtain the vector $(1, -2, 1, 0)$ in Nul A. Another choice, setting $x_3 = 0$ and $x_4 = 1$, might be $(-1, 4, 0, 1)$.

19. No. The vectors cannot be a basis for $\mathbb{R}^3$ because they only span a plane in $\mathbb{R}^3$. Or, point

out that the columns of the matrix $\begin{bmatrix} 3 & 6 \\ -8 & 2 \\ 1 & -5 \end{bmatrix}$ cannot possibly span $\mathbb{R}^3$ because the matrix

cannot have a pivot in every row. So the columns are not a basis for $\mathbb{R}^3$. Be careful *not* to say that the vectors are a basis for $\mathbb{R}^3$. They are not *in* $\mathbb{R}^3$, because they each have three entries.

Warning: $\mathbb{R}^2$ is *not* a subspace of $\mathbb{R}^3$. The notation $\mathbb{R}^2$ refers explicitly to lists of numbers with exactly two entries. $\mathbb{R}^3$ is the set of all lists of three entries from $\mathbb{R}$.

21. Carefully read the definition at the beginning of the section. What is missing?

23. See the paragraph before Example 4.

25. See Theorem 12. The numbers m and n need not be equal.

27. See Example 5.

29. See the first part of the solution of Example 8.

31. $A = \begin{bmatrix} 4 & 5 & 9 & -2 \\ 6 & 5 & 1 & 12 \\ 3 & 4 & 8 & -3 \end{bmatrix} \sim \begin{bmatrix} 1 & 2 & 6 & -5 \\ 0 & 1 & 5 & -6 \\ 0 & 0 & 0 & 0 \end{bmatrix}$. The echelon form identifies columns 1 and 2 as

the pivot columns. A basis for Col A uses columns 1 and 2 of A: $\begin{bmatrix} 4 \\ 6 \\ 3 \end{bmatrix}, \begin{bmatrix} 5 \\ 5 \\ 4 \end{bmatrix}$. This is not the

only choice, but it is the "standard" choice. A *wrong* choice is to select columns 1 and 2 of the echelon form. These columns have zero in the third entry and could not possibly generate the columns displayed in A.

For Nul A, obtain the *reduced* (and augmented) echelon form for $A\mathbf{x} = \mathbf{0}$:

Copyright © 2021 Pearson Education, Inc.

$$\begin{bmatrix} 1 & 0 & -4 & 7 & 0 \\ 0 & 1 & 5 & -6 & 0 \\ 0 & 0 & 0 & 0 & 0 \end{bmatrix}.$$ This corresponds to: $\begin{aligned} x_1 \quad - 4x_3 + 7x_4 &= 0 \\ x_2 + 5x_3 - 6x_4 &= 0. \\ 0 &= 0 \end{aligned}$

Solve for the basic variables and write the solution of $A\mathbf{x} = \mathbf{0}$ in parametric vector form:

$$\begin{bmatrix} x_1 \\ x_2 \\ x_3 \\ x_4 \end{bmatrix} = \begin{bmatrix} 4x_3 - 7x_4 \\ -5x_3 + 6x_4 \\ x_3 \\ x_4 \end{bmatrix} = x_3 \begin{bmatrix} 4 \\ -5 \\ 1 \\ 0 \end{bmatrix} + x_4 \begin{bmatrix} -7 \\ 6 \\ 0 \\ 1 \end{bmatrix}.$$ Basis for Nul A: $\begin{bmatrix} 4 \\ -5 \\ 1 \\ 0 \end{bmatrix}, \begin{bmatrix} -7 \\ 6 \\ 0 \\ 1 \end{bmatrix}$

Note: A basis is a *set* of vectors. For simplicity, the answers here and in the text list the vectors without enclosing the list inside set brackets. Ask your instructor if this format is acceptable.

Warning: A common error is to confuse Col A with Nul A. This happens easily when the definitions of these spaces are not known precisely. Another error is to think that the nonpivot columns of an $m \times n$ matrix A form a basis for Nul A. This is not true in general, even when $m = n$.

33. $A = \begin{bmatrix} 1 & 4 & 8 & -3 & -7 \\ -1 & 2 & 7 & 3 & 4 \\ -2 & 2 & 9 & 5 & 5 \\ 3 & 6 & 9 & -5 & -2 \end{bmatrix} \sim \begin{bmatrix} 1 & 4 & 8 & 0 & 5 \\ 0 & 2 & 5 & 0 & -1 \\ 0 & 0 & 0 & 1 & 4 \\ 0 & 0 & 0 & 0 & 0 \end{bmatrix}.$

Basis for Col A: $\begin{bmatrix} 1 \\ -1 \\ -2 \\ 3 \end{bmatrix}, \begin{bmatrix} 4 \\ 2 \\ 2 \\ 6 \end{bmatrix}, \begin{bmatrix} -3 \\ 3 \\ 5 \\ -5 \end{bmatrix}.$

For Nul A, obtain the *reduced* (and augmented) echelon form for $A\mathbf{x} = \mathbf{0}$:

$$[A \quad \mathbf{0}] \sim \begin{bmatrix} 1 & 0 & -2 & 0 & 7 & 0 \\ 0 & 1 & 2.5 & 0 & -.5 & 0 \\ 0 & 0 & 0 & 1 & 4 & 0 \\ 0 & 0 & 0 & 0 & 0 & 0 \end{bmatrix}. \quad \begin{aligned} x_1 \quad - 2x_3 \quad + 7x_5 &= 0 \\ x_2 + 2.5x_3 \quad - .5x_5 - 0 \\ x_4 + 4x_5 &= 0. \\ 0 &= 0 \end{aligned}$$

Copyright © 2021 Pearson Education, Inc.

The solution in parametric vector form:

$$\begin{bmatrix} x_1 \\ x_2 \\ x_3 \\ x_4 \\ x_5 \end{bmatrix} = \begin{bmatrix} 2x_3 - 7x_5 \\ -2.5x_3 + .5x_5 \\ x_3 \\ -4x_5 \\ x_5 \end{bmatrix} = x_3 \begin{bmatrix} 2 \\ -2.5 \\ 1 \\ 0 \\ 0 \end{bmatrix} + x_5 \begin{bmatrix} -7 \\ .5 \\ 0 \\ -4 \\ 1 \end{bmatrix}.$$

$$\underset{\mathbf{u}}{\uparrow} \qquad \underset{\mathbf{v}}{\uparrow}$$

Basis for Nul A: $\{\mathbf{u}, \mathbf{v}\}$.

Note: This solution illustrates how you can save time on an exam and not copy the ten numbers in the basis vectors for your answer. Just label the basis vectors as $\mathbf{u}$ and $\mathbf{v}$, and write something such as "$\mathbf{u}$ and $\mathbf{v}$ form a basis for Nul A." You might ask your instructor if this is acceptable.

Warning: Do not become too attached to the symbols commonly used for certain ideas. For instance, calling a vector "$\mathbf{b}$" does not imply that it can only be the "right side" of an equation $A\mathbf{x} = \mathbf{b}$. In Exercise 37, you should be looking for a vector $\mathbf{b}$ such that $A\mathbf{b} = \mathbf{0}$.

37. A simple construction is to write any nonzero 3×3 matrix whose columns are obviously linearly dependent, and then make $\mathbf{b}$ a vector of weights that come from a linear dependence relation among the columns. For instance, if the first two columns of A are equal, then $\mathbf{a}_1 - \mathbf{a}_2 + 0\mathbf{a}_3 = \mathbf{0}$. So, $\mathbf{b}$ could be $(1, -1, 0)$.

39. The text has an answer. An answer such as "Nul F is a subset of $\mathbb{R}^5$" says something true, but not much. "Nul F is a subspace of $\mathbb{R}^5$" ought to be good for some partial credit, but this fact does not use the information given about the column space of F. Probably the best possible answer is that Nul F is a nonzero subspace of $\mathbb{R}^5$.

45. Use the command that produces the reduced echelon form in one step (**rref**). By Theorem 13, the pivot columns of A form a basis for Col A.

$$A = \begin{bmatrix} 3 & -5 & 0 & -1 & 3 \\ -7 & 9 & -4 & 9 & -11 \\ -5 & 7 & -2 & 5 & -7 \\ 3 & -7 & -3 & 4 & 0 \end{bmatrix} \sim \begin{bmatrix} 1 & 0 & 2.5 & -4.5 & 3.5 \\ 0 & 1 & 1.5 & -2.5 & 1.5 \\ 0 & 0 & 0 & 0 & 0 \\ 0 & 0 & 0 & 0 & 0 \end{bmatrix}$$

Copyright © 2021 Pearson Education, Inc.

Basis for Col A: $\begin{bmatrix} 3 \\ -7 \\ -5 \\ 3 \end{bmatrix}, \begin{bmatrix} -5 \\ 9 \\ 7 \\ -7 \end{bmatrix}$

For Nul A, obtain the solution of $A\mathbf{x} = \mathbf{0}$ in parametric vector form:

$$x_1 \quad + 2.5x_3 - 4.5x_4 + 3.5x_5 = 0$$
$$x_2 + 1.5x_3 - 2.5x_4 + 1.5x_5 = 0$$

Solution: $\begin{cases} x_1 = -2.5x_3 + 4.5x_4 - 3.5x_5 \\ x_2 = -1.5x_3 + 2.5x_4 - 1.5x_5 \\ x_3, x_4, \text{ and } x_5 \text{ are free} \end{cases}$

$$\mathbf{x} = \begin{bmatrix} x_1 \\ x_2 \\ x_3 \\ x_4 \\ x_5 \end{bmatrix} = \begin{bmatrix} -2.5x_3 + 4.5x_4 - 3.5x_5 \\ -1.5x_3 + 2.5x_4 - 1.5x_5 \\ x_3 \\ x_4 \\ x_5 \end{bmatrix} = x_3 \begin{bmatrix} -2.5 \\ -1.5 \\ 1 \\ 0 \\ 0 \end{bmatrix} + x_4 \begin{bmatrix} 4.5 \\ 2.5 \\ 0 \\ 1 \\ 0 \end{bmatrix} + x_5 \begin{bmatrix} -3.5 \\ -1.5 \\ 0 \\ 0 \\ 1 \end{bmatrix} = x_3\mathbf{u} + x_4\mathbf{v} + x_5\mathbf{w}$$

Basis for Nul A: $\{\mathbf{u}, \mathbf{v}, \mathbf{w}\}$.

Mastering Linear Algebra Concepts: Subspace, Column Space, Null Space, Basis

To form strong mental images of a subspace and the two main types of subspaces (Col A and Nul A), prepare a review sheet that covers the following categories:

- definitions
- equivalent descriptions
- geometric interpretations   Fig. 1, Sentence after Example 1
- special cases
- examples and counterexamples   Examples 1, 2, 3, Exercises 1–4
- typical computations   Example 4, Practice Problems 1 and 2, Exercises 5–14
- contrast between Col A and Nul A

Copyright © 2021 Pearson Education, Inc.

The concept of a basis deserves a separate review sheet. Use the categories below. Also, add notes to your review sheets for Span and Linear Independence that say when a spanning set is a basis for a subspace Span$\{\mathbf{v}_1, \ldots, \mathbf{v}_p\}$ and when a linearly independent set in $\mathbb{R}^n$ is a basis for $\mathbb{R}^n$.

- definition
- geometric interpretations   Fig. 3
- special cases   Example 5
- examples and counterexamples   Warning, Exercises 15–20
- typical computations   Examples 6 and 8, Exercises 31–34

MATLAB   rref

The command **rref(A)** produces the reduced echelon form of A. From that you can write a basis for Col A or write the homogeneous equations that describe Nul A. (Don't forget that A is a coefficient matrix, not an augmented matrix.)

2.9 - Dimension and Rank

This section and Section 2.8 cover the ideas from Chapter 4 that you need for Chapters 5–7. There is no need to read this section if your course covers Chapter 4.

KEY IDEAS

The two fundamental concepts in this section are the dimension of a subspace and the rank of a matrix. Coordinate vectors are used to give an intuitive understanding of dimension and a geometric explanation of why a k-dimensional subspace of $\mathbb{R}^n$ behaves as if it were $\mathbb{R}^k$.

The Basis Theorem ties together the concepts of dimension, subspace, linear independence, span, and basis. So does the Invertible Matrix Theorem. Make sure you know the precise wording of the statements in these theorems.

The following table lists all statements that are in the Invertible Matrix Theorem at this point in the course. They are arranged in the scheme used in Section 2.3 of this *Study Guide*. As before, a few extra statements have been added to make the table more symmetrical.

Copyright © 2021 Pearson Education, Inc.

STATEMENTS FROM THE INVERTIBLE MATRIX THEOREM

Equivalent statements for an $m \times n$ matrix A.	Equivalent statements for an $n \times n$ square matrix A.	Equivalent statements for an $n \times p$ matrix A.
k. There is a matrix D such that $AD = I$. *. A has a pivot position in every row. h. The columns of A span $\mathbb{R}^m$. g. The equation $A\mathbf{x} = \mathbf{b}$ has at least one solution for each $\mathbf{b}$ in $\mathbb{R}^m$. i. The transformation $\mathbf{x} \mapsto A\mathbf{x}$ maps $\mathbb{R}^n$ onto $\mathbb{R}^m$. n. Col $A = \mathbb{R}^m$. o. dim Col $A = m$. *. rank $A = m$.	a. A is an invertible matrix. c. A has n pivot positions. m. The columns of A form a basis for $\mathbb{R}^n$. *. The equation $A\mathbf{x} = \mathbf{b}$ has a unique solution for each $\mathbf{b}$ in $\mathbb{R}^n$. *. The transformation $\mathbf{x} \mapsto A\mathbf{x}$ is invertible. b. A is row equivalent to I. l. A^T is invertible. p. rank $A = n$.	j. There is a matrix C such that $CA = I$. *. A has a pivot position in every column. e. The columns of A are linearly independent. d. The equation $A\mathbf{x} = \mathbf{0}$ has only the trivial solution. f. The transformation $\mathbf{x} \mapsto A\mathbf{x}$ is one-to-one. q. Nul $A = \{\mathbf{0}\}$. r. dim Nul $A = 0$. *. rank $A = p$.

With many concepts to learn in Sections 2.8 and 2.9, you need to be careful not to combine terms in ways that are undefined, even though they may sound reasonable to you. For example, after you finish your work on this section you should recognize that the following phrases (which have appeared on my students' papers) are meaningless: "the basis of a matrix," "the dimension of a basis," and "the rank of a basis."

_____

SOLUTIONS TO EXERCISES

1. If $[\mathbf{x}]_B = \begin{bmatrix} 3 \\ 2 \end{bmatrix}$, then $\mathbf{x}$ is formed from $\mathbf{b}_1$ and $\mathbf{b}_2$ using weights 3 and 2:

$$\mathbf{x} = 3\mathbf{b}_1 + 2\mathbf{b}_2 = 3\begin{bmatrix} 1 \\ 1 \end{bmatrix} + 2\begin{bmatrix} 2 \\ -1 \end{bmatrix} = \begin{bmatrix} 7 \\ 1 \end{bmatrix}$$

7. The figure in the text suggests that $\mathbf{w} = 2\mathbf{b}_1 - \mathbf{b}_2$ and $\mathbf{x} = 1.5\mathbf{b}_1 + .5\mathbf{b}_2$, in which case,

$$[\mathbf{w}]_B = \begin{bmatrix} 2 \\ -1 \end{bmatrix} \text{ and } [\mathbf{x}]_B = \begin{bmatrix} 1.5 \\ .5 \end{bmatrix}.$$

To confirm $[\mathbf{x}]_B$, compute

Copyright © 2021 Pearson Education, Inc.

$$1.5\mathbf{b}_1 + .5\mathbf{b}_2 = 1.5\begin{bmatrix} 3 \\ 0 \end{bmatrix} + .5\begin{bmatrix} -1 \\ 2 \end{bmatrix} = \begin{bmatrix} 4 \\ 1 \end{bmatrix} = \mathbf{x}$$

Note: The figures in the text for Exercises 7 and 8 display what Section 4.4 calls $\mathcal{B}$-graph paper.

Study Tip: Exercises 9–16 make good test questions because they do not require much arithmetic. The problem of finding a basis for a null space is particularly important, because this skill is needed throughout Chapters 5 and 7.

13. The four vectors span the column space H of a matrix that can be reduced to echelon form:

$$\begin{bmatrix} 1 & -3 & 2 & -4 \\ -3 & 9 & -1 & 5 \\ 2 & -6 & 4 & -3 \\ -4 & 12 & 2 & 7 \end{bmatrix} \sim \begin{bmatrix} 1 & -3 & 2 & -4 \\ 0 & 0 & 5 & -7 \\ 0 & 0 & 0 & 5 \\ 0 & 0 & 10 & -9 \end{bmatrix} \sim \begin{bmatrix} 1 & -3 & 2 & -4 \\ 0 & 0 & 5 & -7 \\ 0 & 0 & 0 & 5 \\ 0 & 0 & 0 & 5 \end{bmatrix} \sim \begin{bmatrix} 1 & -3 & 2 & -4 \\ 0 & 0 & 5 & -7 \\ 0 & 0 & 0 & 5 \\ 0 & 0 & 0 & 0 \end{bmatrix}$$

Columns 1, 3, and 4 of the original matrix form a basis for H, so dim $H = 3$.

17. Check the definition of coordinates relative to a basis.

19. Dimension is defined only for a subspace.

21. See the sentence before Example 3.

23. See the Rank Theorem.

25. See the Basis Theorem.

27. The text answer uses the Rank Theorem, which is fine. However, you can also answer Exercises 27–30 without explicit reference to the Rank Theorem. For instance, in Exercise 27, if the null space of a matrix A is three-dimensional, then the equation $A\mathbf{x} = \mathbf{0}$ has three free variables, and three of the columns of A are nonpivot columns. Since a 5×7 matrix has seven columns, A must have four pivot columns (which form a basis of Col A). So rank A = dim Col A = 4.

33. The text has the solution. This question provides a good way to test knowledge of the Basis Theorem.

35. **a**. Start with $B = [\mathbf{b}_1 \ \cdots \ \mathbf{b}_p]$ and $A = [\mathbf{a}_1 \ \cdots \ \mathbf{a}_q]$, where $q > p$. For $j = 1, \dots, q$, the vector $\mathbf{a}_j$ is in W. Since the columns of B span W, the vector $\mathbf{a}_j$ is in the column space of B. That is, $\mathbf{a}_j = B\mathbf{c}_j$ for some vector $\mathbf{c}_j$ of weights. Note that $\mathbf{c}_j$ is in $\mathbb{R}^p$ because B has p columns.

   b. Let $C = [\mathbf{c}_1 \ \cdots \ \mathbf{c}_q]$. Then C is a $p \times q$ matrix because each of the q columns is in $\mathbb{R}^p$. By hypothesis, q is larger than p, so C has more columns than rows. By a theorem, the columns of C are linearly dependent and there exists a nonzero vector $\mathbf{u}$ in $\mathbb{R}^q$ such that $C\mathbf{u} = \mathbf{0}$.

   c. From part (a) and the definition of matrix multiplication

Copyright © 2021 Pearson Education, Inc.

$$A = [\mathbf{a}_1 \ \cdots \ \mathbf{a}_q] = [B\mathbf{c}_1 \ \cdots \ B\mathbf{c}_q] = BC$$

From part (b), $A\mathbf{u} = (BC)\mathbf{u} = B(C\mathbf{u}) = B\mathbf{0} = \mathbf{0}$. Since $\mathbf{u}$ is nonzero, the columns of A are linearly dependent.

Mastering Linear Algebra Concepts: Dimension and Rank

The concepts of dimension and rank are relatively simple, but they are used so often later that they deserve a review sheet. Pay attention to how they are used in the sentences of the Invertible Matrix Theorem. "Dimension" is always attached to a subspace (not a matrix or vector or basis), and "rank" is attached to a matrix (not a subspace or other object).

- definitions
- geometric interpretations   Fig. 1, a coordinate system on a 2-dim subspace
- special cases   Paragraph before Example 2
- examples and counterexamples   Examples 2 and 3
- typical computations   Practice Problem 1, Exercises 9–16 and 27–29

MATLAB   rank

You can use **rref(A)** to check the rank of A, but roundoff error or an extremely small pivot entry can produce an incorrect echelon form. A more reliable command is **rank(A)**.

Chapter 2 - Supplementary Exercises

In this chapter, Exercises 1-15 consist of true/false questions, whose level of difficulty varies. Some are similar to the ones that appear in many sections of the text, in which a word or phrase is sometimes missing or slightly misstated. Some follow fairly easily from a theorem: others may need careful reasoning. A few may require an argument that uses several ideas. In each case, think carefully about the statement and attempt to write a solution. The text provides the true/false answer, but you must supply the justification or counterexample. Careful work on these exercises will help you prepare for an exam over the chapter material.

21. Since $A^{-1}B$ is the solution of $AX = B$, row reduction of $[A \ \ B]$ to $[I \ \ X]$ produces $X = A^{-1}B$.

See Exercise 22 in Section 2.2. In fact, $A^{-1}B = \begin{bmatrix} 10 & -1 \\ 9 & 10 \\ -5 & -3 \end{bmatrix}$.

Copyright © 2021 Pearson Education, Inc.

25. c. When $x_1, \ldots, x_n$ are distinct, the columns of V are linearly independent, by (b). By the Invertible Matrix Theorem, V is invertible and its columns span $\mathbf{R}^n$. So, for every vector $\mathbf{y} = (y_1, \ldots, y_n)$ in $\mathbf{R}^n$, there is a vector $\mathbf{c}$ such that $V\mathbf{c} = \mathbf{y}$. Let p be the polynomial whose coefficients are listed in $\mathbf{c}$. Then, by (a), p is an interpolating polynomial for $(x_1, y_1), \ldots,$ (x_n, y_n).

31. The text has a solution. In addition, note that it is possible that BA is invertible. For example, let C be an invertible 4×4 matrix and construct $A = \begin{bmatrix} C \\ 0 \end{bmatrix}$ and $B = [C^{-1} \ \ 0]$. Then $BA = I_4$, which is invertible.

Chapter 2 - Glossary Checklist

Check your knowledge by attempting to write definitions of the terms below. Then compare your work with the definitions given in the text's Glossary. Ask your instructor which definitions, if any, might appear on a test.

associative law of multiplication: . . . .

basis (for a subspace H of $\mathbb{R}^n$, §2.8):   A set B = $\{\mathbf{v}_1, \ldots, \mathbf{v}_p\}$ in $\mathbb{R}^n$ such that: . . . .

block matrix: *See* partitioned matrix.

block matrix multiplication: The . . . multiplication of . . . as if . . . .

column space (of an $m \times n$ matrix A, §2.8):   The set Col A of . . . .

column sum: The sum of . . . .

commuting matrices: Two matrices A and B such that . . . .

composition of linear transformations: A mapping produced by applying . . . .

conformable for block multiplication: Two partitioned matrices A and B such that . . . .

consumption matrix: A matrix in the . . . model whose columns are . . . .

coordinate vector of x relative to a basis $\mathcal{B} = \{\mathbf{b}_1, \ldots, \mathbf{b}_p\}$ (§2.9):   The vector $[\mathbf{x}]_{\mathcal{B}}$ whose entries c_1, . . ., c_p satisfy . . . .

diagonal entries (in a matrix):   Entries having . . . .

diagonal matrix: A square matrix . . . whose entries are . . . .

dimension (of a subspace, §2.9):   The number . . . .

determinant $\left(of \ A = \begin{bmatrix} a & b \\ c & d \end{bmatrix} \right)$:   The number . . ., denoted by . . . .

distributive laws: (left) . . . (right) . . . .

elementary matrix: An invertible matrix that results by . . . .

Copyright © 2021 Pearson Education, Inc.

final demand vector (or **bill of final demands**):   The vector **d** in the . . . model that lists . . . . The vector **d** can represent . . . .

flexibility matrix: A matrix whose jth column gives . . . of an elastic beam at specified points when . . . is applied at . . . .

Householder reflection: A transformation $\mathbf{x} \mapsto Q\mathbf{x}$, where $Q = \ldots$.

identity matrix: The $n \times n$ matrix I or I_n with . . . .

inner product: A matrix product . . . where **u** and **v** are . . . .

input-output matrix: *See* consumption matrix.

input-output model: *See* Leontief input-output model.

intermediate demands: Demands for goods or services that . . . .

inverse (of an $n \times n$ matrix A)**:**   An $n \times n$ matrix A^{-1} such that . . . .

invertible linear transformation: A linear transformation $T : \mathbb{R}^n \to \mathbb{R}^m$ such that there exists . . . .

invertible matrix: A square matrix that . . . .

ladder network: An electrical network assembled by connecting . . . .

left inverse (of A)**:**   Any rectangular matrix C such that . . . .

Leontief input-output model (or **Leontief production equation**)**:**   The equation . . . , where . . . .

lower triangular matrix: A matrix with . . . .

lower triangular part (of A)**:**   A . . . matrix whose entries on . . . .

LU factorization: The representation of a matrix A in the form $A = LU$, where L is . . . and U is . . . .

main diagonal (of a matrix)**:**   The location of the . . . .

null space (of an $m \times n$ matrix A, §2.8)**:**   The set Nul A of all . . . .

outer product: A matrix product . . . where **u** and **v** are . . . .

partitioned matrix: A matrix whose entries are . . . . Sometimes called . . . .

permuted lower triangular matrix: A matrix such that . . . .

permuted LU factorization: The representation of a matrix A in the form $A = LU$, where L is . . . and U is . . . .

production vector: The vector in the . . . model that lists . . . .

rank (of a matrix A, §2.9)**:**   . . . .

right inverse (of A)**:**   Any rectangular matrix C such that . . . .

row-column rule: The rule for computing a product AB in which . . . .

Schur complement: A certain matrix formed from the blocks of a 2×2 partitioned matrix $A = [A_{ij}]$. If A_{11} is invertible, its Schur complement is given by . . . . If A_{22} is invertible, its Schur complement is given by . . . .

Copyright © 2021 Pearson Education, Inc.

stiffness matrix: The inverse of a . . . matrix. The jth column of a stiffness matrix gives . . . at specified points on an elastic beam in order to produce . . . .

subspace (of $\mathbb{R}^n$, §2.8):   A subset H of $\mathbb{R}^n$ with the properties: . . . .

transfer matrix: A matrix A associated with an electrical circuit having input and output terminals, such that . . . .

transpose (of A):   An $n \times m$ matrix A^T whose . . . are the corresponding . . . of . . . .

unit consumption vector: A column vector in the . . . model that lists . . . .

unit lower triangular matrix: A matrix with . . . .

upper triangular matrix: A matrix U with . . . .

Vandermonde matrix: An $n \times n$ matrix V or its transpose, of the form . . . .

Copyright © 2021 Pearson Education, Inc.

3 Determinants

3.1 - Introduction to Determinants

This section is relatively short and easy. Some exercises provide computational practice and others allow you to discover properties of determinants to be studied in the next section. You will enjoy Section 3.2 more if you finish your work on this section first.

KEY IDEAS

The second paragraph of the section sets the stage for what follows. Read it quickly, without worrying about the details of the row operations. The main idea is that a number related to the determinant of A in a predictable way can be formed by taking the product of the diagonal elements of an echelon form of A, and this number is nonzero if and only if the matrix is invertible. Later (in Section 3.3) you will see why this idea is important. For now, this 3×3 case is only used to motivate the definition of det A.

Determinants are defined here via a cofactor expansion along the first row. Since the cofactors involve determinants of smaller matrices, the definition is said to be *recursive*. For each $n \geq 2$, the determinant of an $n \times n$ matrix is based on the definition of the determinant of an $(n–1) \times (n–1)$ matrix. There are other equivalent definitions of the determinant, but we shall not digress to discuss them.

Study Tip: Watch how parentheses are used in Example 2 to avoid a common mistake. The cofactor expansion puts a minus sign in front of a_{32} because $(-1)^{3+2} = -1$. Since a_{32} happens to be negative, the correct term in the expansion is $-(-2) \det A_{32}$, *not* $-2 \det A_{32}$.

SOLUTIONS TO EXERCISES

1. By definition, det A is computed via a cofactor expansion along the first row:

$$\begin{vmatrix} 3 & 0 & 4 \\ 2 & 3 & 2 \\ 0 & 5 & -1 \end{vmatrix} = 3 \begin{vmatrix} 3 & 2 \\ 5 & -1 \end{vmatrix} - 0 \begin{vmatrix} 2 & 2 \\ 0 & -1 \end{vmatrix} + 4 \begin{vmatrix} 2 & 3 \\ 0 & 5 \end{vmatrix}$$

$$= 3(-3-10) - 0 + 4(10-0) = -39 + 40 = 1$$

For comparison, a cofactor expansion down the second column yields

Copyright © 2021 Pearson Education, Inc.

$$\begin{vmatrix} 3 & 0 & 4 \\ 2 & 3 & 2 \\ 0 & 5 & -1 \end{vmatrix} = (-1)^{1+2} \cdot 0 \begin{vmatrix} 2 & 2 \\ 0 & -1 \end{vmatrix} + (-1)^{2+2} \cdot 3 \begin{vmatrix} 3 & 4 \\ 0 & -1 \end{vmatrix} + (-1)^{3+2} \cdot 5 \begin{vmatrix} 3 & 4 \\ 2 & 2 \end{vmatrix}$$

$$= 0 + 3(-3 - 0) - 5(6 - 8) = -9 + 10 = 1$$

Study Tip: To save time, omit the zero terms in a cofactor expansion, but be careful to use the proper plus or minus signs with the nonzero terms.

7. By definition,

$$\begin{vmatrix} 4 & 3 & 0 \\ 6 & 5 & 2 \\ 9 & 7 & 3 \end{vmatrix} = 4\begin{vmatrix} 5 & 2 \\ 7 & 3 \end{vmatrix} - 3\begin{vmatrix} 6 & 2 \\ 9 & 3 \end{vmatrix} = 4(15 - 14) - 3(18 - 18) = 4$$

Using the second column of A instead,

$$\begin{vmatrix} 4 & 3 & 0 \\ 6 & 5 & 2 \\ 9 & 7 & 3 \end{vmatrix} = -3\begin{vmatrix} 6 & 2 \\ 9 & 3 \end{vmatrix} + 5\begin{vmatrix} 4 & 0 \\ 9 & 3 \end{vmatrix} - 7\begin{vmatrix} 4 & 0 \\ 6 & 2 \end{vmatrix}$$

$$= -3(18 - 18) + 5(12 - 0) - 7(8 - 0) = 0 + 60 - 56 = 4$$

13. Row 2 or column 2 are the best choices because they contain the most zeros. We'll use row 2. Since the only nonzero entry in that row is 2, the determinant is $(-1)^{2+3} \cdot 2 \cdot 2A_{23}$.

$$\det A = (-1)^{2+3} \cdot 2 \begin{vmatrix} 4 & 0 & -7 & 3 & -5 \\ 0 & 0 & 2 & 0 & 0 \\ 7 & 3 & -6 & 4 & -8 \\ 5 & 0 & 5 & 2 & -3 \\ 0 & 0 & 9 & -1 & 2 \end{vmatrix} = (-2) \cdot \begin{vmatrix} 4 & 0 & 3 & -5 \\ 7 & 3 & 4 & -8 \\ 5 & 0 & 2 & -3 \\ 0 & 0 & -1 & 2 \end{vmatrix}$$

The best choice for this 4×4 determinant is to expand down the second column. Notice that the cofactor associated with the 3 in the $(2, 2)$ position is the $(2, 2)$-cofactor of the 4×4 matrix. The original location of the "3" in the 5×5 matrix is irrelevant.

$$\det A = (-2) \cdot (-1)^{2+2}(3) \cdot \begin{vmatrix} 4 & 3 & -5 \\ 5 & 2 & -3 \\ 0 & -1 & 2 \end{vmatrix} = -6 \begin{vmatrix} 4 & 3 & -5 \\ 5 & 2 & -3 \\ 0 & -1 & 2 \end{vmatrix}$$

Finally, use column 1 (although row 3 would work as well).

Copyright © 2021 Pearson Education, Inc.

$$\det A = (-6)\cdot\left(4\begin{vmatrix} 2 & -3 \\ -1 & 2 \end{vmatrix} - 5\begin{vmatrix} 3 & -5 \\ -1 & 2 \end{vmatrix} + 0\right)$$

$$= -6[4(4-3)-5(6-5)] = -6(4-5) = 6$$

Checkpoint: Try to complete the following statement: "If the kth column of the $n\times n$ identity matrix is replaced by a column vector $\mathbf{x}$ whose entries are $x_1, ..., x_n$, then the determinant of the resulting matrix is _____." To discover the answer, compute the determinants of the following matrices:

a. $\begin{bmatrix} 1 & 3 & 0 & 0 \\ 0 & 4 & 0 & 0 \\ 0 & 5 & 1 & 0 \\ 0 & 6 & 0 & 1 \end{bmatrix}$ **b.** $\begin{bmatrix} 1 & 0 & 3 & 0 \\ 0 & 1 & 4 & 0 \\ 0 & 0 & 5 & 0 \\ 0 & 0 & 6 & 1 \end{bmatrix}$ **c.** $\begin{bmatrix} 1 & 0 & 3 & 0 & 0 \\ 0 & 1 & 4 & 0 & 0 \\ 0 & 0 & 5 & 0 & 0 \\ 0 & 0 & 6 & 1 & 0 \\ 0 & 0 & 7 & 0 & 1 \end{bmatrix}$

19. $\det\begin{bmatrix} a & b \\ c & d \end{bmatrix} = ad - bc$, and $\det\begin{bmatrix} c & d \\ a & b \end{bmatrix} = cb - da = -\det\begin{bmatrix} a & b \\ c & d \end{bmatrix}$.

Interchanging two rows reverses the sign of the determinant, at least for the 2×2 case. Do you expect this to be true for larger matrices?

25. The matrix is triangular, so use Theorem 2.

$$\det\begin{bmatrix} 1 & 0 & 0 \\ 0 & 1 & 0 \\ 0 & k & 1 \end{bmatrix} = 1\cdot 1\cdot 1 = 1 \qquad \text{Product of the diagonal entries}$$

31. A 3×3 row replacement matrix has one of the following forms:

$$\begin{bmatrix} 1 & 0 & 0 \\ k & 1 & 0 \\ 0 & 0 & 1 \end{bmatrix}, \begin{bmatrix} 1 & 0 & 0 \\ 0 & 1 & 0 \\ k & 0 & 1 \end{bmatrix}, \begin{bmatrix} 1 & 0 & 0 \\ 0 & 1 & 0 \\ 0 & k & 1 \end{bmatrix},$$

$$\begin{bmatrix} 1 & k & 0 \\ 0 & 1 & 0 \\ 0 & 0 & 1 \end{bmatrix}, \begin{bmatrix} 1 & 0 & k \\ 0 & 1 & 0 \\ 0 & 0 & 1 \end{bmatrix}, \begin{bmatrix} 1 & 0 & 0 \\ 0 & 1 & k \\ 0 & 0 & 1 \end{bmatrix}$$

In each case the matrix is triangular with 1's on the diagonal, so its determinant equals 1. The determinant of a row replacement matrix is 1, at least for the 3×3 case. Do you expect this to be true for larger matrices?

37. $\det A = \det\begin{bmatrix} 3 & 1 \\ 4 & 2 \end{bmatrix} = 3(2)-1(4) = 6-4 = 2$. Since $5A = \begin{bmatrix} 5\cdot 3 & 5\cdot 1 \\ 5\cdot 4 & 5\cdot 2 \end{bmatrix}$,

$$\det 5A = (5\cdot 3)(5\cdot 2)-(5\cdot 1)(5\cdot 4) = 150-100 = 50.$$

Copyright © 2021 Pearson Education, Inc.

So, det $5A \neq 5 \cdot \det A$. Can you see what the true relation between det $5A$ and det A really is, at least for this example? What about det $5A$ for any 2×2 matrix? Try to guess (and perhaps verify) a formula for det rA, where r is any scalar and A is any $n\times n$ matrix.

39. See the paragraph preceding the definition of det A.

41. See the definition of cofactor, preceding Theorem 1.

43. $\det[\mathbf{u}\quad\mathbf{v}] = \det\begin{bmatrix} 3 & 1 \\ 0 & 2 \end{bmatrix} = 6$, $\det[\mathbf{u}\quad\mathbf{x}] = \det\begin{bmatrix} 3 & x \\ 0 & 2 \end{bmatrix} = 6$, and the areas of the parallelograms

determined by $[\mathbf{u}\quad\mathbf{v}]$ and $[\mathbf{u}\quad\mathbf{x}]$ both equal 6. To see why the areas are equal, consider the parallelograms determined by $\mathbf{u} = (3, 0)$ and $\mathbf{v} = (1, 2)$ and by $\mathbf{u}$ and $\mathbf{x} = (x, 2)$:

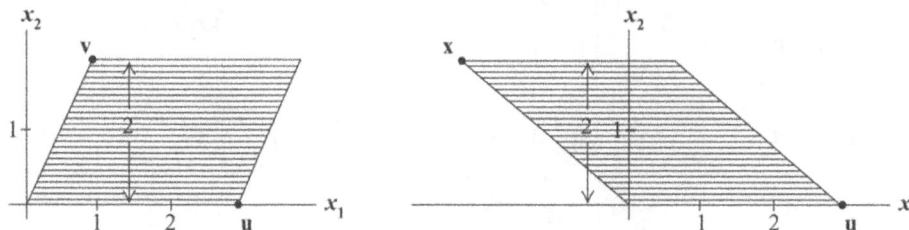

The parallelogram on the left is determined by $\mathbf{u}$ and $\mathbf{v}$ (and the vertices $\mathbf{u} + \mathbf{v}$ and $\mathbf{0}$). Its base is 3 and its altitude is 2, so the area is (base)(altitude) = 6. The parallelogram on the right, determined by $\mathbf{u}$ and $\mathbf{x} = (x, 2)$, has the same base. Also, the altitude is 2 for any value of x, so the area again equals $3 \cdot 2 = 6$.

Answer to Checkpoint: **a.** 4 **b.** 5 **c.** 5 "If the kth column of the $n\times n$ identity matrix is replaced by a column vector $\mathbf{x}$ whose entries are $x_1, ..., x_n$, then the determinant of the resulting matrix is x_k." Can you explain why this is true? You'll learn the answer when you begin Section 3.3.

3.2 - Properties of Determinants

This section presents the main properties of determinants, gives an efficient method of computation, and proves that a matrix is invertible if and only if its determinant is nonzero.

KEY IDEAS

It is not surprising that row operations relate nicely to determinants. After all, we found the definition of a 3×3 determinant by row reducing a 3×3 matrix. Theorem 3 can be rephrased informally as follows:

a. Adding a multiple of one row (or column) of A to another does not change the determinant.

b. Interchanging two rows (or columns) of A reverses the sign of the determinant.

c. A constant may be factored out of one row (or column) of the determinant of A.

The other properties to learn are stated in Theorems 4, 5, and 6, together with the boxed formula for det A labeled as Equation (1). Theorem 4 is sometimes stated as: *A square matrix is*

Copyright © 2021 Pearson Education, Inc.

nonsingular if and only if its determinant is nonzero. Theorems 4 and 6 will be used extensively in Chapter 5.

Your instructor may or may not want you to know the (multi-) linearity property. This property is important in more advanced courses but is not used later in the text. *Warning*: in general, det($A + B$) is *unequal* to det A + det B.

_____

SOLUTIONS TO EXERCISES

1. Rows 1 and 2 are interchanged, so the determinant changes sign.

7.
$$
\begin{vmatrix} 1 & 3 & 0 & 2 \\ -2 & -5 & 7 & 4 \\ 3 & 5 & 2 & 1 \\ 1 & -1 & 2 & -3 \end{vmatrix}
=
\begin{vmatrix} 1 & 3 & 0 & 2 \\ 0 & 1 & 7 & 8 \\ 0 & -4 & 2 & -5 \\ 0 & -4 & 2 & -5 \end{vmatrix}
=
\begin{vmatrix} 1 & 3 & 0 & 2 \\ 0 & 1 & 7 & 8 \\ 0 & 0 & 30 & 27 \\ 0 & 0 & 30 & 27 \end{vmatrix}
=
\begin{vmatrix} 1 & 3 & 0 & 2 \\ 0 & 1 & 7 & 8 \\ 0 & 0 & 30 & 27 \\ 0 & 0 & 0 & 0 \end{vmatrix}
= 0
$$

Note, the second array already shows that the determinant is zero, because two rows are equal, as in Example 3.

Study Tip: In general, computation of a 3×3 determinant by row reduction takes 10 multiplications (and divisions), but cofactor expansion only takes 9 multiplications. At $n = 4$, the advantage switches to row reduction, which requires 23 multiplications, cofactor expansion 40 (9 for four 3×3 determinants, plus 4 multiplications of a_{ij} times det A_{ij}). Often, the best strategy is to combine the two techniques, as in Exercises 11–14.

13. Use row or column operations whenever convenient to create a row or column that has only one nonzero entry. (I recommend using only row operations, because you already have experience with them.) Then use a cofactor expansion to reduce the size of the matrix.

Copyright © 2021 Pearson Education, Inc.

$$\begin{vmatrix} 2 & 5 & 4 & 1 \\ 4 & 7 & 6 & 2 \\ 6 & -2 & -4 & 0 \\ -6 & 7 & 7 & 0 \end{vmatrix} = \begin{vmatrix} 2 & 5 & 4 & 1 \\ 0 & -3 & -2 & 0 \\ 6 & -2 & -4 & 0 \\ -6 & 7 & 7 & 0 \end{vmatrix}$$

Zero created
in column 4

$$= - \begin{vmatrix} 0 & -3 & -2 \\ 6 & -2 & -4 \\ -6 & 7 & 7 \end{vmatrix}$$

Result of cofactor
expansion down column 4

$$= - \begin{vmatrix} 0 & -3 & -2 \\ 6 & -2 & -4 \\ 0 & 5 & 3 \end{vmatrix}$$

Zero created
in column 1

$$= -(-6) \begin{vmatrix} -3 & -2 \\ 5 & 3 \end{vmatrix}$$

Result of cofactor
expansion down column 1

$$= 6 \cdot (-9 + 10) = 6$$

19. $$\begin{vmatrix} a & b & c \\ 2d+a & 2e+b & 2f+c \\ g & h & i \end{vmatrix} = \begin{vmatrix} a & b & c \\ 2d & 2e & 2f \\ g & h & i \end{vmatrix}$$

$(-1) \cdot$ row 1 added
to row 2

$$= 2 \cdot \begin{vmatrix} a & b & c \\ d & e & f \\ g & h & i \end{vmatrix}$$

2 factored
out of row 2

$$= 2 \cdot 7 = 14$$

25. By Theorem 4 and the IMT, the set $\{\mathbf{v}_1, \mathbf{v}_2, \mathbf{v}_3\}$ is linearly independent if and only if $\det[\mathbf{v}_1 \ \ \mathbf{v}_2 \ \ \mathbf{v}_3] \neq 0$. Rather than use row operations on $[\mathbf{v}_1 \ \ \mathbf{v}_2 \ \ \mathbf{v}_3]$, you might choose to expand the determinant by cofactors of the third column:

$$\begin{vmatrix} 7 & -8 & 7 \\ -4 & 5 & 0 \\ -6 & 7 & -5 \end{vmatrix} = 7 \begin{vmatrix} -4 & 5 \\ -6 & 7 \end{vmatrix} + (-5) \begin{vmatrix} 7 & -8 \\ -4 & 5 \end{vmatrix} = 7(-28+30) - 5(35-32)$$

$$= 7(2) - 5(3) = -1$$

The determinant is nonzero, so the vectors are linearly independent.

Study Tip: For 3×3 matrices, some students tend to prefer the special trick suggested for Exercises 15–18 in Section 3.1, even though in general there are 12 multiplications instead of the 9 multiplications needed for cofactor expansion. Note, however, that numbers in the special method can sometimes be large. For comparison, here are those computations for the matrix studied above in Exercise 25:

Copyright © 2021 Pearson Education, Inc.

$$\begin{vmatrix} 7 & -8 & 7 \\ -4 & 5 & 0 \\ -6 & 7 & -5 \end{vmatrix} \begin{matrix} 7 & -8 \\ -4 & 5 \\ -6 & 7 \end{matrix}$$

$$\det\begin{bmatrix} \mathbf{v}_1 & \mathbf{v}_2 & \mathbf{v}_3 \end{bmatrix} = 7(5)(-5) + (-8)(0)(-6) + 7(-4)(7)$$
$$- (-6)(5)(7) - 7(0)(7) - (-5)(-4)(-8)$$

$$= -175 + 0 + (-196) - (-210) - 0 - (-160) = -1$$

27. See Theorem 3.

29. See the remark following Theorem 4.

31. See Theorem 3.

33. See the warning after Example 5.

37. Since the determinant is multiplicative (Theorem 6),

$(\det A)(\det A^{-1}) = \det(AA^{-1}) = \det I = 1$. So $\det A^{-1} = 1/\det A$.

Study Tip: The result of Exercise 37 will be useful in future homework.

39. By Theorem 6 (twice), $\det AB = (\det A)(\det B) = (\det B)(\det A) = \det BA$.

41. By Theorem 5, $\det U^T = \det U$. So, by Theorem 6,

$$\det U^T U = (\det U^T)(\det U) = (\det U)^2$$

If $U^T U = I$, then $(\det U)^2 = \det I = 1$, which implies that $\det U = \pm 1$.

43. The solution is in the text. (The determinant of a triangular matrix is the product of the entries on the main diagonal.)

Study Tip: Exercises 15–26, 45, and 46 make good test questions because they check your knowledge of determinant properties without requiring much computation. Exercise 45(b) is the one most likely to be answered incorrectly. What would be the answer to 45(b) if A were 4×4?

49. Compute $\det A$ by a cofactor expansion down column 3:

$$\det A = (u_1 + v_1) \cdot \det A_{13} - (u_2 + v_2) \cdot \det A_{23} + (u_3 + v_3) \cdot \det A_{33}$$

$$= u_1 \cdot \det A_{13} - u_2 \cdot \det A_{23} + u_3 \cdot \det A_{33} + v_1 \cdot \det A_{13} - v_2 \cdot \det A_{23} + v_3 \cdot \det A_{33}$$

$$= u_1 \cdot \det B_{13} - u_2 \cdot \det B_{23} + u_3 \cdot \det B_{33} + v_1 \cdot \det C_{13} - v_2 \cdot \det C_{23} + v_3 \cdot \det C_{33}$$

$$= \det B + \det C$$

53. Suppose A is $m \times n$ with more columns than rows. Then $A^T A$ is $n \times n$ and must be singular. If A is generated with random entries, then AA^T will be nonsingular (invertible) practically all the time. Try to explain why these statements should be true. (Use the IMT.)

Copyright © 2021 Pearson Education, Inc.

MATLAB   Computing Determinants

To compute det A, set $\mathbf{U} = \mathbf{A}$ and then reduce A step-by-step to an echelon form U, keeping track of how your steps effect the determinant. Then, the determinant of A can be computed from the steps and the command

prod(diag(U))

The command **diag(U)** extracts the diagonal entries of U and places them in a column vector, and **prod** computes the product of those entries.

You can also use **det(A)** to check your work, but the longer sequence of commands helps you think about the *process* of computing det A.

3.3 - Cramer's Rule, Volume, and Linear Transformations

This section will be a valuable reference for students who plan to take a course in multivariable calculus. Mathematics and statistics majors probably will encounter the material here several times. Also, economics students and engineers (particularly electrical engineers) are likely to need Cramer's rule and some of the supplementary exercises in later courses.

KEY IDEAS

The main results of the section are stated in Theorems 7, 8, 9, and 10. The proof of Theorem 7 is simple and yet involves three important ideas: the definition of a matrix product, the multiplicative property of the determinant, and the evaluation of a determinant by cofactors. Check with your instructor about whether you should be able to reproduce the proof of Theorem 7.

A heuristic proof of Theorem 9 for 2×2 matrices is given in an appendix at the end of this section. Theorem 10 provides a key idea in calculus and physics needed for the study of double and triple integrals. The determinant used there in calculus is called a *Jacobian*.

STUDY NOTE

In Exercise 25, you are asked to use Theorem 9 to explain why the determinant of a 3×3 matrix A is zero if and only if A is not invertible. (A similar explanation holds for the 2×2 case.) The answer is in the text, so be sure to work on this before looking at the answer section. *Work on Exercise 25, even if it is not assigned.*

Remember, learning *does* take place when you think hard about an exercise, even when you are unsuccessful, if you try to look at the problem from different angles, browse back through the text, and perhaps look at earlier exercises. *Write* your solution, don't just talk to yourself about what you would write if you had to.

Copyright © 2021 Pearson Education, Inc.

SOLUTIONS TO EXERCISES

1. The system is equivalent to $A\mathbf{x} = \mathbf{b}$, where $A = \begin{bmatrix} 5 & 7 \\ 2 & 4 \end{bmatrix}$ and $\mathbf{b} = \begin{bmatrix} 3 \\ 1 \end{bmatrix}$. Write

$$A_1(\mathbf{b}) = \begin{bmatrix} 3 & 7 \\ 1 & 4 \end{bmatrix}, \quad A_2(\mathbf{b}) = \begin{bmatrix} 5 & 3 \\ 2 & 1 \end{bmatrix}$$
$$\qquad\qquad \uparrow \qquad\qquad\qquad \uparrow$$
$$\qquad\qquad \mathbf{b} \qquad\qquad\qquad \mathbf{b}$$

and compute

$$\det A = 20 - 14 = 6, \quad \det A_1(\mathbf{b}) = 12 - 7 = 5, \det A_2(\mathbf{b}) = 5 - 6 = -1$$

$$x_1 = \frac{\det A_1(\mathbf{b})}{\det A} = \frac{5}{6}, \quad x_2 = \frac{\det A_2(\mathbf{b})}{\det A} = \frac{-1}{6} = -\frac{1}{6}$$

7. The system is equivalent to $A\mathbf{x} = \mathbf{b}$, where $A = \begin{bmatrix} 6s & 4 \\ 9 & 2s \end{bmatrix}$ and $\mathbf{b} = \begin{bmatrix} 5 \\ -2 \end{bmatrix}$. Write

$$A_1(\mathbf{b}) = \begin{bmatrix} 5 & 4 \\ -2 & 2s \end{bmatrix}, \quad A_2(\mathbf{b}) = \begin{bmatrix} 6s & 5 \\ 9 & -2 \end{bmatrix}$$

and compute

$$\det A = 12s^2 - 36 = 12(s^2 - 3) = 12(s - \sqrt{3})(s + \sqrt{3})$$

$$\det A_1(\mathbf{b}) = 10s + 8, \quad \det A_2(\mathbf{b}) = -12s - 45$$

The system has a unique solution when $\det A \neq 0$, that is, when $s \neq \pm\sqrt{3}$. For such a system, the solution is $\mathbf{x} = (x_1, x_2)$, where

$$x_1 = \frac{\det A_1(\mathbf{b})}{\det A} = \frac{10s + 8}{12(s^2 - 3)} = \frac{5s + 4}{6(s^2 - 3)}$$

$$x_2 = \frac{\det A_2(\mathbf{b})}{\det A} = \frac{-12s - 45}{12(s^2 - 3)} = \frac{-4s - 15}{4(s^2 - 3)}$$

13. First, find the cofactors of $A = \begin{bmatrix} 3 & 5 & 4 \\ 1 & 0 & 1 \\ 2 & 1 & 1 \end{bmatrix}$.

Copyright © 2021 Pearson Education, Inc.

$$C_{11} = + \begin{vmatrix} 0 & 1 \\ 1 & 1 \end{vmatrix} = -1, \qquad C_{12} = - \begin{vmatrix} 1 & 1 \\ 2 & 1 \end{vmatrix} = 1, \qquad C_{13} = + \begin{vmatrix} 1 & 0 \\ 2 & 1 \end{vmatrix} = 1$$

$$C_{21} = - \begin{vmatrix} 5 & 4 \\ 1 & 1 \end{vmatrix} = -1, \qquad C_{22} = + \begin{vmatrix} 3 & 4 \\ 2 & 1 \end{vmatrix} = -5, \qquad C_{23} = - \begin{vmatrix} 3 & 5 \\ 2 & 1 \end{vmatrix} = 7$$

$$C_{31} = + \begin{vmatrix} 5 & 4 \\ 0 & 1 \end{vmatrix} = 5, \qquad C_{32} = - \begin{vmatrix} 3 & 4 \\ 1 & 1 \end{vmatrix} = 1, \qquad C_{33} = + \begin{vmatrix} 3 & 5 \\ 1 & 0 \end{vmatrix} = -5$$

Then, arrange the *transpose* of the array of cofactors into the adjugate of A.

$$\operatorname{adj} A = \begin{bmatrix} -1 & -1 & 5 \\ 1 & -5 & 1 \\ 1 & 7 & -5 \end{bmatrix}$$

Were you to compute $\det A$ now, you could write A^{-1}, but you would still need to check whether your calculations are correct. To build in this check, compute

$$A \cdot \operatorname{adj} A = \begin{bmatrix} 3 & 5 & 4 \\ 1 & 0 & 1 \\ 2 & 1 & 1 \end{bmatrix} \begin{bmatrix} -1 & -1 & 5 \\ 1 & -5 & 1 \\ 1 & 7 & -5 \end{bmatrix} = \begin{bmatrix} 6 & 0 & 0 \\ 0 & 6 & 0 \\ 0 & 0 & 6 \end{bmatrix}$$

If any off-diagonal entries in the product are nonzero, or if the diagonal entries are not all the same, then some errors have been made, and you can recheck your cofactor calculations. (One possible mistake is to forget the $\pm$ signs in front of the 2×2 determinants. Another error is to *not* transpose the array of cofactors.) In this case, the calculations above *are* correct and $\det A$ must be 6. So

$$A^{-1} = \frac{1}{\det A} \operatorname{adj} A = \frac{1}{6} \begin{bmatrix} -1 & -1 & 5 \\ 1 & -5 & 1 \\ 1 & 7 & -5 \end{bmatrix}$$

19. The parallelogram with vertices $(0, 0)$, $(5, 2)$, $(6, 4)$, $(11, 6)$ is shown below. If no vertex were zero, we would have to translate the parallelogram to the origin by subtracting one vertex from all four vertices. Since one vertex already is zero, use the two vertices adjacent to the origin to construct the columns of A, and compute $|\det A|$.

$$A = \begin{bmatrix} 5 & 6 \\ 2 & 4 \end{bmatrix}, \quad \begin{pmatrix} \text{area of the} \\ \text{parallelogram} \end{pmatrix} = |\det A| = |20 - 12| = 8$$

Copyright © 2021 Pearson Education, Inc.

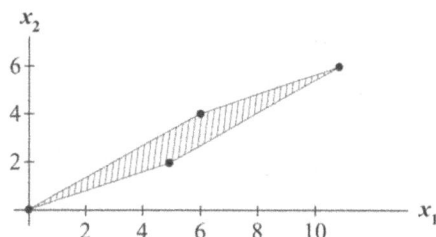

25. The answer is in the text. I hope you took the advice at the beginning of this *Study Guide* section and worked the problem (or at least tried hard to work the problem) before checking the answer section. If you were successful, you should be proud of yourself; you are mastering the material—not only determinants but also linear dependence!

31. Let $\mathbf{x} = \begin{bmatrix} x_1 \\ x_2 \\ x_3 \end{bmatrix}$, $\mathbf{u} = \begin{bmatrix} u_1 \\ u_2 \\ u_3 \end{bmatrix}$, and $A = \begin{bmatrix} a & 0 & 0 \\ 0 & b & 0 \\ 0 & 0 & c \end{bmatrix}$. Also, let S be the unit ball in $\mathbb{R}^3$, whose

bounding surface consists of all vectors $\mathbf{u}$ such that $u_1^2 + u_2^2 + u_3^2 = 1$, and let S' be the image of S under the mapping $\mathbf{u} \mapsto A\mathbf{u}$.

a. If $\mathbf{x}$ is in S', then $\mathbf{x} = A\mathbf{u}$ for some $\mathbf{u}$ in S, and $\mathbf{u} = A^{-1}\mathbf{x} = \begin{bmatrix} x_1/a \\ x_2/b \\ x_3/c \end{bmatrix}$.

The condition on u_1, u_2, u_3 shows that $\left(\dfrac{x_1}{a}\right)^2 + \left(\dfrac{x_2}{b}\right)^2 + \left(\dfrac{x_3}{c}\right)^2 = 1$.

b. Since the volume of the unit ball bounded by S is $4\pi/3$ and the determinant of A is abc, Theorem 10 shows that the volume of the region bounded by S' is $4\pi abc/3$.

35. See Figure 2.

37. See Theorem 8.

Chapter 3 - Supplementary Exercises

In this chapter, Exercises 1-15 consist of true/false questions, whose level of difficulty varies. Some are similar to the ones that appear in many sections of the text, in which a word or phrase is sometimes missing or slightly misstated. Some follow fairly easily from a theorem: others may need careful reasoning. A few may require an argument that uses several ideas. In each case, think carefully about the statement and attempt to write a solution. The text provides the true/false answer, but you must supply the justification or counterexample. Careful work on these exercises will help you prepare for an exam over the chapter material.

Copyright © 2021 Pearson Education, Inc.

19. $\begin{vmatrix} 9 & 1 & 9 & 9 & 9 \\ 9 & 0 & 9 & 9 & 2 \\ 4 & 0 & 0 & 5 & 0 \\ 9 & 0 & 3 & 9 & 0 \\ 6 & 0 & 0 & 7 & 0 \end{vmatrix} = (-1)\begin{vmatrix} 9 & 9 & 9 & 2 \\ 4 & 0 & 5 & 0 \\ 9 & 3 & 9 & 0 \\ 6 & 0 & 7 & 0 \end{vmatrix} = (-1)(-2)\begin{vmatrix} 4 & 0 & 5 \\ 9 & 3 & 9 \\ 6 & 0 & 7 \end{vmatrix}$

$= (-1)(-2)(3)\begin{vmatrix} 4 & 5 \\ 6 & 7 \end{vmatrix} = (-1)(-2)(3)(-2) = -12$

25. To tell if a quadrilateral determined by four points is a parallelogram, first translate one of the vertices to the origin. If we label the vertices of this new quadrilateral as $\mathbf{0}$, $\mathbf{v}_1$, $\mathbf{v}_2$, and $\mathbf{v}_3$, then they will be the vertices of a parallelogram if one of $\mathbf{v}_1$, $\mathbf{v}_2$, or $\mathbf{v}_3$ is the sum of the other two. In this example, subtract $(1, 4)$ from each vertex to get a new parallelogram with vertices $\mathbf{0} = (0, 0)$, $\mathbf{v}_1 = (-2,1)$, $\mathbf{v}_2 = (2,5)$, and $\mathbf{v}_3 = (4,4)$. Since $\mathbf{v}_2 = \mathbf{v}_3 + \mathbf{v}_1$, the quadrilateral is a parallelogram as stated. The translated parallelogram has the same area as the original, and is determined by the columns of $A = \begin{bmatrix} \mathbf{v}_1 & \mathbf{v}_3 \end{bmatrix} = \begin{bmatrix} -2 & 4 \\ 1 & 4 \end{bmatrix}$, so the area of the parallelogram is $|\det A| = |-12| = 12$.

31. First consider the case $n = 2$. In this case

$$\det B = \begin{vmatrix} a-b & b \\ 0 & a \end{vmatrix} = a(a-b), \det C = \begin{vmatrix} b & b \\ b & a \end{vmatrix} = ab - b^2,$$

so

$\det A = \det B + \det C = a(a-b) + ab - b^2 = a^2 - b^2 = (a-b)(a+b) = (a-b)^{2-1}(a+(2-1)b)$, and the formula holds for $n = 2$.

Now assume that the formula holds for all $(k-1) \times (k-1)$ matrices, and let A, B, and C be $k \times k$ matrices. By a cofactor expansion along the first column,

$$\det B = (a-b)\begin{vmatrix} a & b & \dots & b \\ b & a & \dots & b \\ \vdots & \vdots & \ddots & \vdots \\ b & b & \dots & a \end{vmatrix} = (a-b)(a-b)^{k-2}(a+(k-2)b) = (a-b)^{k-1}(a+(k-2)b)$$

since the matrix in the above formula is a $(k-1) \times (k-1)$ matrix. We can perform a series of row operations on C to "zero out" below the first pivot, and produce the following matrix whose determinant is $\det C$: $\begin{bmatrix} b & b & \dots & b \\ 0 & a-b & \dots & 0 \\ \vdots & \vdots & \ddots & \vdots \\ 0 & 0 & \dots & a-b \end{bmatrix}$.   Since this is a triangular matrix,

Copyright © 2021 Pearson Education, Inc.

we have found that $\det C = b(a-b)^{k-1}$. Thus

$$\det A = \det B + \det C = (a-b)^{k-1}(a+(k-2)b) + b(a-b)^{k-1} = (a-b)^{k-1}(a+(k-1)b),$$

which is what was to be shown. Thus the formula has been proven by mathematical induction.

Appendix: A Geometric Proof of a Determinant Property

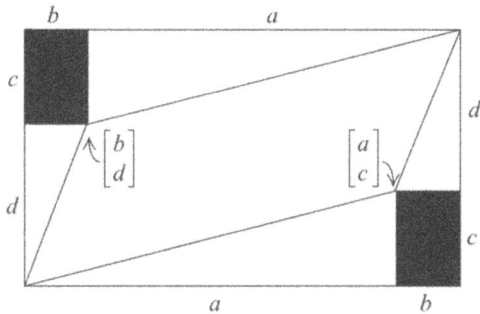

$$\det \begin{bmatrix} a & b \\ c & d \end{bmatrix} = ad - bc$$

$$(a + b)(c + d) = ac + ad + bc + bd$$

$$= \qquad -2bc - bd$$

$$= -ac$$

(Area of the Parallelogram) $=$ $ad - bc$

Chapter 3 - Glossary Checklist

Check your knowledge by attempting to write definitions of the terms below. Then compare your work with the definitions given in the text's Glossary. Ask your instructor which definitions, if any, might appear on a test.

Copyright © 2021 Pearson Education, Inc.

adjugate (or **classical adjoint**):   The matrix adj A formed from a square matrix A by replacing the (i, j)-entry of A by . . . .

cofactor: A number $C_{ij} =$ . . . , called the (i, j)-*cofactor of* A, where A_{ij} is the submatrix formed by deleting . . . .

cofactor expansion: A formula for det A using cofactors associated with one row or one column, such as for row 1: . . . .

Cramer's Rule: A formula for each entry in . . . .

interchange (matrix):   An elementary matrix obtained by interchanging . . . .

row replacement (matrix):   An elementary matrix obtained from the identity matrix by . . . .

scale by *r* (matrix):   An elementary matrix obtained by multiplying . . . .

Copyright © 2021 Pearson Education, Inc.

4 Vector Spaces

4.1 - Vector Spaces and Subspaces

The main focus of the chapter is on $\mathbb{R}^n$ and its subspaces. However, Section 4.1 builds a framework within which the theory for $\mathbb{R}^n$ rests. Most of the exercises in the chapter concern subspaces of $\mathbb{R}^n$, but some are designed to help you learn gradually about other important vector spaces.

KEY IDEA

A *vector space* is any collection of objects that behave as vectors do in $\mathbb{R}^n$. (The precise meaning of "behave" is described by the axioms.) A *vector* is simply any object that belongs to a vector space. Arrows, polynomials, and infinite sequences of numbers are all examples of vectors, in different vector spaces.

The most important vector spaces in this text are subspaces of $\mathbb{R}^n$. Visualize subspaces as lines or planes *through the origin*, the origin by itself, or the entire space $\mathbb{R}^n$. In this section, Theorem 1 is a useful tool to show that a set is a vector space. To show that a set is *not* a subspace, show that one of the properties in the subspace definition is violated. (See Exercises 1–4.)

STUDY NOTES

Parts of this chapter are somewhat more theoretical than the earlier chapters, but that is necessary in order to give a solid foundation for the rest of the course and for careers in the high-tech industry. Learn the key definitions and theorems as they appear in the text (rather than waiting until just before an exam). You need this knowledge to get through the conceptual exercises and to be prepared for later sections.

In Example 5, the concept of a function as a single "vector" in a vector space is difficult to absorb on a first reading, and you should not expect to master it in a few days.

The word subspace should usually be accompanied by a phrase such as "of V," or "of $\mathbb{R}^n$." Without such a phrase, the nature of the elements in the subspace is unknown. For instance, the statement "H is a subspace" does not say whether the elements in H are pairs of numbers, or polynomials, or something else. But the phrase "H is a subspace of P_3" includes the information that each vector in H is a polynomial of degree three or less.

Set Notation: The notation introduced in Example 9 is sometimes used in this chapter as an efficient way to describe a set. The symbols and phrases inside the set brackets describe the set. The part to the left of the colon displays the basic nature of the elements in the set (such as vectors

Copyright © 2021 Pearson Education, Inc.

in $\mathbb{R}^3$) while the part to the right adds any qualifying conditions that must be met in order for an element to belong to the set (such as all vector entries must be positive). For instance, the set in Example 9 can also be written as

$$H = \left\{ \begin{bmatrix} s \\ t \\ u \end{bmatrix} : s \text{ and } t \text{ are real, and } u = 0 \right\}$$

As another example, the set in Example 12 can be written as

$$W = \left\{ \begin{bmatrix} c \\ d \\ a \\ b \end{bmatrix} : c = a - 3b \text{ and } d = b - a, \text{ where } a \text{ and } b \text{ are arbitrary} \right\}$$

Here, and elsewhere, reference to the scalars a and b as "arbitrary" means that the scalars can be any real numbers.

Study Tip: Review Sections 1.3–1.5 and 1.7 before you reach Section 4.3.

SOLUTIONS TO EXERCISES

1. **a.** V is a subset of $\mathbb{R}^2$. The defining property of V is that the entries of every vector in V are nonnegative. So, if **u** and **v** are in V, their entries are nonnegative. Since a sum of nonnegative numbers is nonnegative, the vector **u** + **v** satisfies the condition that defines V. That is, **u** + **v** is in V.

   b. The text's solution gives a specific **u** and c. One specific "counterexample" suffices to show that V is not a vector space. However, you could also simply say, "If any nonzero vector **v** in V is multiplied by a negative scalar c, then c**v** is not in V because at least one of its entries is negative."

7. Most examples in the text involve integers, to make calculations simple, and one can easily overlook the fact that a subspace must be closed under multiplication by *all* real numbers.

Checkpoint: Let H be the set of all points (x, y, z) in $\mathbb{R}^3$ that satisfy the condition $x^2 + y^2 - z^2 = 0$. Is H a subspace of $\mathbb{R}^3$?

13. **a.** **w** is certainly not one of the three vectors in $\{\mathbf{v}_1, \mathbf{v}_2, \mathbf{v}_3\}$.

   b. Span$\{\mathbf{v}_1, \mathbf{v}_2, \mathbf{v}_3\}$ contains infinitely many vectors.

   c. **w** is in the subspace (of $\mathbb{R}^3$) spanned by $\mathbf{v}_1, \mathbf{v}_2, \mathbf{v}_3$ if and only if the equation $x_1\mathbf{v}_1 + x_2\mathbf{v}_2 + x_3\mathbf{v}_3 = \mathbf{w}$ is consistent (has a solution). Row reduce the augmented matrix:

Copyright © 2021 Pearson Education, Inc.

$$\begin{bmatrix} 1 & 2 & 4 & 3 \\ 0 & 1 & 2 & 1 \\ -1 & 3 & 6 & 2 \end{bmatrix} \sim \begin{bmatrix} 1 & 2 & 4 & 3 \\ 0 & 1 & 2 & 1 \\ 0 & 5 & 10 & 5 \end{bmatrix} \sim \begin{bmatrix} 1 & 2 & 4 & 3 \\ 0 & 1 & 2 & 1 \\ 0 & 0 & 0 & 0 \end{bmatrix}$$

There is no pivot in the augmented column, so the vector equation is consistent, and $\mathbf{w}$ is in Span$\{\mathbf{v}_1, \mathbf{v}_2, \mathbf{v}_3\}$.

19. Let H be the set of all functions described in (5). Then H is a subset of the vector space V of all real-valued functions, and H consists of all linear combinations of the functions cos ωt and sin ωt. By Theorem 1, H is a subspace of V, and hence is a vector space.

23. See Example 5.

25. See the definition of a vector.

27. See Exercises 1, 2, or 3.

29. See the paragraph before Example 6.

31. See Example 7.

33. Axiom 4 (plus Axiom 2) shows that $\mathbf{0} + \mathbf{w} = \mathbf{w}$. Exercises 33–38 show how facts that we take for granted in $\mathbb{R}^n$ depend only on a few basic properties of $\mathbb{R}^n$, properties that are now axioms for a general vector space.

39. Let H be a subspace of V that contains the vectors $\mathbf{u}$ and $\mathbf{v}$. Since H is closed under multiplication by scalars, H must contain all scalar multiples of $\mathbf{u}$ and $\mathbf{v}$. Since H is also closed under vector addition, it contains all sums of scalar multiples of $\mathbf{u}$ and $\mathbf{v}$. That is, H contains all vectors in Span$\{\mathbf{u}, \mathbf{v}\}$.

Answer to Checkpoint: H is not a subspace. Counterexample: $(1, 0, 1)$ and $(0, 1, 1)$ are in H and yet their sum, $(1, 1, 2)$, is not in H (because $1^2 + 1^2 - 2^2 \neq 0$). Many counterexamples are possible. Only one is needed.

MATLAB   Graphing Functions

The following commands will graph the function f in Exercise 45:

```
t = linspace(0,2*pi);
f = t.^0 - 8*cos(t).^2 + 8*cos(t).^4;
grid on
plot(t,f)
```

Here, **t** is a vector with 101 entries, the endpoints of 100 equal subintervals of $[0, 2\pi]$; **cos(t)** is a vector whose entries are the cosines of the corresponding entries in **t**, and **t.^0** is a vector of 1's. In general, **t.^k** is a vector whose entries are the kth powers of the corresponding entries in **t**. The command **grid on** (or **off**) turns on (or off) gridlines the next time a display is created. Use **hold on** if you want the next graph to appear on the current display. The command **hold off** makes each graph appear in a separate display. See the topic "Basic Plotting" in MATLAB's **Help** menu for examples of plotting options or use a search engine to look up commands online.

Copyright © 2021 Pearson Education, Inc.

> To take advantage of MATLAB's interactive plotting features see the topic "Plotting Tools in MATLAB's **Help** menu. For example, typing in **plottools** at the command line will summon the plotting tools.

4.2 - Null Spaces, Column Spaces, Row Spaces, and Linear Transformations

Many problems in linear algebra involve a subspace in one way or another. This section provides an opportunity to become comfortable with the concept. The foundation for this section was laid in Sections 1.3 and 1.5. Have you reviewed those sections yet?

KEY IDEAS

Theorems 2 and 3 describe the main types of subspaces. (The proof of Theorem 2 makes a good exam question, because it tests both the definition of Nul A and the definition of a subspace.)

Theorem 3 actually has two conclusions: Col A is a subspace and it is a subspace of $\mathbb{R}^m$. The first phrase tells us that linear combinations of vectors in Col A remain in Col A. The phrase " of $\mathbb{R}^m$ " reminds us that each vector has m entries (because A has m rows). A similar remark applies to the statement from Theorem 2 that Nul A is a subspace of $\mathbb{R}^n$.

The box after Example 4 shows that the statement " Col $A = \mathbb{R}^m$ " can be added to the list of equivalent statements in Theorem 4 of Section 1.4.

STUDY NOTES

In Example 3, the statement "Nul A = Span{**u**, **v**, **w**}" would be an *explicit* description of Nul A (provided you specify what **u**, **v**, and **w** are).

In some applications, it is important to know that for a given $m \times n$ matrix A, every equation $A\mathbf{x} = \mathbf{b}$ has a solution (assuming **b** is in $\mathbb{R}^m$). Yet it may require some effort to determine whether this is the case. If not every equation $A\mathbf{x} = \mathbf{b}$ has a solution, then not every **b** belongs to Col A, and hence Col A is a proper subspace of $\mathbb{R}^m$. One of the goals of the next few sections is to obtain a method for determining when Col $A = \mathbb{R}^m$.

Checkpoint 1:   How many pivot positions does an $m \times n$ matrix A have if Col $A = \mathbb{R}^m$?

Study Tip: Theorems 1, 2, and 3 are the main tools for showing that a set is a vector space (that is, a subspace of some known vector space). Review these theorems now, before starting the exercises.

SOLUTIONS TO EXERCISES

1. Now is the time to learn the *definition* of Nul A. A vector **x** is in Nul A precisely when the product $A\mathbf{x}$ is defined and $A\mathbf{x} = \mathbf{0}$. Given **x**, simply compute $A\mathbf{x}$ to determine whether $A\mathbf{x}$ is zero.

Copyright © 2021 Pearson Education, Inc.

$$A\mathbf{x} = \begin{bmatrix} 3 & -5 & -3 \\ 6 & -2 & 0 \\ -8 & 4 & 1 \end{bmatrix} \begin{bmatrix} 1 \\ 3 \\ -4 \end{bmatrix} = \begin{bmatrix} 0 \\ 0 \\ 0 \end{bmatrix}, \text{ so } \begin{bmatrix} 1 \\ 3 \\ -4 \end{bmatrix} \text{ is in Nul } A.$$

Warning: In Exercises 3–6, writing an equation $\mathbf{x} = c\mathbf{u} + d\mathbf{v}$ is not the same as listing, say, the vectors $\mathbf{u}$ and $\mathbf{v}$ that span the null space. The appropriate answer for these exercises is a list of a small finite number of vectors that span Nul A, not a description of *all* the vectors in Nul A.

Study Tip: Try Practice Problem 1 before you work on Exercises 7–14. If you can't find *two* ways to work the practice problem, reread the first paragraph of Section 4.2 (but don't look at it until you have attempted the practice problem).

7. The set W is a subset of $\mathbb{R}^3$. If W were a vector space (under the standard operations in $\mathbb{R}^3$), it would be a subspace of $\mathbb{R}^3$. But W fails *every* property of a subspace, so it is not a vector space. For instance, the vector $(0, 0, 0)$ does not satisfy the condition $a + b + c = 2$, and so the zero vector is not in W.

13. A typical element of W can be written as follows:

$$\begin{bmatrix} c - 6d \\ d \\ c \end{bmatrix} = \begin{bmatrix} c \\ 0 \\ c \end{bmatrix} + \begin{bmatrix} -6d \\ d \\ 0 \end{bmatrix} = c\underset{\underset{\mathbf{u}}{\uparrow}}{\begin{bmatrix} 1 \\ 0 \\ 1 \end{bmatrix}} + d\underset{\underset{\mathbf{v}}{\uparrow}}{\begin{bmatrix} -6 \\ 1 \\ 0 \end{bmatrix}} = \begin{bmatrix} 1 & -6 \\ 0 & 1 \\ 1 & 0 \end{bmatrix} \begin{bmatrix} c \\ d \end{bmatrix}$$

Since c and d are any real numbers, this calculation shows that W is a subspace of $\mathbb{R}^3$ (and hence is a vector space) by Theorem 3. Alternatively, this calculation shows that W is the same as Span$\{\mathbf{u}, \mathbf{v}\}$, so W is a subspace of $\mathbb{R}^3$, by Theorem 1.

Checkpoint 2: Why is W a subspace "of $\mathbb{R}^3$"?

19. The matrix A is 2×5, so vectors in Nul A must have 5 entries and vectors in Col A must have 2 entries. Thus Nul A is a subspace of $\mathbb{R}^5$ and Col A is a subspace of $\mathbb{R}^2$.

Study Tip: Exercises 17–20 may seem simple, but they will help you in Section 4.5.

25. Check the definition before Example 1.

27. See Theorem 2.

29. See the remark just before Example 4.

31. See the table that contrasts Nul A and Col A.

33. See Fig. 2.

35. See the remark after Theorem 3.

37. See the paragraph before Example 5.

Copyright © 2021 Pearson Education, Inc.

43. The solution in the text shows that T is a linear transformation. If $T(\mathbf{p})$ is the zero vector, then $\mathbf{p}(0) = 0$ and $\mathbf{p}(1) = 0$, by definition of T. One such polynomial is $\mathbf{p}(t) = t(t-1)$. Any other *quadratic* polynomial that vanishes at 0 and 1 must be a multiple of $\mathbf{p}$, so $\mathbf{p}$ spans the kernel of T.

For the range of T, observe that the image of the constant 1 function is $\begin{bmatrix} 1 \\ 1 \end{bmatrix}$, and the image of the polynomial t is $\begin{bmatrix} 0 \\ 1 \end{bmatrix}$. Denote these two images by $\mathbf{u}$ and $\mathbf{v}$, respectively. Since the range of T is a subspace of $\mathbb{R}^2$ that contains $\mathbf{u}$ and $\mathbf{v}$, the range must contain all linear combinations of $\mathbf{u}$ and $\mathbf{v}$. By inspection, $\mathbf{u}$ and $\mathbf{v}$ are linearly independent, so they span $\mathbb{R}^2$. Thus the range of T must contain all of $\mathbb{R}^2$.

49. The vector $\mathbf{w}$ is in Col A because row reduction of the augmented matrix $[A \quad \mathbf{w}]$ shows that the equation $A\mathbf{x} = \mathbf{w}$ is consistent. The vector $\mathbf{w}$ is not in Nul A because $A\mathbf{x}$ is not the zero vector.

Answers to Checkpoints:

1. An $m \times n$ matrix A must have m pivot positions in order for Col A to be all of $\mathbb{R}^m$.

2. W is a subspace "of $\mathbb{R}^3$" because each vector in W has three entries.

Mastering Linear Algebra Concepts: Vector Space, Subspace, and Column and Null Spaces

A vector space is a nonempty set of objects on which are defined two operations, called addition and multiplication by scalars, that satisfy ten axioms. Make one review sheet for vector space and subspace. Include as much of the definition of a vector space as you are required to know. (Ask your instructor.) Include the full definition of a subspace.

Organize what you have learned in Sections 4.1 and 4.2 (together with Sections 1.3–1.5), using the categories listed below. Your examples of vector spaces will be the same as your examples of subspaces. (See the paragraph just before Example 6 in Section 4.1.) The column space and null space of a matrix should be among your examples of subspaces. (The references cited below contain information that will help you to construct your review sheets.)

- definitions of vector space, subspace
- equivalent descriptions   Paragraph before Example 6 in Sec. 4.1
- geometric interpretations   Figs. 6, 7, 9, in Sec. 4.1
- subspaces described explicitly   Theorems 1, 3; Exercises 15–18 in Sec. 4.1
- subspaces described implicitly   Theorem 2; Exercises 20(b), 22 in Sec. 4.1; Example 2 in Sec. 4.2
- examples and counterexamples   Examples and exercises in Sec. 4.1, 4.2

Copyright © 2021 Pearson Education, Inc.

- algorithms and computations   Exercises in Sec. 4.2

- connections with other concepts   Fig. 2 and Example 8 in Sec. 4.2

The references above are not exhaustive. You can find more facts and examples in the exercises for Sections 4.1 and 4.2. In addition to the review sheet for vector spaces, you should make another sheet for column space and null space. List the two basic definitions, equivalent formulations of each definition, and examples of computations. Also, you could attach a copy of the table that contrasts the two types of subspaces. Many students say that the table is quite helpful.

4.3 - Linearly Independent Sets; Bases

The definition of linear independence carries over from $\mathbb{R}^n$ to any vector space. The geometric interpretations in Chapter 1 of linearly independent and dependent sets should help you visualize these concepts here.

KEY IDEAS

In general, you cannot use an ordinary matrix equation $A\mathbf{x} = \mathbf{0}$ to study linear dependence of $\{\mathbf{v}_1, ..., \mathbf{v}_p\}$. You have to work with the vector equation $c_1\mathbf{v}_1 + \cdots + c_p\mathbf{v}_p = \mathbf{0}$, or with Theorem 4, unless the vectors happen to be n-tuples of numbers.

A set $\{\mathbf{v}\}$ with $\mathbf{v} \neq \mathbf{0}$ is linearly independent because the vector equation $x_1\mathbf{v} = \mathbf{0}$ has only the trivial solution. (See Exercise 38 in Section 4.1.) The set $\{\mathbf{0}\}$ is linearly dependent, because the equation $x_1\mathbf{0} = \mathbf{0}$ has many nontrivial solutions.

Theorem 6 is important for later work, but its proof is rather subtle. Study Examples 8 and 9 carefully, as well as the proof of the theorem.

A basis for a vector space V is a set in V that is large enough to span V and small enough to be linearly independent. See also the subsection *Two Views of a Basis*. The plural of *basis* is *bases*.

Warning: If a set in V does not span V, the set may or may not be linearly dependent.

SOLUTIONS TO EXERCISES

1. The complete solution is in the text. For a general set of n vectors in $\mathbb{R}^n$, row operations on a matrix will usually be needed to determine if the matrix has n pivot positions.

7. Again, the solution is in the text. Any set in $\mathbb{R}^n$ with fewer than n vectors cannot span $\mathbb{R}^n$ and therefore cannot be a basis for $\mathbb{R}^n$. Such a set may or may not be linearly independent. What similar statement can you make about a set in $\mathbb{R}^n$ with more than n vectors? See Exercise 8 for ideas.

Study Tip: Theorem 4 in Section 1.4 may help you decide whether a set of vectors spans $\mathbb{R}^n$.

Copyright © 2021 Pearson Education, Inc.

4-8 Chapter 4 • Vector Spaces

13. The matrix B is in echelon form and displays the pivot columns. A basis for Col A consists of columns 1 and 2 of A: $\begin{bmatrix} -2 \\ 2 \\ -3 \end{bmatrix}, \begin{bmatrix} 4 \\ -6 \\ 8 \end{bmatrix}$. This is not the only correct choice, but it is the "standard" choice. A *wrong* choice would be columns 1 and 2 of B. See the Warning after Theorem 6.

For the null space, solve $A\mathbf{x} = \mathbf{0}$:

$$[A \quad \mathbf{0}] \sim [B \quad \mathbf{0}] = \begin{bmatrix} 1 & 0 & 6 & 5 & 0 \\ 0 & 2 & 5 & 3 & 0 \\ 0 & 0 & 0 & 0 & 0 \end{bmatrix} \sim \begin{bmatrix} 1 & 0 & 6 & 5 & 0 \\ 0 & 1 & 5/2 & 3/2 & 0 \\ 0 & 0 & 0 & 0 & 0 \end{bmatrix}$$

Then $x_1 = -6x_3 - 5x_4$, $x_2 = -(5/2)x_3 - (3/2)x_4$, with x_3 and x_4 free. The general solution is

$$\mathbf{x} = \begin{bmatrix} x_1 \\ x_2 \\ x_3 \\ x_4 \end{bmatrix} = \begin{bmatrix} -6x_3 & - & 5x_4 \\ -(5/2)x_3 & - & (3/2)x_4 \\ x_3 & & \\ & & x_4 \end{bmatrix} = x_3 \begin{bmatrix} -6 \\ -5/2 \\ 1 \\ 0 \end{bmatrix} + x_4 \begin{bmatrix} -5 \\ -3/2 \\ 0 \\ 1 \end{bmatrix}$$

This equation describes *all* vectors in Nul A, not just a basis for Nul A. For a basis, the "standard" choice is $(-6, -5/2, 1, 0)$ and $(-5, -3/2, 0, 1)$. Another choice is $(-12, -5, 2, 0)$ and $(-10, -3, 0, 2)$, which avoids fractions.

A basis for Row A can be taken from the pivot rows of B: $\left\{ \begin{bmatrix} 1 & 0 & 6 & 5 \end{bmatrix}, \begin{bmatrix} 0 & 2 & 5 & 3 \end{bmatrix} \right\}$.

Warning: You really need to know the definition of Nul A and the definition of Col A, not just the procedures for finding bases for these spaces. The definitions will help you avoid the fairly common mistake of attempting to use the null space procedure to find a basis for a column space, or vice-versa.

19. You can solve the equation $4\mathbf{v}_1 + 5\mathbf{v}_2 - 3\mathbf{v}_3 = \mathbf{0}$ for any one of the three vectors in terms of the others. By the Spanning Set Theorem, the set spanned by all three vectors is the same as the set spanned by any two of the vectors—any one of the three vectors can be discarded. If you discard $\mathbf{v}_3$, then $\mathbf{v}_1$ and $\mathbf{v}_2$ span H and are obviously linearly independent. Hence $\{\mathbf{v}_1, \mathbf{v}_2\}$ is a basis for H. The same reasoning applies to $\{\mathbf{v}_1, \mathbf{v}_3\}$ and $\{\mathbf{v}_2, \mathbf{v}_3\}$. These are the most likely answers. But, once you realize that $\{\mathbf{v}_1, \mathbf{v}_2\}$ is a basis for H, you can make others, such as $\{\mathbf{v}_1, \mathbf{v}_2 + r\mathbf{v}_1\}$ for any scalar r. Showing that this set *is* a basis for H does take some work, however. You might try to do this. In some courses that spend time on theoretical questions, this problem could appear on a test.

21. See the paragraph preceding Theorem 4.

23. See the definition of a basis.

25. See Example 3.

Copyright © 2021 Pearson Education, Inc.

27. See the subsection *Two Views of a Basis*.

29. See the box before Example 9.

31. See the footnote for Example 10.

33. Let $A = [\mathbf{v}_1 \quad \mathbf{v}_2 \quad \mathbf{v}_3 \quad \mathbf{v}_4]$. Since A is square and its columns span $\mathbb{R}^4$, the columns of A must be linearly independent, by the Invertible Matrix Theorem. So $\{\mathbf{v}_1, \mathbf{v}_2, \mathbf{v}_3, \mathbf{v}_4\}$ is a basis for $\mathbb{R}^4$.

Checkpoint: Suppose $\{\mathbf{v}_1, \ldots, \mathbf{v}_n\}$ is a basis for $\mathbb{R}^n$ and A is an invertible $n \times n$ matrix. Explain why $\{A\mathbf{v}_1, \ldots, A\mathbf{v}_n\}$ is a basis for $\mathbb{R}^n$.

35. The displayed equation shows only that Span $\{\mathbf{v}_1, \mathbf{v}_2, \mathbf{v}_3\}$ *contains H*. In fact, the vectors $\mathbf{v}_1, \mathbf{v}_2, \mathbf{v}_3$ are not all in H, so Span$\{\mathbf{v}_1, \mathbf{v}_2, \mathbf{v}_3\}$ cannot be H. Therefore $\{\mathbf{v}_1, \mathbf{v}_2, \mathbf{v}_3\}$ cannot be a basis for H. (It is easy to check that $\{\mathbf{v}_1, \mathbf{v}_2, \mathbf{v}_3\}$ is a basis for $\mathbb{R}^3$.)

41. (This generalizes Exercise 39 in Section 1.8.) Suppose $\{\mathbf{v}_1, \ldots, \mathbf{v}_p\}$ is linearly dependent. Then there exist $c_1, \ldots, c_p$, not all zero, such that

$$c_1\mathbf{v}_1 + \cdots + c_p\mathbf{v}_p = \mathbf{0}$$

Then, since T is linear,

$$T(c_1\mathbf{v}_1 + \cdots + c_p\mathbf{v}_p) = T(\mathbf{0}) = \mathbf{0} \qquad \text{Section 1.8, after the definition of linear transformation}$$

and

$$c_1T(\mathbf{v}_1) + \cdots + c_pT(\mathbf{v}_p) = \mathbf{0}$$

Since not all the c_i are zero, $\{T(\mathbf{v}_1), \ldots, T(\mathbf{v}_p)\}$ is linearly dependent.

Study Tip: The solution of Exercise 41 illustrates how to use linear dependence: If $\{\mathbf{v}_1, \mathbf{v}_2, \ldots, \mathbf{v}_p\}$ is known to be linearly dependent, then you can write $c_1\mathbf{v}_1 + \cdots + c_p\mathbf{v}_p = \mathbf{0}$, assume that not all the c_k are zero, and use this equation in some way.

Answer to Checkpoint: Let $B = [\mathbf{v}_1 \quad \cdots \quad \mathbf{v}_n]$. Then B is invertible because $\{\mathbf{v}_1, \mathbf{v}_2, \ldots, \mathbf{v}_n\}$ is a basis for $\mathbb{R}^n$. Since A is also an invertible $n \times n$ matrix, so is AB; hence the columns of AB, namely $A\mathbf{v}_1, \ldots, A\mathbf{v}_n$, form a basis for $\mathbb{R}^n$.

Mastering Linear Algebra Concepts: Basis

To the review sheet(s) you have on linear independence, add Examples 1, 2, and 6 from this section. The definition and geometric interpretations are unchanged. Add a note about not using the matrix equation $A\mathbf{x} = \mathbf{0}$ in the general case.

Start a separate review sheet for "basis," even though it involves two other concepts (span and linear independence) already being reviewed.

Copyright © 2021 Pearson Education, Inc.

- definition
- geometric interpretation   Fig. 1
- special cases   Standard bases for $\mathbb{R}^n$ and $\mathbb{P}_n$
- examples and counterexamples   Example 10, Exercise 35
- algorithms and computations   Examples 7 and 9; Example 3 in Sec. 4.2
- connections with other concepts   Invertible Matrix Theorem
                                        Unique Representation Theorem (in Sec. 4.4)

MATLAB   rref and **cos**

The command **rref(A)** produces the reduced echelon form of A. From that you can write a basis for Col A or write the homogeneous equations that describe Nul A. (Don't forget that A is a coefficient matrix, not an augmented matrix.)

In some cases, roundoff error or an extremely small pivot entry can cause **rref** to produce an incorrect echelon form. The more reliable singular value decomposition (see Section 7.4) can produce bases for Col A and Nul A, but **rref** is satisfactory for our purposes.

For Exercise 48, see the MATLAB box for Section 4.1.

4.4 - Coordinate Systems

This section contains a variety of geometric and algebraic explanations of the idea of a coordinate system for a vector space.

KEY IDEAS

The coordinate mapping from a vector space V (with a basis of n elements) onto $\mathbb{R}^n$ is a rule for giving " $\mathbb{R}^n$ -names" to vectors in V in such a way that the vector space structure of V is still visible in $\mathbb{R}^n$. Every vector space calculation in V is precisely mirrored by the same calculation in $\mathbb{R}^n$.

An important special case is when V is itself $\mathbb{R}^n$, and each vector $\mathbf{x}$ and its coordinate vector $[\mathbf{x}]_\mathcal{B}$ are related by a matrix equation $\mathbf{x} = P_\mathcal{B}[\mathbf{x}]_\mathcal{B}$.

Everything in the section depends on the Unique Representation Theorem. The proof of that theorem could appear on an exam because it shows precisely why the two properties of a basis $\mathcal{B}$ are important, and it illustrates how linear independence can be used in an argument:

Any vector in V has "coordinates" because $\mathcal{B}$ spans V, and the coordinates are uniquely determined because $\mathcal{B}$ is linearly independent.

If you are asked to prove the theorem, make sure your proof shows exactly where each property of a basis is needed in the proof. Also, be careful *not* to use a matrix in the proof. The vectors

Copyright © 2021 Pearson Education, Inc.

$\mathbf{v}_1, \ldots, \mathbf{v}_n$ cannot be arranged as the columns of an ordinary matrix when the vectors are in some abstract vector space.

Checkpoint: Let B = {$\mathbf{b}_1$, ..., $\mathbf{b}_n$} be a basis for $\mathbb{R}^n$. Apply the Invertible Matrix Theorem to the matrix $A = [\mathbf{b}_1 \ \cdots \ \mathbf{b}_n]$ and deduce the Unique Representation Theorem for the case when $V = \mathbb{R}^n$

.

STUDY NOTES

Be careful to distinguish between $\mathbf{x}$ and $[\mathbf{x}]_{\mathcal{B}}$. They are *not equal* in general, even if $\mathbf{x}$ itself is in $\mathbb{R}^n$ (unless $\mathcal{B}$ is the standard basis for $\mathbb{R}^n$).

Theorem 9 and Exercises 29 and 30 show that the coordinate mapping translates vector space statements or calculations in V into equivalent (and familiar) calculations in $\mathbb{R}^n$. The table below lists some examples of typical linear algebra statements.

CORRESPONDING STATEMENTS IN ISOMORPHIC VECTOR SPACES

Linear Algebra in V	Matrix Algebra in $\mathbb{R}^n$
a. $\mathbf{u}$, $\mathbf{v}$, and $\mathbf{w}$ are in V	$[\mathbf{u}]_{\mathcal{B}}$, $[\mathbf{v}]_{\mathcal{B}}$, and $[\mathbf{w}]_{\mathcal{B}}$ are in $\mathbb{R}^n$
b. $\mathbf{w}$ is in Span{$\mathbf{u}$, $\mathbf{v}$}, or $\mathbf{w}$ is in the subspace of V spanned by $\mathbf{u}$ and $\mathbf{v}$	$[\mathbf{w}]_{\mathcal{B}}$ is in Span{$[\mathbf{u}]_{\mathcal{B}}$, $[\mathbf{v}]_{\mathcal{B}}$}, or $[\mathbf{w}]_{\mathcal{B}}$ is in the subspace of $\mathbb{R}^n$ spanned by $[\mathbf{u}]_{\mathcal{B}}$ and $[\mathbf{v}]_{\mathcal{B}}$
c. $\mathbf{w} = c\mathbf{u} + d\mathbf{v}$	$[\mathbf{w}]_{\mathcal{B}} = c[\mathbf{u}]_{\mathcal{B}} + d[\mathbf{v}]_{\mathcal{B}}$
d. {$\mathbf{v}_1$, ..., $\mathbf{v}_p$} is lin. indep.	{$[\mathbf{v}_1]_{\mathcal{B}}$, ..., $[\mathbf{v}_p]_{\mathcal{B}}$} is lin. indep.
e. {$\mathbf{v}_1$, ..., $\mathbf{v}_p$} spans V	{$[\mathbf{v}_1]_{\mathcal{B}}$, ..., $[\mathbf{v}_p]_{\mathcal{B}}$} spans $\mathbb{R}^n$
f. {$\mathbf{v}_1$, ..., $\mathbf{v}_n$} is a basis for V	{$[\mathbf{v}_1]_{\mathcal{B}}$, ..., $[\mathbf{v}_n]_{\mathcal{B}}$} is a basis for $\mathbb{R}^n$

SOLUTIONS TO EXERCISES

1. Since $[\mathbf{x}]_{\mathcal{B}} = \begin{bmatrix} 5 \\ 3 \end{bmatrix}$, we have $\mathbf{x} = 5\mathbf{b}_1 + 3\mathbf{b}_2 = 5\begin{bmatrix} 3 \\ -5 \end{bmatrix} + 3\begin{bmatrix} -4 \\ 6 \end{bmatrix} = \begin{bmatrix} 3 \\ -7 \end{bmatrix}$.

7. The B-coordinates of $\mathbf{x}$ are scalars c_1, c_2, c_3 that satisfy $c_1\mathbf{b}_1 + c_2\mathbf{b}_2 + c_3\mathbf{b}_3 = \mathbf{x}$. To solve this vector equation, row reduce the augmented matrix:

Copyright © 2021 Pearson Education, Inc.

$$\begin{bmatrix} 1 & -3 & 2 & 8 \\ -1 & 4 & -2 & -9 \\ -3 & 9 & 4 & 6 \end{bmatrix} \sim \begin{bmatrix} 1 & -3 & 2 & 8 \\ 0 & 1 & 0 & -1 \\ 0 & 0 & 10 & 30 \end{bmatrix} \sim \begin{bmatrix} 1 & -3 & 0 & 2 \\ 0 & 1 & 0 & -1 \\ 0 & 0 & 1 & 3 \end{bmatrix} \sim \begin{bmatrix} 1 & 0 & 0 & -1 \\ 0 & 1 & 0 & -1 \\ 0 & 0 & 1 & 3 \end{bmatrix}$$

$$\uparrow \quad \uparrow \quad \uparrow \quad \uparrow$$
$$\mathbf{b}_1 \quad \mathbf{b}_2 \quad \mathbf{b}_3 \quad \mathbf{x}$$

So $[\mathbf{x}]_{\mathcal{B}} = \begin{bmatrix} -1 \\ -1 \\ 3 \end{bmatrix}$.

13. The $\mathcal{B}$–coordinates of $\mathbf{p}$ are scalars c_1, c_2, c_3 that satisfy

$$c_1(1+t^2) + c_2(t+t^2) + c_3(1+2t+t^2) = \mathbf{p}(t) = 1 + 4t + 7t^2 \tag{1}$$

Multiply out terms on the left:

$$c_1 + c_1 t^2 + c_2 t + c_2 t^2 + c_3 + 2c_3 t + c_3 t^2 = 1 + 4t + 7t^2$$

On the left, group the constant terms, the terms involving t, and the terms involving t^2:

$$(c_1 + c_3) + (c_2 + 2c_3)t + (c_1 + c_2 + c_3)t^2 = 1 + 4t + 7t^2$$

Equate coefficients of like powers of t to obtain the system of equations:

$$c_1 \quad + \quad c_3 = 1$$
$$c_2 + 2c_3 = 4$$
$$c_1 + c_2 + c_3 = 7$$

Row reduce the augmented matrix to obtain

$$\begin{bmatrix} 1 & 0 & 1 & 1 \\ 0 & 1 & 2 & 4 \\ 1 & 1 & 1 & 7 \end{bmatrix} \sim \begin{bmatrix} 1 & 0 & 1 & 1 \\ 0 & 1 & 2 & 4 \\ 0 & 0 & -2 & 2 \end{bmatrix} \sim \begin{bmatrix} 1 & 0 & 0 & 2 \\ 0 & 1 & 0 & 6 \\ 0 & 0 & 1 & -1 \end{bmatrix}. \text{ Thus, } [\mathbf{p}]_{\mathcal{B}} = \begin{bmatrix} 2 \\ 6 \\ -1 \end{bmatrix} \tag{2}$$

Perhaps you can skip writing the second and third displayed equations and mentally go from (1) directly to the systems of equations. That will save writing time, but mistakes can occur.

A shorter solution uses Theorem 9 and the fact that a calculation in P_2 can be done instead with coordinate vectors relative to the standard basis $\{1, t, t^2\}$. Using this idea, you go directly from equation (1) to the equivalent equation using coordinate vectors:

$$c_1 \begin{bmatrix} 1 \\ 0 \\ 1 \end{bmatrix} + c_2 \begin{bmatrix} 0 \\ 1 \\ 1 \end{bmatrix} + c_3 \begin{bmatrix} 1 \\ 2 \\ 1 \end{bmatrix} = \begin{bmatrix} 1 \\ 4 \\ 7 \end{bmatrix} \tag{3}$$

This vector equation is, of course, equivalent to the system of equations above, and you solve it by row reducing the augmented matrix as in (2).

Warning: The second solution for Exercise 13 is faster, but students can easily forget what their calculations mean. I expect my students to *write* about what they are doing, to show that they

Copyright © 2021 Pearson Education, Inc.

understand what they are calculating. For an acceptable solution to Exercise 13, write that the $\mathcal{B}$–coordinates of **p** are scalars c_1, c_2, c_3 that satisfy equation (1), write something about using coordinate vectors to express (1) in the equivalent form (3), and then solve (3) by the calculations in (2).

15. See the definition of a $\mathcal{B}$-coordinate vector.
17. See equation (4).
19. See Example 5.

23. The set S spans V because every **x** in V has a representation as a (unique) linear combination of elements of S. To show linear independence, suppose that $S = \{\mathbf{v}_1, \ldots, \mathbf{v}_n\}$ and $c_1\mathbf{v}_1 + \cdots + c_n\mathbf{v}_n = \mathbf{0}$ for some scalars $c_1, \ldots, c_n$. The case when $c_1 = \cdots = c_n = 0$ is one possibility. By hypothesis, this is the *only* possible representation of the zero vector as a linear combination of the elements of S. So S is linearly independent and hence is a basis for V.

27. Suppose that $[\mathbf{u}]_{\mathcal{B}} = [\mathbf{w}]_{\mathcal{B}}$ for some **u** and **w** in V, and denote the entries in this coordinate vector by $c_1, \ldots, c_n$. By definition of the coordinate vectors,

$$\mathbf{u} = c_1\mathbf{b}_1 + \cdots + c_n\mathbf{b}_n \text{ and } \mathbf{w} = c_1\mathbf{b}_1 + \cdots + c_n\mathbf{b}_n$$

which shows that $\mathbf{u} = \mathbf{w}$. Since **u** and **w** were arbitrary elements of V, this shows that the coordinate mapping is one-to-one.

29. Since the coordinate mapping is one-to-one, the following equations have the same solutions, $c_1, \ldots, c_p$:

$$c_1\mathbf{u}_1 + \cdots + c_p\mathbf{u}_p = \mathbf{0} \qquad \text{(the zero vector in } V) \qquad (1)$$

$$[c_1\mathbf{u}_1 + \cdots + c_p\mathbf{u}_p]_{\mathcal{B}} = [\mathbf{0}]_{\mathcal{B}} \qquad \text{(the zero vector in } \mathbb{R}^n) \qquad (2)$$

Since the coordinate mapping is linear, (2) is equivalent to

$$c_1[\mathbf{u}_1]_{\mathcal{B}} + \cdots + c_p[\mathbf{u}_p]_{\mathcal{B}} = \begin{bmatrix} 0 \\ \vdots \\ 0 \end{bmatrix} \qquad (3)$$

Hence $c_1, \ldots, c_p$ satisfy (1) if and only if they satisfy (3). So (1) has only the trivial solution if and only if (3) has only the trivial solution. It follows that $\{\mathbf{u}_1, \ldots, \mathbf{u}_p\}$ is linearly independent if and only if $\{[\mathbf{u}_1]_{\mathcal{B}}, \ldots, [\mathbf{u}_p]_{\mathcal{B}}\}$ is linearly independent. (This fact is also an immediate consequence of Exercises 41 and 42 in Section 4.3.)

Study Tip: Exercises 31–35, 37, and 38 tell you to explain your work and justify your conclusions. This requirement is to help you think about your calculations. Check with your instructor. A similar requirement is likely to appear on exam questions.

Warning: The standard mistake in Exercises 31–35, 37, and 38 is to write the coordinate vectors as the *rows* of a matrix and then to check the linear independence of the *columns*. This is completely wrong! Since we mainly work with column vectors, it is wise to write the coordinate vectors first

Copyright © 2021 Pearson Education, Inc.

(as columns) and afterwards write a matrix that can be row reduced to check for linear independence. See the second solution of Exercise 13 and the sample solution for Exercise 31, below

31. The coordinate vectors of the polynomials are

$$\mathbf{v}_1 = \begin{bmatrix} 1 \\ 0 \\ 0 \\ 2 \end{bmatrix}, \mathbf{v}_2 = \begin{bmatrix} 2 \\ 1 \\ -3 \\ 0 \end{bmatrix}, \mathbf{v}_3 = \begin{bmatrix} 0 \\ -1 \\ 2 \\ -1 \end{bmatrix} \text{ (relative to the standard basis).}$$

To check linear independence of these vectors in $\mathbb{R}^4$, compute

$$\begin{bmatrix} 1 & 2 & 0 & 0 \\ 0 & 1 & -1 & 0 \\ 0 & -3 & 2 & 0 \\ 2 & 0 & -1 & 0 \end{bmatrix} \sim \begin{bmatrix} 1 & 2 & 0 & 0 \\ 0 & 1 & -1 & 0 \\ 0 & -3 & 2 & 0 \\ 0 & -4 & -1 & 0 \end{bmatrix} \sim \begin{bmatrix} 1 & 2 & 0 & 0 \\ 0 & 1 & -1 & 0 \\ 0 & 0 & -1 & 0 \\ 0 & 0 & -5 & 0 \end{bmatrix} \sim \begin{bmatrix} 1 & 2 & 0 & 0 \\ 0 & 1 & -1 & 0 \\ 0 & 0 & 1 & 0 \\ 0 & 0 & 0 & 0 \end{bmatrix}$$

The three coordinate vectors are linearly independent in $\mathbb{R}^4$, because the equation $x_1\mathbf{v}_1 + x_2\mathbf{v}_2 + x_3\mathbf{v}_3 = \mathbf{0}$ has only the trivial solution. By the isomorphism with P_3, the three polynomials are linearly independent in P_3.

Note: When you write that an equation has only the trivial solution, you must indicate in some way what equation you have in mind.

Study Tip: Exercises 31–34 could be expanded to ask whether the given polynomials form a basis for P_3.

35. In each part, place the coordinate vectors of the polynomials into the columns of a matrix and reduce the matrix to echelon form:

a. $\begin{bmatrix} 1 & -3 & -4 & 1 \\ -3 & 5 & 5 & 0 \\ 5 & -7 & -6 & -1 \end{bmatrix} \sim \begin{bmatrix} 1 & -3 & -4 & 1 \\ 0 & -4 & -7 & 3 \\ 0 & 8 & 14 & -6 \end{bmatrix} \sim \begin{bmatrix} 1 & -3 & -4 & 1 \\ 0 & -4 & -7 & 3 \\ 0 & 0 & 0 & 0 \end{bmatrix}$. The four coordinate

vectors do not span $\mathbb{R}^3$ because there is no pivot in row 3. Because of the isomorphism between $\mathbb{R}^3$ and P_2, the corresponding polynomials do not span P_2.

Copyright © 2021 Pearson Education, Inc.

$$\textbf{b.} \begin{bmatrix} 0 & 1 & -3 & 2 \\ 5 & -8 & 4 & -3 \\ 1 & -2 & 2 & 0 \end{bmatrix} \sim \begin{bmatrix} 1 & -2 & 2 & 0 \\ 5 & -8 & 4 & -3 \\ 0 & 1 & -3 & 2 \end{bmatrix} \sim \begin{bmatrix} 1 & -2 & 2 & 0 \\ 0 & 2 & -6 & -3 \\ 0 & 1 & -3 & 2 \end{bmatrix} \sim \begin{bmatrix} 1 & -2 & 2 & 0 \\ 0 & 2 & -6 & -3 \\ 0 & 0 & 0 & 3.5 \end{bmatrix}.$$

The four coordinate vectors span $\mathbb{R}^3$ because there is a pivot in each row. Because of the isomorphism between $\mathbb{R}^3$ and P_2, the corresponding polynomials span P_2.

Study Tip: Carefully study the solutions in Exercise 35, because some student papers have a discussion that is far removed from a correct answer. After creating the matrix from the coordinate vectors and row reducing, a student might write something such as "They do not span because there is not a pivot in each row."

Error #1: "They" is a pronoun with no mention of what "they" is. Does the student mean the polynomials, the coordinate vectors, or the columns of the matrix?

Error #2: Span what? The verb "span" requires an object, such as P_2 or $\mathbb{R}^3$.

Error #3. There is no mention of "isomorphism." The phrase "not a pivot in each row" has relevance only to the fact that the coordinate vectors do not span $\mathbb{R}^3$. The only way to get from this fact to the polynomials in the exercise is to use the isomorphism between P_2 and $\mathbb{R}^3$. (Some instructors may permit the term "correspondence" instead of the more precise "isomorphism."

Answer to Checkpoint: The columns of the matrix $A = [\mathbf{b}_1 \; \cdots \; \mathbf{b}_n]$ form a basis for $\mathbb{R}^n$, so A is invertible, by the Invertible Matrix Theorem. By Theorem 5 in Section 2.2, for each $\mathbf{x}$ in $\mathbb{R}^n$ there exists a unique vector $\mathbf{c} = (c_1, \ldots, c_n)$ such that $\mathbf{x} = A\mathbf{c}$, that is, $\mathbf{x} = c_1\mathbf{b}_1 + \cdots + c_n\mathbf{b}_n$.

**MATLAB   The Backslash Operator **

If an equation $A\mathbf{x} = \mathbf{b}$ has a unique solution, MATLAB will automatically produce $\mathbf{x}$ if you use the command

x = A\b

In this section, the equation probably will have the form $P\mathbf{u} = \mathbf{x}$, with $\mathbf{u}$ the $\mathcal{B}$-coordinate vector of $\mathbf{x}$, and the command will be $\mathbf{u} = P\backslash\mathbf{x}$.

The "backslash" command works in two different ways. When A is square, the command **A\b** causes MATLAB to create an LU factorization of A (see Section 2.5); if A is invertible, the factorization is used to produce the unique solution to $A\mathbf{x} = \mathbf{b}$; and if A is not invertible, MATLAB gives the error message "*matrix is singular*" (even if the system $A\mathbf{x} = \mathbf{b}$ has a solution). When A is not square, **A\b** creates a least-squares solution (see Section 6.5).

4.5 - The Dimension of a Vector Space

This short section provides a convenient way to compare the "sizes" of various subspaces of a vector space. The notion of dimension will be used frequently throughout the rest of the text.

Copyright © 2021 Pearson Education, Inc.

KEY IDEAS

Theorem 11 shows that the dimension of a finite-dimensional vector space does not depend on the particular basis for the space. Example 4 shows how to visualize subspaces of various dimensions. Theorems 10 and 13 are important for later theory and applications. You might remember Theorem 10 more easily in this form:

> In an n-dimensional vector space, any set of more than n vectors must be linearly dependent.

Theorem 13, the Basis Theorem, may be restated as follows:

> If V is a p-dimensional vector space, with $p \geq 1$, and if S is a subset of V that contains exactly p elements, then S is linearly independent if and only if S spans V.

Warning: Theorem 12 shows that any basis of a subspace H of a finite-dimensional space V can be extended to a basis of V. But it is *not* true that any basis of V can be cut down to a basis for H. That is, if S is a basis for V, it is not likely that a subset of S is a basis for H. For instance, suppose S is the standard basis for $\mathbb{R}^3$ and H is a plane in $\mathbb{R}^3$ that contains the origin but none of the coordinate axes. Then no subset of S can be a basis for H.

SOLUTIONS TO EXERCISES

1. Since $\begin{bmatrix} s-2t \\ s+t \\ 3t \end{bmatrix} = s\begin{bmatrix} 1 \\ 1 \\ 0 \end{bmatrix} + t\begin{bmatrix} -2 \\ 1 \\ 3 \end{bmatrix}$ for all s, t, the set $\left\{ \begin{bmatrix} 1 \\ 1 \\ 0 \end{bmatrix}, \begin{bmatrix} -2 \\ 1 \\ 3 \end{bmatrix} \right\}$ certainly spans the subspace,

   call it H. Also, the set is obviously linearly independent (because the vectors are not multiples), so the set is a basis for H. Hence, dim $H = 2$.

3. The given subspace, call it H, is the set of all linear combinations of the vectors

   $$\mathbf{v}_1 = \begin{bmatrix} 0 \\ 1 \\ 0 \\ 1 \end{bmatrix}, \mathbf{v}_2 = \begin{bmatrix} 0 \\ -1 \\ 1 \\ 2 \end{bmatrix}, \mathbf{v}_3 = \begin{bmatrix} 2 \\ 0 \\ -3 \\ 0 \end{bmatrix}$$

   First determine if $\{\mathbf{v}_1, \mathbf{v}_2, \mathbf{v}_3\}$ is linearly independent. One way to do this is to row reduce the augmented matrix $[\mathbf{v}_1 \ \ \mathbf{v}_2 \ \ \mathbf{v}_3 \ \ \mathbf{0}]$. A faster way is to use Theorem 4 in Section 4.3.

   Clearly, $\mathbf{v}_1 \neq \mathbf{0}$, $\mathbf{v}_2$ is not a multiple of $\mathbf{v}_1$, and $\mathbf{v}_3$ is not a linear combination of the vectors $\mathbf{v}_1$, $\mathbf{v}_2$ that precede it, because the first entry in $\mathbf{v}_3$ is not zero. Hence $\{\mathbf{v}_1, \mathbf{v}_2, \mathbf{v}_3\}$ is linearly independent and thus is a basis for the space H it spans. Thus dim $H = 3$.

7. Standard calculations show that the set of solutions of the homogeneous system consists of only the trivial solution. So the subspace is $\{\mathbf{0}\}$, and it has no basis. (The vector $\mathbf{0}$ spans the space, but $\{\mathbf{0}\}$ is a linearly dependent set.) By definition, the dimension is zero. [*Note:* Instructors who want every subspace to have a basis often define the empty set to be a basis for $\{\mathbf{0}\}$. The number of vectors in this basis is zero, so the dimension of $\{\mathbf{0}\}$ is still zero.]

Copyright © 2021 Pearson Education, Inc.

13. A has two pivot columns, so dim Col $A = 2$. There are two columns without pivot positions, so the equation $A\mathbf{x} = \mathbf{0}$ has two free variables, and nullity $A = 2$.

17. See the proof of Theorem 14.

19. Read Example 4 carefully.

21. Create a linearly independent set of signals with more than 10 vectors.

23. See the warning prior to Theorem 7 in Section 4.3.

25. See Theorem 13.

27. Form the matrix whose columns are the coordinate vectors of the Hermite polynomials, relative to the standard basis $\{1, t, t^2, t^3\}$:

$$A = \begin{bmatrix} 1 & 0 & -2 & 0 \\ 0 & 2 & 0 & -12 \\ 0 & 0 & 4 & 0 \\ 0 & 0 & 0 & 8 \end{bmatrix}$$

The matrix has four pivots and hence is invertible. So its columns, the coordinate vectors, are linearly independent. Hence the Hermite polynomials themselves are linearly independent in P_3. Since there are four Hermite polynomials, and dim $P_3 = 4$, we conclude from The Basis Theorem that the Hermite polynomials form a *basis* for P_3.

Note: You could, of course, say that the columns of the matrix A span $\mathbb{R}^4$. But you cannot stop with that assertion, because you need the polynomials to span P_3. You have to go on and point out that because of the isomorphism between P_3 and $\mathbb{R}^4$, a set of vectors spans P_3 if and only if the set of coordinate vectors (the columns of A) spans $\mathbb{R}^4$. So the solution is shorter if you appeal to the Basis Theorem.

31. Note that $n \geq 1$, because S cannot have fewer than 0 vectors. If dim $V = n \geq 1$, then $V \neq \{0\}$. If S spans V, then a subset U of S is a basis for V, by the Spanning Set Theorem. But if S has fewer than n vectors, then U also has fewer than n vectors. This is impossible, by Theorem 10, because dim $V = n$. So S cannot span V.

37. Since rank A + nullity A = number of columns of A, rank $A + 4 = 6$, thus rank $A = 2$, which is also the dimension of the row and columns spaces of A.

41. If P were finite-dimensional, then Theorem 12 would imply that $n + 1 = \dim P_n \leq \dim P$ for each n, because each P_n is a subspace of P. This is impossible, so P must be infinite-dimensional.

43. True. Apply the Spanning Set Theorem to the set $\{\mathbf{v}_1, \ldots, \mathbf{v}_p\}$ and produce a basis for V. This basis will have no more than p elements in it, so dim V must be no more than p.

45. True. By Theorem 11, $\{\mathbf{v}_1, \ldots, \mathbf{v}_p\}$ can be expanded to a basis for V. The basis will have at least p elements in it, so dim V must be at least p.

47. True. Take any basis (of p vectors) for V and adjoin the zero vector. Spanning sets can be arbitrarily large. The dimension of V being p only keeps spanning sets from having *fewer* than p elements.

Copyright © 2021 Pearson Education, Inc.

51. Since H is a nonzero subspace of a finite-dimensional space, H is finite-dimensional and has a basis, say, $\mathbf{v}_1, \ldots, \mathbf{v}_p$. Any vector in $T(H)$ has the form $T(\mathbf{y})$ for some $\mathbf{y}$ in H. Since $\{\mathbf{v}_1, \ldots, \mathbf{v}_p\}$ spans H, there exist scalars $c_1, \ldots, c_p$ such that $\mathbf{y} = c_1\mathbf{v}_1 + \cdots + c_p\mathbf{v}_p$. Since T is linear, $T(\mathbf{y}) = c_1 T(\mathbf{v}_1) + \cdots + c_p T(\mathbf{v}_p)$. This shows that $\{T(\mathbf{v}_1), \ldots, T(\mathbf{v}_p)\}$ spans $T(H)$ and hence dim $T(H) \leq p = $ dim H.

Second proof: Let $k = $ dim $T(H)$. If $k = 0$, then $k < $ dim H. Otherwise, $T(H)$ has a basis, which can be written in the form $T(\mathbf{v}_1), \ldots, T(\mathbf{v}_k)$ for some vectors $\mathbf{v}_1, \ldots, \mathbf{v}_k$ in H. Since $\{T(\mathbf{v}_1), \ldots, T(\mathbf{v}_k)\}$ is linearly independent, so is $\{\mathbf{v}_1, \ldots, \mathbf{v}_k\}$, by Exercise 41 in Section 4.3. Since $\mathbf{v}_1, \ldots, \mathbf{v}_k$ are in H, the dimension of H must be at least k.

KEY IDEAS

The Rank Theorem is the main result. By definition, rank $A = $ dim Col A. But because rank A is also the dimension of Row A, the displayed equation in the theorem leads to the equation: dim Row $A + $ dim Nul $A = n$.

Equivalent Descriptions of Rank

The rank of an $m \times n$ matrix A may be described in several ways:

- the dimension of the column space of A, (our definition)
- the number of pivot positions in A, (from Theorem 6)
- the maximum number of linearly independent columns in A,
- the dimension of the row space of A, (from the Rank Theorem)
- the maximum number of linearly independent rows in A,
- the number of nonzero rows in an echelon form of A,
- the maximum number of columns in an invertible submatrix of A. (Supplementary Exercise 35 at the end of the chapter)

Pay attention to how Theorem 13 differs from the results in Section 4.3 about Col A: If you are interested in *rows* of A, use the nonzero rows of an echelon form B as a basis for Row A; if you are interested in the *columns* of A, only use B to obtain *information* about A (namely, to identify the pivot columns), and use the pivot columns of A as a basis for Col A. For Nul A, it is important to use the *reduced* echelon form of A.

When a matrix A is changed into a matrix B by one or more elementary row operations, the row space, null space, and column space of A may or may not be the same as the corresponding subspaces for B. The following table summarizes what can happen in this situation.

Effects of Elementary Row Operations

Copyright © 2021 Pearson Education, Inc.

- Row operations do not affect the linear dependence relations among the columns. (That is, the columns of A have exactly the same linear dependence relations as the columns of any matrix that is row-equivalent to A.)
- Row operations usually change the column space.
- Row operations never change the row space.
- Row operations never change the null space.

The four subspaces shown in Figure 1 in the text are called the *fundamental subspaces* determined by A. The main difficulty here is to avoid confusion between Row A, Nul A, and Col A. The fourth subspace will appear again in Sections 6.1 and 7.4.

The following table lists all statements that are in the Invertible Matrix Theorem at this point in the course, arranged in the scheme used in Section 2.3 of this *Study Guide*. The statements in all three columns are equivalent when A is square ($m = n = p$). As before, a few extra statements have been added to make the table more symmetrical.

STATEMENTS FROM THE INVERTIBLE MATRIX THEOREM

Equivalent statements for an $m{\times}n$ matrix A.	Equivalent statements for an $n{\times}n$ square matrix A.	Equivalent statements for any $n{\times}p$ matrix A.
k. There is a matrix D such that $AD = I$.	a. A is an invertible matrix.	j. There is a matrix C such that $CA = I$.
*. A has a pivot position in every row.	c. A has n pivot positions.	*. A has a pivot position in every column.
h. The columns of A span $\mathbb{R}^m$.	m. The columns of A form a basis for $\mathbb{R}^n$	e. The columns of A are linearly independent.
g. The equation $A\mathbf{x} = \mathbf{b}$ has at least one solution for each $\mathbf{b}$ in $\mathbb{R}^m$.	*. The equation $A\mathbf{x} = \mathbf{b}$ has a unique solution for each $\mathbf{b}$ in $\mathbb{R}^n$.	d. The equation $A\mathbf{x} = \mathbf{0}$ has only the trivial solution.
i. The transformation $\mathbf{x} \mapsto A\mathbf{x}$ maps $\mathbb{R}^n$ onto $\mathbb{R}^m$.	*. The transformation $\mathbf{x} \mapsto A\mathbf{x}$ is invertible.	f. The transformation $\mathbf{x} \mapsto A\mathbf{x}$ is one-to-one.
n. Col $A = \mathbb{R}^m$.	b. A is row equivalent to I.	q. Nul $A = \{\mathbf{0}\}$.
o. dim Col $A = m$.	l. A^T is invertible.	r. dim Nul $A = 0$.
*. rank $A = m$.	p. rank $A = n$.	*. rank $A = p$.

With so many concepts in your linear algebra vocabulary, you need to be careful not to combine terms in ways that are undefined, even though they may sound reasonable to you. For example,

Copyright © 2021 Pearson Education, Inc.

after you finish your work on this section, you should recognize that the following phrases (which have appeared on my students' papers) are meaningless: "the basis of a matrix," "the dimension of a basis," and "the rank of a basis."

Mastering Linear Algebra Concepts: Eight Basic Ideas

Sometime between now and when you finish the chapter, you should do a major review of the eight key concepts introduced in this chapter: vector space, subspace, column space, null space, basis, coordinate vector, dimension, and rank. (The row space of A is not really a separate concept; it is just the column space of A^T.) Study your old review sheets, and prepare new summary sheets for coordinate vector, dimension, and rank. Use as many of the standard categories (special cases, examples, algorithms, etc.) as possible. The tables in this section will be helpful. Also, add cross-references about dimension and rank to other sheets (subspace, column space, etc.) and update your summary sheet for the Invertible Matrix Theorem.

MATLAB   rref and rank

In this course, you can use either **rref(A)** or **rank(A)** to check the rank of A. In practical work, you should use the more reliable command **rank(A)**, based on the singular value decomposition (Section 7.4).

4.6 - Change of Basis

This section will help you better understand coordinate systems. A review of Section 4.4 now is strongly recommended.

KEY IDEAS

Figure 1 and the accompanying discussion will help you visualize the main idea of the section. Imagine superimposing the C-graph paper (Figure 1-b) on the B-graph paper (Figure 1-a). Can you see where b_1 will lie on the C-coordinate system? Four units in the c_1-direction and one unit in the c_2-direction. That is the geometric interpretation of the equation $[b_2]_C = \begin{bmatrix} 4 \\ 1 \end{bmatrix}$ in Example 1.

Similarly, since $[b_2]_C = \begin{bmatrix} -6 \\ 1 \end{bmatrix}$, b_2 lies six units in the negative c_1-direction and one unit in the c_2-direction.

Copyright © 2021 Pearson Education, Inc.

In general, the locations of $\mathbf{b}_1$ and $\mathbf{b}_2$ on the c-graph paper are precisely what you must find in order to build the columns of the change-of-coordinates matrix:

$$\underset{C \leftarrow B}{P} = \left[\, [\mathbf{b}_1]_C \quad [\mathbf{b}_2]_C \,\right]$$

The notation for this matrix should help you remember the basic equation for changing B-coordinates into C-coordinates:

$$[\mathbf{x}]_C = \underset{C \leftarrow B}{P} = [\mathbf{x}]_B$$

The calculations are simple when B and C are bases for $\mathbb{R}^n$. The box after Example 2 illustrates the algorithm for computing the change-of-coordinates matrix. In general,

$$[c_1 \quad \cdots \quad c_n \mid \mathbf{b}_1 \quad \cdots \quad \mathbf{b}_n] \sim [I \mid \underset{C \leftarrow B}{P}]$$

Equivalently, using the notation of Section 4.4,

$$[P_C \quad P_B] \sim [I \quad \underset{C \leftarrow B}{P}]$$

where P_B is the matrix $[\mathbf{b}_1 \quad \cdots \quad \mathbf{b}_n]$ that changes B-coordinates to *standard coordinates*, and P_C is similarly defined. If you refer back to Exercise 25 of Section 2.2, you will see that $\underset{C \leftarrow B}{P}$ is the same as $(P_C)^{-1} P_B$. Since $(P_C)^{-1}$ changes standard coordinates to C-coordinates, you can obtain $[\mathbf{x}]_C$ from $[\mathbf{x}]_B$ as follows:

$$[\mathbf{x}]_B \xrightarrow{} P_B[\mathbf{x}]_B \xrightarrow{} (P_C)^{-1} P_B[\mathbf{x}]_B = \underset{C \leftarrow B}{P}[\mathbf{x}]_B = [\mathbf{x}]_C$$

$$\underset{B\text{-coordinates}}{} \quad \underset{\substack{\text{standard} \\ \text{coordinates}}}{} \quad \underset{C\text{-coordinates}}{}$$

This diagram provides another way of viewing the change of coordinates.

SOLUTIONS TO EXERCISES

1. a. From $\mathbf{b}_1 = 6\mathbf{c}_1 - 2\mathbf{c}_2$ and $\mathbf{b}_2 = 9\mathbf{c}_1 - 4\mathbf{c}_2$, write

$$[\mathbf{b}_1]_c = \begin{bmatrix} 6 \\ -2 \end{bmatrix}, \mathbf{b}_2 = \begin{bmatrix} 9 \\ -4 \end{bmatrix}, \text{ and } \underset{C \leftarrow B}{P} = \begin{bmatrix} 6 & 9 \\ -2 & -4 \end{bmatrix}$$

b. Since $\mathbf{x} = -3\mathbf{b}_1 + 2\mathbf{b}_2$,

$$[\mathbf{x}]_B = \begin{bmatrix} -3 \\ 2 \end{bmatrix} \text{ and } [\mathbf{x}]_C = \begin{bmatrix} 6 & 9 \\ -2 & -4 \end{bmatrix}\begin{bmatrix} -3 \\ 2 \end{bmatrix} = \begin{bmatrix} 0 \\ -2 \end{bmatrix}$$

7. Unlike Exercise 1, you do not have direct information from which you can write $[\mathbf{b}_1]_C$ and $[\mathbf{b}_2]_C$. Rather than compute these two coordinate vectors separately, use the algorithm from Example 2:

Copyright © 2021 Pearson Education, Inc.

$$[\mathbf{c}_1 \quad \mathbf{c}_2 \quad \mathbf{b}_1 \quad \mathbf{b}_2] = \begin{bmatrix} 1 & -2 & 7 & -3 \\ -5 & 2 & 5 & -1 \end{bmatrix} \sim \begin{bmatrix} 1 & -2 & 7 & -3 \\ 0 & -8 & 40 & -16 \end{bmatrix}$$

$$\sim \begin{bmatrix} 1 & -2 & 7 & -3 \\ 0 & 1 & -5 & 2 \end{bmatrix} \sim \begin{bmatrix} 1 & 0 & -3 & 1 \\ 0 & 1 & -5 & 2 \end{bmatrix}$$

Thus $\underset{C \leftarrow B}{P} = \begin{bmatrix} -3 & 1 \\ -5 & 2 \end{bmatrix}$. The change-of-coordinates matrix from C to B is

$$\underset{B \leftarrow C}{P} = (\underset{C \leftarrow B}{P})^{-1} = \begin{bmatrix} -3 & 1 \\ -5 & 2 \end{bmatrix}^{-1} = \frac{1}{-1}\begin{bmatrix} 2 & -1 \\ 5 & -3 \end{bmatrix} = \begin{bmatrix} -2 & 1 \\ -5 & 3 \end{bmatrix}$$

11. See Theorem 15.

13. See the first paragraph in the subsection on *Change of Basis* in $\mathbb{R}^n$.

15. Let $\mathbf{b}_1$ represent the polynomial $1 - 2t + t^2$, let $\mathbf{b}_2$ be $3 - 5t + 4t^2$, let $\mathbf{b}_3$ be $2t + 3t^2$, and let C be the standard basis $\{1, t, t^2\}$ for $\mathbb{P}_2$. The C-coordinate vectors of the vectors $\mathbf{b}_1, \mathbf{b}_2, \mathbf{b}_3$ are

$$[\mathbf{b}_1]_C = \begin{bmatrix} 1 \\ -2 \\ 1 \end{bmatrix}, [\mathbf{b}_2]_C = \begin{bmatrix} 3 \\ -5 \\ 4 \end{bmatrix}, [\mathbf{b}_3]_C = \begin{bmatrix} 0 \\ 2 \\ 3 \end{bmatrix}, \text{ and}$$

$$\underset{C \leftarrow B}{P} = \begin{bmatrix} 1 & 3 & 0 \\ -2 & -5 & 2 \\ 1 & 4 & 3 \end{bmatrix}$$

The coordinate vector $[-1 + 2t]_B$ satisfies

$$\underset{C \leftarrow B}{P}[-1 + 2t]_B = [-1 + 2t]_C = \begin{bmatrix} -1 \\ 2 \\ 0 \end{bmatrix}$$

This equation can be solved by row reduction:

$$\begin{bmatrix} 1 & 3 & 0 & -1 \\ -2 & -5 & 2 & 2 \\ 1 & 4 & 3 & 0 \end{bmatrix} \sim \cdots \sim \begin{bmatrix} 1 & 0 & 0 & 5 \\ 0 & 1 & 0 & -2 \\ 0 & 0 & 1 & 1 \end{bmatrix}; \quad [-1 + 2t]_B = \begin{bmatrix} 5 \\ -2 \\ 1 \end{bmatrix}$$

Copyright © 2021 Pearson Education, Inc.

19. **a.** Since we found P in Exercise 54 of Section 4.5, we can calculate that

$$P^{-1} = \frac{1}{32} \begin{bmatrix} 32 & 0 & 16 & 0 & 12 & 0 & 10 \\ 0 & 32 & 0 & 24 & 0 & 20 & 0 \\ 0 & 0 & 16 & 0 & 16 & 0 & 15 \\ 0 & 0 & 0 & 8 & 0 & 10 & 0 \\ 0 & 0 & 0 & 0 & 4 & 0 & 6 \\ 0 & 0 & 0 & 0 & 0 & 2 & 0 \\ 0 & 0 & 0 & 0 & 0 & 0 & 1 \end{bmatrix}.$$

b. Since P is the change-of-coordinates matrix from C to $\mathcal{B}$, P^{-1} will be the change-of-coordinates matrix from $\mathcal{B}$ to C. By Theorem 15, the columns of P^{-1} will be the C-coordinate vectors of the basis vectors in $\mathcal{B}$. Thus

$$\cos^2 t = \frac{1}{2}(1 + \cos 2t)$$

$$\cos^3 t = \frac{1}{4}(3\cos t + \cos 3t)$$

$$\cos^4 t = \frac{1}{8}(3 + 4\cos 2t + \cos 4t)$$

$$\cos^5 t = \frac{1}{16}(10\cos t + 5\cos 3t + \cos 5t)$$

$$\cos^6 t = \frac{1}{32}(10 + 15\cos 2t + 6\cos 4t + \cos 6t)$$

21. **a.** If P is to be the change-of-coordinates matrix from $\{\mathbf{u}_1, \mathbf{u}_2, \mathbf{u}_3\}$ to $\{\mathbf{v}_1, \mathbf{v}_2, \mathbf{v}_3\}$, then the columns of P should be C-coordinate vectors, where $C = \{\mathbf{v}_1, \mathbf{v}_2, \mathbf{v}_3\}$. That is, the columns of P should be $[\mathbf{u}_1]_C$, $[\mathbf{u}_2]_C$, and $[\mathbf{u}_3]_C$. You know P, but you do not know $\mathbf{u}_1$, $\mathbf{u}_2$, or $\mathbf{u}_3$. Ask yourself, for example, what is the meaning of $[\mathbf{u}_1]_C$? By definition, a C-coordinate vector tells how to build a vector out of the C-basis vectors, $\mathbf{v}_1$, $\mathbf{v}_2$, and $\mathbf{v}_3$. So, for $j = 1, 2, 3$,

$$\mathbf{u}_j = [\mathbf{v}_1 \quad \mathbf{v}_2 \quad \mathbf{v}_3][\mathbf{u}_j]_C = V[\mathbf{u}_j]_C$$

where $V = [\mathbf{v}_1 \quad \mathbf{v}_2 \quad \mathbf{v}_3]$. Then, by the definition of matrix multiplication,

$$[\mathbf{u}_1 \quad \mathbf{u}_2 \quad \mathbf{u}_3] = [V[\mathbf{u}_j]_C \quad V[\mathbf{u}_j]_C \quad V[\mathbf{u}_j]_C] = V[[\mathbf{u}_j]_C \quad [\mathbf{u}_j]_C \quad [\mathbf{u}_j]_C] = VP$$

You know V and P, so you can compute $\mathbf{u}_1$, $\mathbf{u}_2$, and $\mathbf{u}_3$

$$VP = \begin{bmatrix} -2 & -8 & -7 \\ 2 & 5 & 2 \\ 3 & 2 & 6 \end{bmatrix} \begin{bmatrix} 1 & 2 & -1 \\ -3 & -5 & 0 \\ 4 & 6 & 1 \end{bmatrix} = \begin{bmatrix} -6 & -6 & -5 \\ -5 & -9 & 0 \\ 21 & 32 & 3 \end{bmatrix}$$

Copyright © 2021 Pearson Education, Inc.

Thus, $\mathbf{u}_1 = \begin{bmatrix} -6 \\ -5 \\ 21 \end{bmatrix}$, $\mathbf{u}_2 = \begin{bmatrix} -6 \\ -9 \\ 32 \end{bmatrix}$, $\mathbf{u}_3 = \begin{bmatrix} -5 \\ 0 \\ 3 \end{bmatrix}$.

b. If you were not able to work part (a) by yourself and you have read the solution, then you should try part (b) by yourself. Here are the steps of the solution:

(i) Write in symbols what the columns of P should be.

(ii) Decide how these columns are related to the matrix $W = [\mathbf{w}_1 \quad \mathbf{w}_2 \quad \mathbf{w}_3]$.

(iii) Obtain a matrix equation that involves W in some way.

(iv) Compute W, and list its columns as the answer to the problem.

Study the solution to (a), close the *Study Guide*, and work on (b) as if it were a new problem. The solution of (b) is at the end of the solutions for Section 4.8.

MATLAB   Change-of-Coordinates Matrix

The command **rref(M)** row reduces a matrix such as $[\mathbf{c}_1 \quad \mathbf{c}_2 \quad \mathbf{b}_1 \quad \mathbf{b}_2]$ to the desired form.

4.7 - Digital Signal Processing

When I asked my colleague in electrical engineering for advice on curriculum in linear algebra, he immediately mentioned that signals are a vector space and filters are just linear transformations. The material in this section is foundational for understanding the underlying structure of signals and our digital world in general. Talking with mathematicians in industry, I also heard repeatedly the need for students to see the power of abstract reasoning and understand how to use it.

You should be able to

- add and subtract signals

- establish whether or not a subset of the signals is a subspace

- show that a function is an LTI

- discuss the range and null space of an LTI

KEY IDEAS

Signals are just vectors. Signal processors are just linear transformations. These two facts allow you to use and rely on most of the techniques and theory you have learned so far in this course. For infinite signal spaces, we can no longer use matrices to represent linear transformations.

Copyright © 2021 Pearson Education, Inc.

Storage is also an issue, so in this section we discuss creating needed signals using linear transformations, rather than storing them.

1. $\chi + \alpha = (..., 0, 2, 0, 2, 0, 2, 0, ...)$.

For Exercise 7, recall that $I(\{x_k\}) = \{x_k\}$ and $S(\{x_k\}) = \{x_{k-1}\}$. Apply the transformation to each signal from the table and observe which signals get mapped to the zero signal.

7. $(I - cS)(\varepsilon_c) = I(\varepsilon_c) - cS(\varepsilon_c) = \{(c)^k\} - c\{(c)^{k-1}\} = \{0\}$, and ε_c is the only signal from Table 1 that gets mapped to the zero signal by $I - cS$.

13. Apply T to any signal to get a signal in the range of T. For example, $T(\delta) = (..., 0, 0, 0, 1, -1, 0, 0, ...)$, is the range of T.

15. See Theorem 17.

17. See Theorem 17.

19. Compare the definition of an LTI to the definition of a linear transformation.

21. Recall that the signals are a vector space.

Guess and check or working backwards through the solution to Practice Problem 3 are two good ways to find solutions to Exercises 23 and 24.

25. Set $r = 0$ to see that the zero signal is in W. Notice that the sum of two signals is in W, as is any scalar multiple of a signal in W.

27. A typical element in W looks like $r(..., 1, 0, 1, 0, 1, 0, 1, ...)$, so $\{\chi - \alpha\}$ is a basis for W. This subspace has dimension 1.

31. A typical element in W looks like $\sum_{m=-\infty}^{\infty} r_{2m-1} S^{2m-1}(\delta)$, so $\beta = \{S^{2m-1}(\delta) |$ with m any integer$\}$ is an infinite linearly independent subset of W, hence W is infinite dimensional. Note that β is not a basis, since not every element in W can be generated using a finite sum of elements from β.

Copyright © 2021 Pearson Education, Inc.

4.8 - Applications to Difference Equations

Difference equations are the discrete analogues of differential equations. Both are important in science and engineering. The discrete and continuous theories are remarkably parallel, and linear algebra is applied in similar ways, although the calculations are somewhat easier for difference equations. A variety of examples and exercises here illustrate some difference equations you may encounter later in your work.

KEY IDEAS

Each signal in $\mathbb{S}$ is an infinite list of numbers. Linear independence of a set of signals can often be demonstrated by looking at short segments of the signals, that is, by showing that a Casorati matrix is invertible. A Casorati matrix cannot be used in general to demonstrate linear *dependence* of a set. However, see the appendix at the end of this *Study Guide* section. The main focus of the section is on difference equations. Given a homogeneous difference equation, you should be able to:

- determine whether a specified signal is a solution of the equation;

- find solutions of the equation, using the auxiliary equation;

- give the general solution (when the auxiliary equation has no multiple roots and no complex roots).

Theorem 20 is the key result that enables you to write the general solution. Just finding some specific solutions is not enough; you must show that they *span* the set of all solutions. But Theorem 20 and the Basis Theorem in Section 4.5 together show that for an *n*th order equation, you only need to find *n linearly independent* solutions. (See Example 4.) The same principle applies to an *n*th order differential equation (discussed later in Section 5.7). This principle is one of the most powerful applications of linear algebra in the text.

The general principle of nonhomogeneous equations is illustrated in Figure 2:

$$\left\{ \begin{array}{c} \text{General solution of} \\ \text{nonhomogeneous eqn.} \end{array} \right\} = \left\{ \begin{array}{c} \text{Particular solution of} \\ \text{nonhomogeneous eqn.} \end{array} \right\} + \left\{ \begin{array}{c} \text{General solution of} \\ \text{homogeneous eqn.} \end{array} \right\}$$

The final subsection shows the modern way to study an *n*th order linear difference equation, rewriting it as a first order system $\mathbf{x}_{k+1} = A\mathbf{x}_k \ (k = 1, 2, \ldots)$. Such systems were introduced in Section 1.10 and they will be discussed further in Sections 5.6 and 5.9.

SOLUTIONS TO EXERCISES

1. If $y_k = (-4)^k$, then $y_{k+1} = (-4)^{k+1}$ and $y_{k+2} = (-4)^{k+2}$. Substitute these formulas into the left side of the equation:

Copyright © 2021 Pearson Education, Inc.

$$y_{k+2} + 2y_{k+1} - 8y_k = (-4)^{k+2} + 2(-4)^{k+1} - 8(-4)^k$$

$$= (-4)^k[(-4)^2 + 2(-4) - 8]$$

$$= (-4)^k[16 - 8 - 8] = 0 \text{ for all } k$$

Since the difference equation holds for all k, $(-4)^k$ is a solution. The text answer displays the similar calculations for $y_k = 2^k$.

7. Compute the Casorati matrix for the signals 1^k, 2^k, and $(-2)^k$, setting $k = 0$ for convenience:

$$\begin{bmatrix} 1^0 & 2^0 & (-2)^0 \\ 1^1 & 2^1 & (-2)^1 \\ 1^2 & 2^2 & (-2)^2 \end{bmatrix} = \begin{bmatrix} 1 & 1 & 1 \\ 1 & 2 & -2 \\ 1 & 4 & 4 \end{bmatrix} \sim \begin{bmatrix} 1 & 1 & 1 \\ 0 & 1 & -3 \\ 0 & 3 & 3 \end{bmatrix} \sim \begin{bmatrix} 1 & 1 & 1 \\ 0 & 1 & -3 \\ 0 & 0 & 12 \end{bmatrix}$$

This Casorati matrix has three pivots and hence is invertible, by the IMT. Hence the set of signals $\{1^k, 2^k, (-2)^k\}$ is linearly independent in S. We know (from the text) that these signals are in the solution space H of a third-order difference equation. By Theorem 20, dim $H = 3$. Since the three signals are linearly independent, they form a basis for H, by the Basis Theorem in Section 4.5.

Warning: Many student papers for Exercise 7 suffer from a lack of precision, often confusing linear independence of the columns of the Casorati matrix with linear independence of the signals in S. There is no need to discuss the columns of the Casorati matrix—just observe that the matrix is invertible. But you must point out that the three *signals* are linearly independent, in order to apply the Basis Theorem to the vector space H of solutions to the difference equation.

13. The auxiliary equation for $y_{k+2} - y_{k+1} + \frac{2}{9}y_k = 0$ is $r^2 - r + \frac{2}{9} = 0$. By the quadratic formula,

$$r = \frac{1 \pm \sqrt{1 - 8/9}}{2} = \frac{1 \pm 1/3}{2} = \frac{2}{3} \text{ or } \frac{1}{3}$$

Two solutions of the difference equation are $\left(\frac{2}{3}\right)^k$ and $\left(\frac{1}{3}\right)^k$. These signals are obviously linearly independent because neither is a multiple of the other. Since the solution space is two-dimensional (Theorem 17), the two signals form a basis for the solution space, by the Basis Theorem.

Study Tip: I think Exercises 7–19 are good questions because they illustrate how important Theorem 20 and the Basis Theorem really are. Probably, you do not have to remember the specific number of Theorem, but your discussion should show that you have it in mind and know how to use it with the Basis Theorem. (Check with your instructor.)

19. Letting $a = .9$ and $b = 4/9$ gives the difference equation $Y_{k+2} - 1.3Y_{k+1} + .4Y_k = 1$. First we find a particular solution $Y_k = T$ of this equation, where T is a constant. The solution of the

Copyright © 2021 Pearson Education, Inc.

equation $T - 1.3T + .4T = 1$ is $T = 10$, so 10 is a particular solution to
$Y_{k+2} - 1.3Y_{k+1} + .4Y_k = 1$. Next we solve the homogeneous difference equation
$Y_{k+2} - 1.3Y_{k+1} + .4Y_k = 0$. The auxiliary equation for this difference equation is
$r^2 - 1.3r + .4 = 0$. By the quadratic formula (or factoring), $r = .8$ or $r = .5$, so two solutions of
the homogeneous difference equation are $.8^k$ and $.5^k$. The signals $(.8)^k$ and $(.5)^k$ are
linearly independent because neither is a multiple of the other. By Theorem 17, the solution
space is two-dimensional, so the two linearly independent signals $(.8)^k$ and $(.5)^k$ form a
basis for the solution space of the homogeneous difference equation by the Basis Theorem.
Translating the solution space of the homogeneous difference equation by the particular
solution 10 of the nonhomogeneous difference equation gives us the general solution of
$Y_{k+2} - 1.3Y_{k+1} + .4Y_k = 1$: $Y_k = c_1(.8)^k + c_2(.5)^k + 10$. As k increases the first two terms in the
solution approach 0, so Y_k approaches 10.

21. The auxiliary equation for $y_{k+2} + 4y_{k+1} + y_k = 0$ is $r^2 + 4r + 1 = 0$. By the quadratic formula,

$$r = \frac{-4 \pm \sqrt{16-4}}{2} = \frac{-4 \pm 2\sqrt{3}}{2} = -2 \pm \sqrt{3}$$

Two solutions of the difference equation are $(-2+\sqrt{3})^k$ and $(-2-\sqrt{3})^k$. They are obviously
linearly independent because neither is a multiple of the other. Since the solution space is
two-dimensional (Theorem 17), the two signals form a fundamental set of solutions by the
Basis Theorem, and the general solution has the form $c_1(-2+\sqrt{3})^k + c_2(-2-\sqrt{3})^k$.

27. To prove that $y_k = k^2$ is a solution of

$$y_{k+2} + 3y_{k+1} - 4y_k = 10k + 7 \tag{1}$$

show that when k^2, $(k+1)^2$, and $(k+2)^2$ are substituted for y_k, y_{k+1}, and y_{k+2}, respectively, the
resulting equation is true for all k:

$$(k+2)^2 + 3(k+1)^2 - 4k^2 = (k^2 + 4k + 4) + 3(k^2 + 2k + 1) - 4k^2$$

$$= (1+3-4)k^2 + (4+6)k + (4+3)$$

$$= 10k + 7 \text{ for all } k$$

So k^2 is a solution of (1). The auxiliary equation for the homogeneous difference equation

$$y_{k+2} + 3y_{k+1} - 4y_k = 0 \text{ for all } k \tag{2}$$

is $r^2 + 3r - 4 = 0$, which factors as $(r-1)(r+4) = 0$, so $r = 1, -4$. Thus 1^k and $(-4)^k$ are
solutions of (2). The signals are linearly independent (for neither is a multiple of the other),
so they form a basis for the two-dimensional solution space. The general solution of (2) is
$c_1 \cdot 1^k + c_2(-4)^k$. Add this to a particular solution of (1) and obtain the general solution
$k^2 + c_1 + c_2(-4)^k$ of (1).

Copyright © 2021 Pearson Education, Inc.

31. Let $\mathbf{x}_k = \begin{bmatrix} y_k \\ y_{k+1} \\ y_{k+2} \\ y_{k+3} \end{bmatrix}$. Then $\mathbf{x}_{k+1} = \begin{bmatrix} y_{k+1} \\ y_{k+2} \\ y_{k+3} \\ y_{k+4} \end{bmatrix} = \begin{bmatrix} 0 & 1 & 0 & 0 \\ 0 & 0 & 1 & 0 \\ 0 & 0 & 0 & 1 \\ 9 & -6 & -8 & 6 \end{bmatrix} \begin{bmatrix} y_k \\ y_{k+1} \\ y_{k+2} \\ y_{k+3} \end{bmatrix} = A\mathbf{x}_k.$

21. b. (*This solution is for Section 4.7.*) The columns of a change-of-coordinates matrix P from $\{\mathbf{v}_1, \mathbf{v}_2, \mathbf{v}_3\}$ to $\{\mathbf{w}_1, \mathbf{w}_2, \mathbf{w}_3\}$ are $\mathcal{D}$-coordinate vectors, where $\mathcal{D} = \{\mathbf{w}_1, \mathbf{w}_2, \mathbf{w}_3\}$. That is, the columns of P are $[\mathbf{v}_1]_{\mathcal{D}}$, $[\mathbf{v}_2]_{\mathcal{D}}$, and $[\mathbf{v}_3]_{\mathcal{D}}$. How are these columns related to the matrix $W = [\mathbf{w}_1 \quad \mathbf{w}_2 \quad \mathbf{w}_3]$? A $\mathcal{D}$-coordinate vector tells how to build a vector out of the $\mathcal{D}$-basis vectors (the columns of the matrix W). For $j = 1, 2, 3$,

$$\mathbf{v}_j = [\mathbf{w}_1 \quad \mathbf{w}_2 \quad \mathbf{w}_3][\mathbf{v}_j]_{\mathcal{D}} = W \,[\mathbf{v}_j]_{\mathcal{D}}$$

By definition of matrix multiplication,

$$V = [\mathbf{v}_1 \quad \mathbf{v}_2 \quad \mathbf{v}_3] = [W\,[\mathbf{v}_j]_{\mathcal{D}} \quad W\,[\mathbf{v}_j]_{\mathcal{D}} \quad W\,[\mathbf{v}_j]_{\mathcal{D}}] = W\,[[\mathbf{v}_j]_{\mathcal{D}} \quad [\mathbf{v}_j]_{\mathcal{D}} \quad [\mathbf{v}_j]_{\mathcal{D}}] = WP$$

You know V and P, so compute W from $VP^{-1} = W$. Use MATLAB or other matrix program to compute P^{-1}. Then

$$W = VP^{-1} = \begin{bmatrix} -2 & -8 & -7 \\ 2 & 5 & 2 \\ 3 & 2 & 6 \end{bmatrix} \begin{bmatrix} 5 & 8 & 5 \\ -3 & -5 & -3 \\ -2 & -2 & -1 \end{bmatrix} = \begin{bmatrix} 28 & 38 & 21 \\ -9 & -13 & -7 \\ -3 & 2 & 3 \end{bmatrix}$$

Thus, $\mathbf{w}_1 = \begin{bmatrix} 28 \\ -9 \\ -3 \end{bmatrix}$, $\mathbf{w}_2 = \begin{bmatrix} 38 \\ -13 \\ 2 \end{bmatrix}$, $\mathbf{w}_3 = \begin{bmatrix} 21 \\ -7 \\ 3 \end{bmatrix}$.

Chapter 4 - Supplementary Exercises

In this chapter, Exercises 1-19 consist of true/false questions, whose level of difficulty varies. Some are similar to the ones that appear in many sections of the text, in which a word or phrase is sometimes missing or slightly misstated. Some follow from a theorem: others may need careful reasoning. A few may require an argument that uses several ideas. In each case, think carefully about the statement and attempt to write a solution. The text provides the true/false answer, but you must supply the justification or counterexample.

25. You would have to know that the solution set of the homogeneous system is spanned by two solutions. In this case, the null space of the 18×20 coefficient matrix A is at most two-dimensional. By the Rank Theorem, rank $A = 20 -$ nullity $A \geq 20 - 2 = 18$. Since Col A is a subspace of $\mathbb{R}^{18}$, Col $A = \mathbb{R}^{18}$. Thus $A\mathbf{x} = \mathbf{b}$ has a solution for every $\mathbf{b}$ in $\mathbb{R}^{18}$.

Copyright © 2021 Pearson Education, Inc.

31. By Exercise 30, rank $PA \leq$ rank A, and rank $A = \operatorname{rank}(P^{-1}P)A = \operatorname{rank} P^{-1}(PA) \leq \operatorname{rank} PA$, so rank $PA = $ rank A.

37. To determine if the matrix pair (A, B) is controllable, we compute the rank of the matrix $\begin{bmatrix} B & AB & A^2B \end{bmatrix}$. To find the rank, we row reduce:

$$\begin{bmatrix} B & AB & A^2B \end{bmatrix} = \begin{bmatrix} 0 & 1 & 0 \\ 1 & -.9 & .81 \\ 1 & .5 & .25 \end{bmatrix} \sim \begin{bmatrix} 1 & 0 & 0 \\ 0 & 1 & 0 \\ 0 & 0 & 1 \end{bmatrix}.$$

The rank of the matrix is 3, and the pair (A, B) is controllable.

Appendix: The Casorati Test

Let $\{\mathbf{y}_1, \ldots, \mathbf{y}_n\}$ be a set of signals in $\mathbb{S}$. For $j = 1, \ldots, n$ and for any k, let $\mathbf{y}_j(k)$ denote the kth entry in the signal $\mathbf{y}_j$ and let $C(k) = \begin{bmatrix} \mathbf{y}_1(k) & \cdots & \mathbf{y}_n(k) \\ \mathbf{y}_1(k+1) & \cdots & \mathbf{y}_n(k+1) \\ \vdots & & \vdots \\ \mathbf{y}_1(k+n-1) & \cdots & \mathbf{y}_n(k+n-1) \end{bmatrix}$ The Casorati matrix

a. If $C(k)$ is invertible for some k, $\{\mathbf{y}_1, \ldots, \mathbf{y}_n\}$ is linearly independent.
b. If $\mathbf{y}_1, \ldots, \mathbf{y}_n$ all satisfy a homogeneous difference equation of order n,

$$y_{k+n} + a_1 y_{k+n-1} + \cdots + a_n y_k = 0 \text{ for all } k \qquad (*)$$

(with $a_n \neq 0$), and if the Casorati matrix $C(k)$ is not invertible for some k, then $\{\mathbf{y}_1, \ldots, \mathbf{y}_n\}$ is linearly dependent in $\mathbb{S}$, and for all k, $C(k)$ is not invertible.

Proof. (a) The argument given in the text for a set of three signals generalizes immediately to n signals. (b) Suppose that $\mathbf{y}_1, \ldots, \mathbf{y}_n$ are in the set H of solutions of (*) and $C(k_o)$ is not invertible for some k_0. It is readily verified that if $T: H \rightarrow \mathbb{R}^n$ is defined by $T(\mathbf{y}) = \begin{bmatrix} \mathbf{y}(k_0) \\ \mathbf{y}(k_0 + 1) \\ \vdots \\ \mathbf{y}(k_0 + n-1) \end{bmatrix}$

then T is a linear transformation. The proof of Theorem 19 is easily modified to show that (*) has a unique solution $\mathbf{y}$ whenever $\mathbf{y}(k_0), \ldots, \mathbf{y}(k_0 + n - 1)$ are specified. This means that T is a one-to-one mapping of H onto $\mathbb{R}^n$. Furthermore, the images $T(\mathbf{y}_1), \ldots, T(\mathbf{y}_n)$ form the columns of the Casorati matrix $C(k_0)$ and hence are linearly dependent, because $C(k_0)$ is not invertible. Since T is one-to-one, $\{\mathbf{y}_1, \ldots, \mathbf{y}_n\}$ is linearly dependent. This proves the first statement in (b). The second

Copyright © 2021 Pearson Education, Inc.

statement follows immediately from part (a), because if $C(k)$ were invertible for some k, then $\{\mathbf{y}_1, ..., \mathbf{y}_n\}$ would be linearly independent, which is not true. So $C(k)$ is not invertible for each k.

MATLAB roots
    The MATLAB command **roots(p)** produces a column vector whose entries are the roots of the polynomial described by **p**.

Chapter 4 - Glossary Checklist

Check your knowledge by attempting to write definitions of the terms below. Then compare your work with the definitions given in the text's Glossary. Ask your instructor which definitions, if any, might appear on a test.

auxiliary equation: A polynomial equation in a variable r, created from . . . .

basis (for a nonzero subspace H):   A set $\mathcal{B} = \{\mathbf{v}_1, ..., \mathbf{v}_p\}$ in V such that: . . . .

$\mathcal{B}$-coordinates of x: *See* coordinates of **x** relative to the basis $\mathcal{B}$.

change-of-coordinates matrix (from a basis $\mathcal{B}$ to a basis C):   A matrix $\underset{C \leftarrow \mathcal{B}}{P}$ that transforms . . . ,

      namely, (equation). . . .

column space (of an $m \times n$ matrix A):   The set Col A of . . . . In set notation, Col A = { :   }.

controllable (pair of matrices):   A matrix pair (A, B) where A is $n \times n$, B has n rows, and . . . .

coordinate mapping (determined by an ordered basis B in a vector space V):   A mapping that

      associates to each . . . .

coordinates of x relative to the basis $\mathcal{B} = \{\mathbf{b}_1, ..., \mathbf{b}_n\}$:   . . . .

coordinate vector of x relative to $\mathcal{B}$:   The vector $[\mathbf{x}]_\mathcal{B}$ whose entries . . . .

dimension (of a vector space V):   The number . . . .

explicit description (of a subspace W of $\mathbb{R}^n$):   A parametric representation of W as the set of . . . .

finite-dimensional (vector space):   A vector space that is . . . .

full rank (matrix):   An $m \times n$ matrix whose rank is . . . .

fundamental set of solutions: A . . . for the set of solutions of . . . .

fundamental subspaces (determined by A):   The . . . of A, ...

implicit description (of a subspace W of $\mathbb{R}^n$):   A set of one or more . . . .

infinite-dimensional (vector space):   A nonzero vector space V that . . . .

isomorphism: A . . . mapping from one vector space . . . .

Copyright © 2021 Pearson Education, Inc.

kernel (of a linear transformation $T : V \rightarrow W$):   The set of . . . such that . . . .

linear combination: A sum . . . .

linear dependence relation: A . . . vector equation where . . . .

linear filter: A . . . equation used to transform . . . .

linearly dependent (vectors):   A set $\{\mathbf{v}_1, ..., \mathbf{v}_p\}$ with the property that . . . .

linearly independent (vectors):   A set $\{\mathbf{v}_1, ..., \mathbf{v}_p\}$ with the property . . . .

linear transformation T (from a vector space V into a vector space W):   A rule $T : V \rightarrow W$ that to each vector $\mathbf{x}$ in V assigns a unique vector $T(\mathbf{x})$ in W, such that: . . . .

Markov chain: A sequence of . . . vectors $\mathbf{v}_0, \mathbf{v}_1, \mathbf{v}_2, \ldots$, together with a . . . matrix P such that . . . .

maximal linearly independent set (in V):   A linearly independent set B in V such that if . . . , then . . . .

minimal spanning set (for a subspace H):   A set B that spans H and has the property that if . . . , then . . . .

null space (of an $m \times n$ matrix A):   The set Nul A of all . . . . In set notation, Nul $A = \{ \ : \quad \}$.

probability vector: A vector in $\mathbb{R}^n$ whose entries . . . .

proper subspace: Any subspace of a vector space V . . . .

range (of a linear transformation $T : V \ \square \ W$):   The set of all vectors . . . .

rank (of a matrix A):   . . . .

regular stochastic matrix: A stochastic matrix P such that . . . .

row space (of a matrix A):   The set Row A of all . . . ; also denoted by . . . .

signal (or **discrete-time signal**):   An element in the space S of all . . . .

Span$\{\mathbf{v}_1, ..., \mathbf{v}_p\}$:   The set of . . . . Also, the . . . *spanned* . . . .

spanning set (for a subspace H):   Any set $\{\mathbf{v}_1, ..., \mathbf{v}_p\}$ . . . such that . . . .

standard basis: The basis . . . for $\mathbb{R}^n$ consisting of . . . , or the basis . . . for P$_n$.

state vector: A . . . vector. In general, a vector that describes . . . , often in connection with a difference equation . . . .

steady-state vector (for a stochastic matrix P):   A . . . vector $\mathbf{v}$ such that . . . .

stochastic matrix: A . . . matrix whose columns . . . .

submatrix (of A):   Any matrix obtained by . . . .

subspace: A subset H of some vector space V such that H has these properties . . . .

vector space: A set of objects, called vectors, on which . . . .

zero subspace: The subspace . . . consisting of . . . .

Copyright © 2021 Pearson Education, Inc.

5 Eigenvalues and Eigenvectors

5.1 – Eigenvalues and Eigenvectors

This section introduces eigenvectors and eigenvalues. A hint about the connection with dynamical systems appears at the end of the section.

KEY IDEAS

In words, a nonzero vector $\mathbf{v}$ is an eigenvector of a matrix A if and only if the transformed vector $A\mathbf{x}$ points in the same or opposite direction of $\mathbf{v}$.

Notice that while an eigenvalue λ might be zero, an eigenvector is never zero (by definition). An eigenspace contains eigenvectors together with the zero vector.

The two equations $A\mathbf{x} = \lambda\mathbf{x}$ and $(A - \lambda I)\mathbf{x} = \mathbf{0}$ are equivalent. See Example 3. The first equation is useful for understanding what eigenvalues and eigenvectors are, and it shows the geometric effect of the linear transformation $\mathbf{x} \mapsto A\mathbf{x}$ on an eigenvector. The second equation shows that the eigenspace is a subspace (because it is the null space of the matrix $A - \lambda I$), and the equation is used to find a basis for the eigenspace, when λ is a known eigenvalue. The second equation will be used again in Section 5.2 for another purpose.

SOLUTIONS TO EXERCISES

1. The number 2 is an eigenvalue of A if and only if the equation $A\mathbf{x} = 2\mathbf{x}$ has a nontrivial solution. This equation is equivalent to $(A - 2I)\mathbf{x} = \mathbf{0}$. Compute

$$A - 2I = \begin{bmatrix} 3 & 2 \\ 3 & 8 \end{bmatrix} - \begin{bmatrix} 2 & 0 \\ 0 & 2 \end{bmatrix} = \begin{bmatrix} 1 & 2 \\ 3 & 6 \end{bmatrix}$$

The columns of A are obviously linearly dependent, so $(A - 2I)\mathbf{x} = \mathbf{0}$ has a nontrivial solution, and so 2 is an eigenvalue of A.

7. Proceed as in Exercise 1:

$$A - 4I = \begin{bmatrix} 3 & 0 & -1 \\ 2 & 3 & 1 \\ -3 & 4 & 5 \end{bmatrix} - \begin{bmatrix} 4 & 0 & 0 \\ 0 & 4 & 0 \\ 0 & 0 & 4 \end{bmatrix} = \begin{bmatrix} -1 & 0 & -1 \\ 2 & -1 & 1 \\ -3 & 4 & 1 \end{bmatrix}$$

Copyright © 2021 Pearson Education, Inc.

You need to know whether $A - 4I$ is invertible. This could be checked in several ways, but since you are asked for an eigenvector, in the event that one exists, the best strategy is to row reduce the augmented matrix for $(A - 4I)\mathbf{x} = \mathbf{0}$:

$$\begin{bmatrix} -1 & 0 & -1 & 0 \\ 2 & -1 & 1 & 0 \\ -3 & 4 & 1 & 0 \end{bmatrix} \sim \begin{bmatrix} -1 & 0 & -1 & 0 \\ 0 & -1 & -1 & 0 \\ 0 & 4 & 4 & 0 \end{bmatrix} \sim \begin{bmatrix} 1 & 0 & 1 & 0 \\ 0 & 1 & 1 & 0 \\ 0 & 0 & 0 & 0 \end{bmatrix}$$

Now it is clear that 4 is an eigenvalue of A [because $(A - 4I)\mathbf{x} = \mathbf{0}$ has a nontrivial solution]. The coordinates of an eigenvector satisfy $-x_1 - x_3 = 0$ and $-x_2 - x_3 = 0$. The general solution is not requested, so take any nonzero value for x_3 to produce an eigenvector. If $x_3 = 1$, then $\mathbf{x} = (-1, -1, 1)$.

Checkpoint 1:   The answer in the text is different, namely, $(1, 1, -1)$. Why is this also correct?

Helpful Hint: Suppose you think that 4 is an eigenvalue of a matrix, as in Exercise 7, and you row reduce the augmented matrix for $(A - 4I)\mathbf{x} = \mathbf{0}$. If you discover that there are no free variables, then there are only two possibilities: (1) 4 is *not* an eigenvalue of A, or (2) you have made an arithmetic error.

13. <u>For $\lambda = 1$:</u>

$$A - 1I = \begin{bmatrix} 4 & 0 & 1 \\ -2 & 1 & 0 \\ -2 & 0 & 1 \end{bmatrix} - \begin{bmatrix} 1 & 0 & 0 \\ 0 & 1 & 0 \\ 0 & 0 & 1 \end{bmatrix} = \begin{bmatrix} 3 & 0 & 1 \\ -2 & 0 & 0 \\ -2 & 0 & 0 \end{bmatrix}$$

The equations for $(A - I)\mathbf{x} = \mathbf{0}$ are easy to solve: $\left\{ \begin{array}{rcl} 3x_1 + x_3 & = & 0 \\ -2x_1 & = & 0 \end{array} \right\}$

Row operations hardly seem necessary. Obviously x_1 is zero, and hence x_3 is also zero. There are three-variables, so x_2 is free. The general solution of $(A - I)\mathbf{x} = \mathbf{0}$ is $x_2\mathbf{e}_2$, where $\mathbf{e}_2 = (0, 1, 0)$, and so $\mathbf{e}_2$ provides a basis for the eigenspace.

<u>For $\lambda = 2$:</u>

$$A - 2I = \begin{bmatrix} 4 & 0 & 1 \\ -2 & 1 & 0 \\ -2 & 0 & 1 \end{bmatrix} - \begin{bmatrix} 2 & 0 & 0 \\ 0 & 2 & 0 \\ 0 & 0 & 2 \end{bmatrix} = \begin{bmatrix} 2 & 0 & 1 \\ -2 & -1 & 0 \\ -2 & 0 & -1 \end{bmatrix}$$

$$[(A - 2I) \quad \mathbf{0}] = \begin{bmatrix} 2 & 0 & 1 & 0 \\ -2 & -1 & 0 & 0 \\ -2 & 0 & -1 & 0 \end{bmatrix} \sim \begin{bmatrix} 2 & 0 & 1 & 0 \\ 0 & -1 & 1 & 0 \\ 0 & 0 & 0 & 0 \end{bmatrix} \sim \begin{bmatrix} ① & 0 & 1/2 & 0 \\ 0 & ① & -1 & 0 \\ 0 & 0 & 0 & 0 \end{bmatrix}$$

Copyright © 2021 Pearson Education, Inc.

So $x_1 = -(1/2)x_3$, $x_2 = x_3$, with x_3 free. The general solution of $(A - 2I)\mathbf{x} = \mathbf{0}$ is $x_3 \begin{bmatrix} -1/2 \\ 1 \\ 1 \end{bmatrix}$. A

nice basis vector for the eigenspace is $\begin{bmatrix} -1 \\ 2 \\ 2 \end{bmatrix}$.

For $\lambda = 3$:

$$A - 3I = \begin{bmatrix} 4 & 0 & 1 \\ -2 & 1 & 0 \\ -2 & 0 & 1 \end{bmatrix} - \begin{bmatrix} 3 & 0 & 0 \\ 0 & 3 & 0 \\ 0 & 0 & 3 \end{bmatrix} = \begin{bmatrix} 1 & 0 & 1 \\ -2 & -2 & 0 \\ -2 & 0 & -2 \end{bmatrix}$$

$$[(A - 3I) \quad \mathbf{0}] = \begin{bmatrix} 1 & 0 & 1 & 0 \\ -2 & -2 & 0 & 0 \\ -2 & 0 & -2 & 0 \end{bmatrix} \sim \begin{bmatrix} 1 & 0 & 1 & 0 \\ 0 & -2 & 2 & 0 \\ 0 & 0 & 0 & 0 \end{bmatrix} \sim \begin{bmatrix} ① & 0 & 1 & 0 \\ 0 & ① & -1 & 0 \\ 0 & 0 & 0 & 0 \end{bmatrix}$$

So $x_1 = -x_3$, $x_2 = x_3$, with x_3 free. A basis vector for the eigenspace is $\begin{bmatrix} -1 \\ 1 \\ 1 \end{bmatrix}$.

Study Tip: The text's answer to Exercise 15 is likely to be the same as yours, but there are many answers. What should you do if your vectors differ from those in the answer key? Example 2 gives the answer. Whenever you compute an **x** that you think is an eigenvector of A, you can check this simply by computing $A\mathbf{x}$. There is a little more to do in Exercise 15, however. The answer shows a basis of two eigenvectors, which means that the eigenspace is two-dimensional. So your answer must consist of two linearly independent eigenvectors. You can check that they are indeed eigenvectors, and then their linear independence can be checked by inspection.

19. The matrix $\begin{bmatrix} 1 & 2 & 3 \\ 1 & 2 & 3 \\ 1 & 2 & 3 \end{bmatrix}$ is not invertible because its columns are linearly dependent. So the

number 0 is an eigenvalue of the matrix. See the discussion following Example 5.

21. Carefully read the definition of an eigenvalue.

23. See the paragraph after Example 5

25. See the Reasonable Answers discussion.

27. See Example 4.

29. See the discussion with equation (3).

31. If a 2×2 matrix A had three distinct eigenvalues, then by Theorem 2 there would correspond three linearly independent eigenvectors (one for each eigenvalue). This is impossible because the vectors all belong to a two-dimensional vector space in which any set of three

Copyright © 2021 Pearson Education, Inc.

vectors is linearly dependent. See Theorem 8 in Section 1.6. In general, if an $n \times n$ matrix has p distinct eigenvalues, then by Theorem 2 there would be a linearly independent set of p eigenvectors (one for each eigenvalue). Since these vectors belong to an n-dimensional vector space, p cannot exceed n.

33. Let $\mathbf{x}$ be a nonzero vector such that $A\mathbf{x} = \lambda\mathbf{x}$. Then $A^{-1}A\mathbf{x} = A^{-1}(\lambda\mathbf{x})$, and $\mathbf{x} = \lambda(A^{-1}\mathbf{x})$. Since $\mathbf{x} \neq \mathbf{0}$ (and since A is invertible), λ cannot be zero. Then $\lambda^{-1}\mathbf{x} = A^{-1}\mathbf{x}$, which shows that λ^{-1} is an eigenvalue of A^{-1}, because $\mathbf{x} \neq \mathbf{0}$.

Note: The relation between the eigenvalues of A and A^{-1} is important in the so-called *inverse power* method for estimating an eigenvalue of a matrix. (See Section 5.8.)

35. For any λ, $(A - \lambda I)^T = A^T - (\lambda I)^T = A^T - \lambda I$. Since $(A - \lambda I)^T$ is invertible if and only if $A - \lambda I$ is invertible (by Theorem 6(c) in Section 2.2), we conclude that $A^T - \lambda I$ is *not* invertible if and only if $A - \lambda I$ is *not* invertible. That is, λ is an eigenvalue of A^T if and only if λ is an eigenvalue of A.

Answer to Checkpoint: The answer in the text is also correct because it is a nonzero multiple of the eigenvector found in the solution to Exercise 7, and any nonzero multiple of an eigenvector is another eigenvector. (The eigenspace is a *subspace* and so is closed under scalar multiplication.)

MATLAB   Finding Eigenvectors

When you know an eigenvalue, MATLAB can help you find a basis for the corresponding eigenspace. In general, **eye(k)** is the $k \times k$ identity matrix. For example, if A is a 5×5 matrix with an eigenvalue 7, then the commands

    C = A – 7*eye(5)
    rref(C)

allows you to read off a basis from the eigenspace. The command **[X,Y]=eig(A)** returns two matrices X and Y. The eigenvalues are on the diagonal of X and the corresponding eigenvectors appear as the columns of Y.

   If the numbers in the basis matrix Y are messy, use **format rat; B**, which will display the entries in B as rational numbers. (All eigenvectors calculated in this section have rational entries, so the rational format introduces no error.) To return to the usual decimal number display, enter **format short**.

Copyright © 2021 Pearson Education, Inc.

5.2 - The Characteristic Equation

There are several equivalent definitions of the determinant of a matrix. The definition here in terms of the pivots in an echelon form has the advantage that it is easy to state and understand, and in most cases it provides the most efficient way to compute a determinant.

KEY IDEAS

When A is 3×3, the geometric interpretation of det A as a volume explains why det $A = 0$ if and only if A is not invertible:

> The determinant of A is zero.
> <=> The parallelepiped determined by the columns of A has zero volume.
> <=> One column of A is in the subspace spanned by the other columns.
> <=> The columns of A are linearly dependent.
> <=> The matrix A is not invertible.

If A is $n \times n$, then $\det(A - \lambda I) = 0$ if and only if $A - \lambda I$ is not invertible, and this happens if and only if λ is an eigenvalue of A.

Exercises 1–14 are designed only to provide some basic familiarity with characteristic polynomials. The main use of $\det(A - \lambda I)$ is as a tool for *studying* eigenvalues rather than computing them.

Sometimes the characteristic polynomial is defined as $\det(\lambda I - A)$. A property of determinants implies that $\det(\lambda I - A) = (-1)^n \det(A - \lambda I)$, when A is $n \times n$, so the two polynomials are either the same (when n is even) or they are negatives of one another. The use of $\det(A - \lambda I)$ tends to make hand calculations easier and less prone to copying errors.

SOLUTIONS TO EXERCISES

1. $A = \begin{bmatrix} 2 & 7 \\ 7 & 2 \end{bmatrix}$, $A - \lambda I = \begin{bmatrix} 2 & 7 \\ 7 & 2 \end{bmatrix} - \begin{bmatrix} \lambda & 0 \\ 0 & \lambda \end{bmatrix} = \begin{bmatrix} 2-\lambda & 7 \\ 7 & 2-\lambda \end{bmatrix}$, the characteristic polynomial is

$$\det(A - \lambda I) = (2 - \lambda)^2 - 7^2 = 4 - 4\lambda + \lambda^2 - 49 = \lambda^2 - 4\lambda - 45$$

In factored form, the characteristic equation is $(\lambda - 9)(\lambda + 5) = 0$, so the eigenvalues of A are 9 and –5.

Warning: Don't row reduce a matrix A to find its eigenvalues. Row reduction preserves the null space of A but not the eigenvalues of A.

7. $A = \begin{bmatrix} 5 & 3 \\ -4 & 4 \end{bmatrix}$, $A - \lambda I = \begin{bmatrix} 5-\lambda & 3 \\ -4 & 4-\lambda \end{bmatrix}$, the characteristic polynomial is

$$\det(A - \lambda I) = (5 - \lambda)(4 - \lambda) - (3)(-4) = 20 - 9\lambda + \lambda^2 + 12$$
$$= \lambda^2 - 9\lambda + 32$$

Copyright © 2021 Pearson Education, Inc.

The characteristic polynomial does not factor easily, but the quadratic formula provides the solutions of $\lambda^2 - 9\lambda + 32 = 0$.

$$\lambda = \frac{+9 \pm \sqrt{81 - 4(32)}}{2}$$

These values for λ are not real numbers, so A has no real eigenvalues. There is no nonzero vector $\mathbf{x}$ in $\mathbb{R}^2$ such that $A\mathbf{x} = \lambda\mathbf{x}$ for such a λ. (For any $\mathbf{x} \neq \mathbf{0}$ in $\mathbb{R}^2$, the vector $A\mathbf{x}$ has only real entries and thus could not equal a complex multiple of $\mathbf{x}$.)

Study Tip: If you are asked to work some of Exercises 9–14, you may be tested on them. This is one way of finding out if you know what the characteristic polynomial is and how it is connected with eigenvalues. Also, you can show that you know some elementary properties of determinants.

13. $A - \lambda I = \begin{bmatrix} 6-\lambda & -2 & 0 \\ -2 & 9-\lambda & 0 \\ 5 & 8 & 3-\lambda \end{bmatrix}$

The method using the special 3×3 formula will produce the characteristic polynomial $-\lambda^3 + 18\lambda^2 - 95\lambda + 150$. Factoring such a polynomial to find the eigenvalues requires a little experience. (See the appendix at the end of the exercise solutions.). However, if you use a cofactor expansion down the third column (see Section 3.1), you immediately obtain

$$\det(A - \lambda I) = (3 - \lambda) \cdot \det \begin{bmatrix} 6-\lambda & -2 \\ -2 & 9-\lambda \end{bmatrix}$$

$$= (3 - \lambda)[(6 - \lambda)(9 - \lambda) - 4]$$

$$= (3 - \lambda)(\lambda^2 - 15\lambda + 50)$$

The characteristic polynomial is already partially factored, and the remaining quadratic factor is itself easily factored. The factored characteristic polynomial is $(3 - \lambda)(\lambda - 10)(\lambda - 5)$ or, equivalently, $-(\lambda - 3)(\lambda - 5)(\lambda - 10)$.

Note: The solutions of Exercises 11–14 are similar to that of Exercise 13. These matrices have the property that if a cofactor expansion is chosen along a column or row that contains two zeros, then the characteristic polynomial appears in a partially factored form.

19. Since the equation $\det(A - \lambda I) = (\lambda_1 - \lambda)(\lambda_2 - \lambda) \cdots (\lambda_n - \lambda)$ holds for all λ, set $\lambda = 0$ and conclude that $\det A = \lambda_1 \lambda_2 \cdots \lambda_n$.

21. See the new addition to the IMT.

23. See Theorem 3.

25. See the boxed expression before Example 3.

27. See the boxed expression before Example 3.

Copyright © 2021 Pearson Education, Inc.

29. See the paragraph before Example 4.

31. If $A = QR$, with Q invertible, and if $A_1 = RQ$, then $A_1 = Q^{-1}QRQ = Q^{-1}AQ$, which shows that A_1 is similar to A.

Appendix: Factoring a Polynomial

In general it is difficult to factor a polynomial of degree 3 or higher (unless you have one of several powerful computer programs available). Fortunately, textbook examples and exercises tend to have small integer solutions. The following observation is helpful.

Let $p(\lambda)$ be a polynomial with integer coefficients. If $p(c) = 0$ for some integer c, then $\lambda - c$ is a factor of $p(\lambda)$ and c is a divisor of the constant term of $p(\lambda)$.

EXAMPLE   Find the eigenvalues of the matrix A whose characteristic polynomial is $p(\lambda) = -\lambda^3 + 18\lambda^2 - 96\lambda + 160$.

Solution   By the observation above, any integer eigenvalue of A must be a divisor of the constant term 160 in the characteristic polynomial. There are twenty-four such divisors: 1, 2, 4, 5, 8, 10, 16, 20, 32, 40, 80, and 160, together with the negatives of these numbers. We let c be one of these numbers and try to divide $\lambda - c$ into the polynomial by long division. The terms $\lambda \pm 1$ and $\lambda \pm 2$ don't work, and the first successful division is

$$\begin{array}{r} -\lambda^2 + 14\lambda - 40 \\ \lambda - 4 \overline{) -\lambda^3 + 18\lambda^2 - 96\lambda + 160} \\ \underline{-\lambda^3 + 4\lambda^2} \\ 14\lambda^2 - 96\lambda \\ \underline{14\lambda^2 - 56\lambda} \\ -40\lambda + 160 \\ \underline{-40\lambda + 160} \end{array}$$

Thus the characteristic polynomial is $(\lambda - 4)(-\lambda^2 + 14\lambda - 40)$. The quadratic polynomial factors easily, and the characteristic equation is $(\lambda - 4)(\lambda - 4)(10 - \lambda) = 0$. The eigenvalues of A are 4 (with multiplicity two) and 10.

MATLAB   poly, plot

You can use the MATLAB command **poly(A)** to check your answers in Exercises 9–14. Note that if A is $n \times n$, this command lists the coefficients of the characteristic polynomial of A in order of decreasing powers of λ, beginning with λ^n. If the polynomial is of odd degree, the coefficients are multiplied by -1, to make $+1$ the coefficient of λ^n. This corresponds to using $\det(\lambda I - A)$ instead of $\det(A - \lambda I)$.

Use **rand(4)** to create a 4×4 matrix with random entries. To get a matrix with integer entries between 1 and 100 enter **randi(100,4)**, which you can use for exercises 33 and 34.,

Copyright © 2021 Pearson Education, Inc.

The following commands will produce the graph of the characteristic polynomial of the matrix A in Exercise 35 (with $a = 32$).

x = linspace(0,3);	Choose 100 points between 0 and 3
grid on	Include a grid on the display
hold on	Add the next graph to the display
A(3,2)= 32; p = poly(A);	Compute the characteristic polynomial
v = polyval(p,x);	Evaluate it at the points in **x**
plot(x,v,'b')	Plot the graph in blue.

Edit line 4 to change the value of a (from 32 to another value). Edit line 6 to change the color of the graph. When the commands are run again, the old graph(s) will remain visible. If you do not specify the color of the graph, MATLAB will automatically cycle through a set of colors, one for each graph on the display. To create a fresh display, enter **hold off** and **clf**.

To take advantage of MATLAB's interactive plotting features, see the topic "Plotting Tools" in MATLAB's **Help** menu.

5.3 - Diagonalization

The factorization $A = PDP^{-1}$ is used to compute powers of A, decouple dynamical systems in Sections 5.6 and 5.7, and study symmetric matrices and quadratic forms in Chapter 7.

KEY IDEAS

Example 3 gives the algorithm for diagonalizing a matrix A. After you construct P and D, check your calculations:

1. Compute AP and PD, and check that $AP = PD$.
2. Make sure the columns of P are linearly independent. Use Theorem 7 to save time. You only have to verify that for each eigenvalue, the corresponding eigenvectors are linearly independent. That's easy if the dimension of the eigenspace is 2 or 1.

The key equation in this section is $AP = PD$. It will help you to keep the order of the factors correct when you write $A = PDP^{-1}$, and it also leads immediately to $P^{-1}AP = D$, which you may need occasionally. Possible test question: If $AP = PD$, explain why the first column of P is an eigenvector of A (if the column is nonzero). (Study the proof of the Diagonalization Theorem.)

Warning: Do not confuse the property of being diagonalizable with the property of being invertible. They are not connected. The matrix in Example 5 is diagonalizable, but it is not invertible because 0 is an eigenvalue. The matrix $\begin{bmatrix} 1 & -2 \\ 0 & 1 \end{bmatrix}$ is invertible, but it is not diagonalizable because the eigenspace for $\lambda = 1$ is only one-dimensional.

Copyright © 2021 Pearson Education, Inc.

SOLUTIONS TO EXERCISES

1. $P = \begin{bmatrix} 5 & 7 \\ 2 & 3 \end{bmatrix}$, $D = \begin{bmatrix} 2 & 0 \\ 0 & 1 \end{bmatrix}$, $A = PDP^{-1}$, and $A^4 = PD^4P^{-1}$. Next compute $P^{-1} = \dfrac{1}{1}\begin{bmatrix} 3 & -7 \\ -2 & 5 \end{bmatrix}$,

$D^4 = \begin{bmatrix} 2^4 & 0 \\ 0 & 1 \end{bmatrix} = \begin{bmatrix} 16 & 0 \\ 0 & 1 \end{bmatrix}$. Putting this together,

$A^4 = \begin{bmatrix} 5 & 7 \\ 2 & 3 \end{bmatrix}\begin{bmatrix} 16 & 0 \\ 0 & 1 \end{bmatrix}\begin{bmatrix} 3 & -7 \\ -2 & 5 \end{bmatrix} = \begin{bmatrix} 80 & 7 \\ 32 & 3 \end{bmatrix}\begin{bmatrix} 3 & -7 \\ -2 & 5 \end{bmatrix} = \begin{bmatrix} 226 & -525 \\ 90 & -209 \end{bmatrix}$

7. $A = \begin{bmatrix} 1 & 0 \\ 6 & -1 \end{bmatrix}$. The eigenvalues are obviously ±1 (since A is triangular).

<u>For $\lambda = 1$</u>: $A - 1I = \begin{bmatrix} 0 & 0 \\ 6 & -2 \end{bmatrix}$. The equation $(A - I)\mathbf{x} = \mathbf{0}$ amounts to $6x_1 - 2x_2 = 0$. So $x_1 = (1/3)x_2$, with x_2 free. The general solution is $x_2 \begin{bmatrix} 1/3 \\ 1 \end{bmatrix}$. A nice basis vector for the eigenspace is $\mathbf{u}_1 = \begin{bmatrix} 1 \\ 3 \end{bmatrix}$.

<u>For $\lambda = -1$</u>: $A - (-1)I = \begin{bmatrix} 2 & 0 \\ 6 & 0 \end{bmatrix}$. The equation $(A + I)\mathbf{x} = \mathbf{0}$ amounts to $2x_1 = 0$, with x_2 free. The general solution is $x_2 \begin{bmatrix} 0 \\ 1 \end{bmatrix}$. Take $\mathbf{u}_2 = \begin{bmatrix} 0 \\ 1 \end{bmatrix}$ as a basis vector for the eigenspace.

From $\mathbf{u}_1$ and $\mathbf{u}_2$, construct $P = [\mathbf{u}_1 \quad \mathbf{u}_2] = \begin{bmatrix} 1 & 0 \\ 3 & 1 \end{bmatrix}$. Then set $D = \begin{bmatrix} 1 & 0 \\ 0 & -1 \end{bmatrix}$, where the eigenvalues in D correspond to $\mathbf{u}_1$ and $\mathbf{u}_2$, respectively.

Warning: The 3×3 matrices in Exercises 11–18 may or may not be diagonalizable since each matrix has only two distinct eigenvalues. (Theorem 6 gives only a *sufficient* condition for diagonalizability.) You have to check for three linearly independent eigenvectors.

13. $A = \begin{bmatrix} 2 & 2 & -1 \\ 1 & 3 & -1 \\ -1 & -2 & 2 \end{bmatrix}$. The eigenvalues 5 and 1 are given. Because A is 3×3, you need three linearly independent eigenvectors.

<u>For $\lambda = 5$</u>: Solve $(A - 5I)\mathbf{x} = \mathbf{0}$. Form $A - 5I = \begin{bmatrix} -3 & 2 & -1 \\ 1 & -2 & -1 \\ -1 & -2 & -3 \end{bmatrix}$ and compute

Copyright © 2021 Pearson Education, Inc.

$$\begin{bmatrix} -3 & 2 & -1 & 0 \\ 1 & -2 & -1 & 0 \\ -1 & -2 & -3 & 0 \end{bmatrix} \sim \begin{bmatrix} 1 & -2 & -1 & 0 \\ -3 & 2 & -1 & 0 \\ -1 & -2 & -3 & 0 \end{bmatrix} \sim \cdots \sim \begin{bmatrix} 1 & 0 & 1 & 0 \\ 0 & 1 & 1 & 0 \\ 0 & 0 & 0 & 0 \end{bmatrix}$$

So $x_1 = -x_3$, $x_2 = -x_3$, with x_3 free. Take $\mathbf{v}_1 = \begin{bmatrix} -1 \\ -1 \\ 1 \end{bmatrix}$, for instance, as a basis vector for the

eigenspace. At this point, you don't know if A is diagonalizable. Your only hope is to find two linearly independent eigenvectors inside the eigenspace for $\lambda = 1$, because there are no other eigenspaces in which to look.

<u>For $\lambda = 1$</u>: Form $A - I = \begin{bmatrix} 1 & 2 & -1 \\ 1 & 2 & -1 \\ -1 & -2 & 1 \end{bmatrix}$. The equation $(A - I)\mathbf{x} = \mathbf{0}$ reduces to $x_1 + 2x_2 - x_3 = 0$.

So $x_1 = -2x_2 + x_3$, with x_2 and x_3 free. At this point you know that the eigenspace is two-dimensional (because there are two free variables). So there are the necessary two linearly independent eigenvectors, and hence A is diagonalizable. To produce the eigenvectors, write the solution of $(A - I)\mathbf{x} = \mathbf{0}$ in the form

$$\begin{bmatrix} x_1 \\ x_2 \\ x_3 \end{bmatrix} = \begin{bmatrix} -2x_2 + x_3 \\ x_2 \\ x_3 \end{bmatrix} = x_2 \begin{bmatrix} -2 \\ 1 \\ 0 \end{bmatrix} + x_3 \begin{bmatrix} 1 \\ 0 \\ 1 \end{bmatrix}$$

Set $\mathbf{v}_2 = \begin{bmatrix} -2 \\ 1 \\ 0 \end{bmatrix}$, $\mathbf{v}_3 = \begin{bmatrix} 1 \\ 0 \\ 1 \end{bmatrix}$, and $P = [\mathbf{v}_1 \ \ \mathbf{v}_2 \ \ \mathbf{v}_3] = \begin{bmatrix} -1 & -2 & 1 \\ -1 & 1 & 0 \\ 1 & 0 & 1 \end{bmatrix}$.

The columns of P are linearly independent, by Theorem 7, because the eigenvectors form bases for their respective eigenspaces. So P is invertible. Since the first column of P corresponds to $\lambda = 5$, the first diagonal entry in D must be 5, which means that

$$D = \begin{bmatrix} 5 & 0 & 0 \\ 0 & 1 & 0 \\ 0 & 0 & 1 \end{bmatrix}$$

19. $A = \begin{bmatrix} 5 & -3 & 0 & 9 \\ 0 & 3 & 1 & -2 \\ 0 & 0 & 2 & 0 \\ 0 & 0 & 0 & 2 \end{bmatrix}$. The eigenvalues are obviously 5, 3, and 2. (Why?)

<u>For $\lambda = 2$</u>: To solve $(A - 2I)\mathbf{x} = \mathbf{0}$, completely reduce $[(A - 2I) \ \ \mathbf{0}]$:

Copyright © 2021 Pearson Education, Inc.

$$\begin{bmatrix} 3 & -3 & 0 & 9 & 0 \\ 0 & 1 & 1 & -2 & 0 \\ 0 & 0 & 0 & 0 & 0 \\ 0 & 0 & 0 & 0 & 0 \end{bmatrix} \sim \begin{bmatrix} 3 & 0 & 3 & 3 & 0 \\ 0 & 1 & 1 & -2 & 0 \\ 0 & 0 & 0 & 0 & 0 \\ 0 & 0 & 0 & 0 & 0 \end{bmatrix} \sim \begin{bmatrix} 1 & 0 & 1 & 1 & 0 \\ 0 & 1 & 1 & -2 & 0 \\ 0 & 0 & 0 & 0 & 0 \\ 0 & 0 & 0 & 0 & 0 \end{bmatrix}$$

So $x_1 = -x_3 - x_4$, $x_2 = -x_3 + 2x_4$, with x_3 and x_4 free. The usual calculations produce a basis for the eigenspace:

$$\begin{bmatrix} x_1 \\ x_2 \\ x_3 \\ x_4 \end{bmatrix} = \begin{bmatrix} -x_3 - x_4 \\ -x_3 + 2x_4 \\ x_3 \\ x_4 \end{bmatrix} = x_3 \begin{bmatrix} -1 \\ -1 \\ 1 \\ 0 \end{bmatrix} + x_4 \begin{bmatrix} -1 \\ 2 \\ 0 \\ 1 \end{bmatrix}. \quad \text{Basis: } \mathbf{v}_1 = \begin{bmatrix} -1 \\ -1 \\ 1 \\ 0 \end{bmatrix}, \mathbf{v}_2 = \begin{bmatrix} -1 \\ 2 \\ 0 \\ 1 \end{bmatrix}$$

Checkpoint: If you happened to choose $\lambda = 2$ first, as in this solution, would you have enough information at this point to determine whether A is diagonalizable?

<u>For $\lambda = 3$</u>: To solve $(A - 3I)\mathbf{x} = \mathbf{0}$, completely reduce $[(A - 3I) \quad \mathbf{0}]$:

$$\begin{bmatrix} 2 & -3 & 0 & 9 & 0 \\ 0 & 0 & 1 & -2 & 0 \\ 0 & 0 & -1 & 0 & 0 \\ 0 & 0 & 0 & -1 & 0 \end{bmatrix} \sim \cdots \sim \begin{bmatrix} 1 & -3/2 & 0 & 0 & 0 \\ 0 & 0 & 1 & 0 & 0 \\ 0 & 0 & 0 & 1 & 0 \\ 0 & 0 & 0 & 0 & 0 \end{bmatrix}, \quad \begin{cases} x_1 = (3/2)x_2 \\ x_2 \text{ is free} \\ x_3 = 0 \\ x_4 = 0 \end{cases}$$

Choosing $x_2 = 2$ produces the eigenvector $\mathbf{v}_3 = (3, 2, 0, 0)$.

<u>For $\lambda = 5$</u>: To solve $(A - 5I)\mathbf{x} = \mathbf{0}$, completely reduce $[(A - 5I) \quad \mathbf{0}]$:

$$\begin{bmatrix} 0 & -3 & 0 & 9 & 0 \\ 0 & -2 & 1 & -2 & 0 \\ 0 & 0 & -3 & 0 & 0 \\ 0 & 0 & 0 & -3 & 0 \end{bmatrix} \sim \cdots \sim \begin{bmatrix} 0 & 1 & 0 & 0 & 0 \\ 0 & 0 & 1 & 0 & 0 \\ 0 & 0 & 0 & 1 & 0 \\ 0 & 0 & 0 & 0 & 0 \end{bmatrix}, \quad \begin{cases} x_1 \text{ is free} \\ x_2 = 0 \\ x_3 = 0 \\ x_4 = 0 \end{cases}$$

A basis vector for the eigenspace is $\mathbf{v}_4 = (1, 0, 0, 0)$. Set

$$P = [\mathbf{v}_1 \quad \mathbf{v}_2 \quad \mathbf{v}_3 \quad \mathbf{v}_4] = \begin{bmatrix} -1 & -1 & 3 & 1 \\ -1 & 2 & 2 & 0 \\ 1 & 0 & 0 & 0 \\ 0 & 1 & 0 & 0 \end{bmatrix}, \quad D = \begin{bmatrix} 2 & 0 & 0 & 0 \\ 0 & 2 & 0 & 0 \\ 0 & 0 & 3 & 0 \\ 0 & 0 & 0 & 5 \end{bmatrix}$$

This answer differs from that in the text. There, $P = [\mathbf{v}_4 \ \mathbf{v}_3 \ \mathbf{v}_1 \ \mathbf{v}_2]$, and the entries in D are rearranged to match the new order of the eigenvectors. According to the Diagonalization Theorem, both answers are correct.

21. The symbol D does not automatically denote a diagonal matrix.

23. Read Theorem 5 carefully.

25. Read Theorem 5 carefully.

Copyright © 2021 Pearson Education, Inc.

27. See Theorem 5.

29. A is diagonalizable because you know that five linearly independent eigenvectors exist: three in the three-dimensional eigenspace and two in the two-dimensional eigenspace. Theorem 7 guarantees that the set of all five eigenvectors is linearly independent.

31. Let $\{\mathbf{v}_1\}$ be a basis for the one-dimensional eigenspace, let $\mathbf{v}_2$ and $\mathbf{v}_3$ form a basis for the two-dimensional eigenspace, and let $\mathbf{v}_4$ be any eigenvector in the remaining eigenspace. By Theorem 7, $\{\mathbf{v}_1, \mathbf{v}_2, \mathbf{v}_3, \mathbf{v}_4\}$ is linearly independent. Since A is 4×4, the Diagonalization Theorem shows that A is diagonalizable.

33. If A is diagonalizable, then $A = PDP^{-1}$ for some invertible P and diagonal D. Since A is invertible, 0 is not an eigenvalue of A. So the diagonal entries in D (which are eigenvalues of A) are not zero, and D is invertible. By the theorem on the inverse of a product, $A^{-1} = (PDP^{-1})^{-1} = (P^{-1})^{-1}D^{-1}P^{-1} = PD^{-1}P^{-1}.$ Since D^{-1} is obviously diagonal, A^{-1} is diagonalizable.

A Second Proof: If A is $n \times n$, it has n linearly independent eigenvectors, say, $\mathbf{v}_1, \ldots, \mathbf{v}_n$. Then $\mathbf{v}_1, \ldots, \mathbf{v}_n$ are also eigenvectors of A^{-1}, by the solution of Exercise 33 in Section 5.1. Hence A^{-1} is diagonalizable, by the Diagonalization Theorem.

35. The diagonal entries in D_1 are reversed from those in D. So interchange the (eigenvector) columns of P to make them correspond properly to the eigenvalues in D_1. In this case,

$P_1 = \begin{bmatrix} 1 & 1 \\ -2 & -1 \end{bmatrix}$ and $D_1 = \begin{bmatrix} 3 & 0 \\ 0 & 5 \end{bmatrix}.$ Although the first column of P must be an eigenvector

corresponding to the eigenvalue 3, there is nothing to prevent us from selecting some multiple of $\begin{bmatrix} 1 \\ -2 \end{bmatrix}$, say $\begin{bmatrix} -3 \\ 6 \end{bmatrix}$, and letting $P_2 = \begin{bmatrix} -3 & 1 \\ 6 & -1 \end{bmatrix}$. We now have three different

factorizations or "diagonalizations" of A: $A = PDP^{-1} = P_1 D_1 P_1^{-1} = P_2 D_1 P_2^{-1}$

Answer to Checkpoint: Yes. In Exercise 19, the fact that the eigenspace for $\lambda = 2$ is two-dimensional guarantees that A is diagonalizable, because each of the other two eigenvalues will produce at least one eigenvector, and the resulting set of four eigenvectors will be linearly independent, by Theorem 7. So A is diagonalizable, by the Diagonalization Theorem. (Note that since one eigenspace is two-dimensional, the other eigenspaces must be one-dimensional, because there could not possibly be *more* than four linearly independent eigenvectors in $\mathbb{R}^4$.)

Mastering Linear Algebra Concepts: Eigenvalue, Eigenvector, Eigenspace

I suggest that you take time now to prepare review sheets for the three terms listed above. Begin with their definitions, of course.

Eigenvalue:

Copyright © 2021 Pearson Education, Inc.

- equivalent description
- geometric interpretation
- special cases
- typical computations
- connections with other concepts
-

Sec. 5.1: Equation (3);
Sec. 5.1: Fig. 2
Theorems 1, 6
Sec. 5.1: Exer. 7, 19; Sec. 5.2: Exer. 7, 15
The IMT; Theorem 4; Sec. 5.2: Exer. 19

Eigenvector:
- special cases
- typical computations
- connections with other concepts

Theorem 2
Sec. 5.1: Examples 2, 4, Exer. 5, 15
Sec. 5.2: Example 5
Sec. 5.3: Theorem 5, Exer. 5, 18

Eigenspace:
- equivalent description
- geometric interpretation
- typical computations
- connections with other concepts

Sec. 5.1: Equation (3)
Sec. 5.1: Fig. 2, 3
Sec. 5.1: Example 4
Theorem 7

MATLAB   Diagonalization
To practice the diagonalization procedure in this section, you should use **[D,P]=eig(A)** to produce eigenvalues and eigenvectors. See the MATLAB box for Section 5.1.

5.4 - Eigenvalues and Linear Transformations

This section introduces the matrix of a linear transformation relative to specified bases for vector spaces, and then uses this concept to give a new interpretation of the matrix factorization $A = PDP^{-1}$.

STUDY NOTES

The exercises will help you learn the definition of the matrix representation of a transformation relative to specified bases, say $\mathcal{B}$ and C.

Algorithm for Finding the $\mathcal{B}$-matrix of $T : V \rightarrow V$

1. Compute the images of the basis vectors:

$$T(\mathbf{b}_1), \ldots, T(\mathbf{b}_n)$$

2. Convert these images into $\mathcal{B}$-coordinate vectors:

$$[T(\mathbf{b}_1)]_\mathcal{B}, \ldots, [T(\mathbf{b}_n)]_\mathcal{B}$$

3. Place the $\mathcal{B}$-coordinate vectors into the columns of $[T]_\mathcal{B}$.

Copyright © 2021 Pearson Education, Inc.

The sentence before Theorem 8 summarizes the main idea of the theorem. Studying the proof should help you understand the theorem and review important concepts. The subsection on Similarity of Matrix Representations could have contained more ideas. Take notes carefully if your instructor decides to expand this material somewhat.

SOLUTIONS TO EXERCISES

1. The coefficients of $T(\mathbf{b}_1)$ form the first column of $[T]_\mathcal{B}$, the coefficients of $T(\mathbf{b}_2)$ form the second column of $[T]_\mathcal{B}$, and so on. $[T]_\mathcal{B} = \begin{bmatrix} 3 & -1 & 0 \\ -5 & 6 & 4 \\ 0 & 0 & 9 \end{bmatrix}$

7. If $P = \begin{bmatrix} \mathbf{b}_1 & \mathbf{b}_2 \end{bmatrix} = \begin{bmatrix} 2 & 1 \\ -1 & 2 \end{bmatrix}$, then the $\mathcal{B}$-matrix is

$$P^{-1}AP = \frac{1}{5}\begin{bmatrix} 2 & -1 \\ 1 & 2 \end{bmatrix}\begin{bmatrix} 3 & 4 \\ -1 & -1 \end{bmatrix}\begin{bmatrix} 2 & 1 \\ -1 & 2 \end{bmatrix} = \begin{bmatrix} 1 & 5 \\ 0 & 1 \end{bmatrix}$$

13. **a.** We compute that $A\mathbf{b}_1 = \begin{bmatrix} 1 & 1 \\ -1 & 3 \end{bmatrix}\begin{bmatrix} 1 \\ 1 \end{bmatrix} = \begin{bmatrix} 2 \\ 2 \end{bmatrix} = 2\mathbf{b}_1.$ so $\mathbf{b}_1$ is an eigenvector of A corresponding to the eigenvalue 2. The characteristic polynomial of A is $\lambda^2 - 4\lambda + 4 = (\lambda - 2)^2$, so 2 is the only eigenvalue for A. Now $A - 2I = \begin{bmatrix} -1 & 1 \\ -1 & 1 \end{bmatrix}$, which implies that the eigenspace corresponding to the eigenvalue 2 is one-dimensional. Thus the matrix A is not diagonalizable.

   b. Following Example 4, if $P = \begin{bmatrix} \mathbf{b}_1 & \mathbf{b}_2 \end{bmatrix}$, then the $\mathcal{B}$-matrix for T is

$$P^{-1}AP = \begin{bmatrix} -4 & 5 \\ 1 & -1 \end{bmatrix}\begin{bmatrix} 1 & 1 \\ -1 & 3 \end{bmatrix}\begin{bmatrix} 1 & 5 \\ 1 & 4 \end{bmatrix} = \begin{bmatrix} 1 & 5 \\ 1 & 4 \end{bmatrix} = \begin{bmatrix} 2 & -1 \\ 0 & 2 \end{bmatrix}.$$

15. **a.** $T(\mathbf{p}) = 3 + 3t + 3t^2 = 3\mathbf{p}$, thus $\mathbf{p}$ is an eigenvector with eigenvalue 3.

   b. $T(\mathbf{p}) = -t - t^2 - t^3$, $-\mathbf{p}$, which is not a multiple of $\mathbf{p}$, so it is not an eigenvector.

17. True. See Theorem 4 from Section 5.2.

18. False. See Theorem 8.

19. False. See Example 1.

20. True. See the definition prior to Example 1.

Copyright © 2021 Pearson Education, Inc.

21. If A is similar to B, then there exists an invertible matrix P such that $P^{-1}AP = B$. Thus B is invertible because it is the product of invertible matrices. By a theorem about inverses of products, $B^{-1} = P^{-1}A^{-1}(P^{-1})^{-1} = P^{-1}A^{-1}P$, which shows that A^{-1} is similar to B^{-1}.

25. If $A\mathbf{x} = \lambda\mathbf{x}$, $\mathbf{x} \neq \mathbf{0}$, then $P^{-1}A\mathbf{x} = \lambda P^{-1}\mathbf{x}$. If $B = P^{-1}AP$, then

$$B(P^{-1}\mathbf{x}) = P^{-1}AP(P^{-1}\mathbf{x}) = P^{-1}A\mathbf{x} = \lambda P^{-1}\mathbf{x} \qquad (*)$$

by the first calculation. Note that $P^{-1}\mathbf{x} \neq \mathbf{0}$, because $\mathbf{x} \neq \mathbf{0}$ and P^{-1} is invertible. Hence (*) shows that $P^{-1}\mathbf{x}$ is an eigenvector of B corresponding to λ. (Of course, λ is an eigenvalue of both A and B because the matrices are similar, by Theorem 4 in Section 5.2.)

Study Tip: It can be shown that for *any* square matrix A, the trace of A is the sum of the eigenvalues of A, counted according to multiplicities. You can use this fact to provide a quick check on your eigenvalue calculations. The *sum* of the eigenvalues must match the *sum* of the diagonal entries in A (even if A is not diagonalizable).

29. Since S shift each of the 1's χ one position to the right, $S(\chi) = \chi$, so χ is an eigenvector with eigenvalue 1.

5.5 - Complex Eigenvalues

If the characteristic equation of an $n \times n$ real matrix A has a complex eigenvalue λ and if $\mathbf{v}$ is a nonzero vector in $\mathbb{C}^n$ such that $A\mathbf{v} = \lambda\mathbf{v}$, then both λ and $\mathbf{v}$ provide useful information about A.

STUDY NOTES

Only matrices with real entries are considered here. If λ is a complex eigenvalue of A, with $\mathbf{v}$ a corresponding eigenvector, then $\bar{\lambda}$ is also an eigenvalue of A, with $\bar{\mathbf{v}}$ an eigenvector. Find out if you should know how to prove this fact.

Example 6 describes the prototype for all 2×2 matrices with a complex eigenvalue λ. Only λ is needed if you want to know the angle φ of rotation and the scale factor $|\lambda|$, but an associated eigenvector $\mathbf{v}$ is also needed if you want to factor A as PCP^{-1}, as in Example 7.

SOLUTIONS TO EXERCISES

1. $A = \begin{bmatrix} 1 & -2 \\ 1 & 3 \end{bmatrix}$, $A - \lambda I = \begin{bmatrix} 1-\lambda & -2 \\ 1 & 3-\lambda \end{bmatrix}$

$\det(A - \lambda I) = (1 - \lambda)(3 - \lambda) - (-2) = \lambda^2 - 4\lambda + 5$

Use the quadratic formula to find the eigenvalues: $\lambda = \dfrac{4 \pm \sqrt{16 - 20}}{2} = 2 \pm i$. Example 2 gives a shortcut for finding one eigenvector, and Example 5 shows how to write the other eigenvector with no effort.

Copyright © 2021 Pearson Education, Inc.

<u>For $\lambda = 2 + i$:</u> $A - (2+i)I = \begin{bmatrix} -1-i & -2 \\ 1 & 1-i \end{bmatrix}$. The equation $(A - \lambda I)\mathbf{x} = \mathbf{0}$ gives

$$(-1-i)x_1 - \quad 2x_2 = 0$$
$$x_1 + (1-i)x_2 = 0$$

As in Example 2, the two equations are equivalent—each determines the same relation between x_1 and x_2. So use the second equation to obtain $x_1 = -(1-i)x_2$, with x_2 free. The general solution is $x_2 \begin{bmatrix} -1+i \\ 1 \end{bmatrix}$, and the vector $\mathbf{v}_1 = \begin{bmatrix} -1+i \\ 1 \end{bmatrix}$ provides a basis for the eigenspace.

<u>For $\lambda = 2 - i$:</u> Let $\mathbf{v}_2 = \overline{\mathbf{v}}_1 = \begin{bmatrix} -1-i \\ 1 \end{bmatrix}$. The remark prior to Example 5 shows that $\mathbf{v}_2$ is automatically an eigenvector for $\overline{2+i}$. In fact, calculations similar to those above would show that $\{\mathbf{v}_2\}$ is a basis for the eigenspace. (In general, for a real matrix A, it can be shown that the set of complex conjugates of the vectors in a basis of the eigenspace for λ is a basis of the eigenspace for $\overline{\lambda}$.)

7. $A = \begin{bmatrix} \sqrt{3} & -1 \\ 1 & \sqrt{3} \end{bmatrix}$. The eigenvalues are $\sqrt{3} \pm i$. Ask your instructor if you are permitted to write down the eigenvalues of $\begin{bmatrix} a & -b \\ b & a \end{bmatrix}$ from memory, or if you are expected to find them via the characteristic equation. Note that the eigenvectors are easy to remember, too. See the Practice Problem.

The scale factor associated with the transformation $\mathbf{x} \mapsto A\mathbf{x}$ is simply $r = |\lambda| = \left((\sqrt{3})^2 + 1^2\right)^{1/2} = 2$. For the angle of the rotation, plot the point $(a, b) = (\sqrt{3}, 1)$ in the xy-plane and use trigonometry.

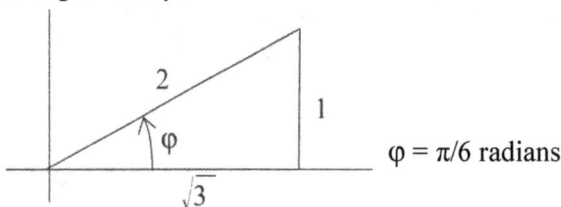

$\varphi = \pi/6$ radians

13. From Exercise 1, $\lambda = 2 \pm i$, and the eigenvector $\mathbf{v} = \begin{bmatrix} -1-i \\ 1 \end{bmatrix}$ corresponds to $\lambda = 2 - i$. Since

Re $\mathbf{v} = \begin{bmatrix} -1 \\ 1 \end{bmatrix}$ and Im $\mathbf{v} = \begin{bmatrix} -1 \\ 0 \end{bmatrix}$, take $P = \begin{bmatrix} -1 & -1 \\ 1 & 0 \end{bmatrix}$. Then compute

$$C = P^{-1}AP = \begin{bmatrix} 0 & 1 \\ -1 & -1 \end{bmatrix}\begin{bmatrix} 1 & -2 \\ 1 & 3 \end{bmatrix}\begin{bmatrix} -1 & -1 \\ 1 & 0 \end{bmatrix} = \begin{bmatrix} 0 & 1 \\ -1 & -1 \end{bmatrix}\begin{bmatrix} 3 & -1 \\ 2 & -1 \end{bmatrix} = \begin{bmatrix} 2 & -1 \\ 1 & 2 \end{bmatrix}$$

Copyright © 2021 Pearson Education, Inc.

Actually, Theorem 9 gives the formula for C. Note that the eigenvector $\mathbf{v}$ corresponds to $a - bi$ instead of $a + bi$. If, for instance, you use the eigenvector for $2 + i$, your C will be $\begin{bmatrix} 2 & 1 \\ -1 & 2 \end{bmatrix}$. The imaginary part of the eigenvalue is the $(1, 2)$-entry in C.

So, there are two possible choices for C (depending on the vector used to produce P). On an exam, if you are not sure of the form of C, you can always compute it quickly from the formula $C = P^{-1}AP$, as in the solution above.

Note: Because there are two possibilities for C in the factorization of a 2×2 matrix as in Exercise 13, the measure of rotation φ associated with the transformation $\mathbf{x} \mapsto A\mathbf{x}$ is determined only up to a change of sign. The "orientation" of the angle is determined by the change of variable $\mathbf{x} = P\mathbf{u}$. See Fig. 4 in the text.

19. $A = \begin{bmatrix} 1.52 & -.7 \\ .56 & .4 \end{bmatrix}$, $\det(A - \lambda I) = \lambda^2 - 1.92\lambda + 1$

Use the quadratic formula to solve $\lambda^2 - 1.92\lambda + 1 = 0$:
$$\lambda = \frac{1.92 \pm \sqrt{-.3136}}{2} = .96 \pm .28i$$

To find the eigenvector for $\lambda = .96 - .28i$, solve
$$(1.52 - .96 + .28i)x_1 \qquad\qquad -.7x_2 = 0$$
$$.56x_1 + (.4 - .96 + .28i)x_2 = 0$$

There is a nonzero solution (because $.96 - .28i$ is an eigenvalue), so you can use either equation to find the solution. From the second equation,
$$x_1 = \frac{.56 - .28i}{.56}x_2 = (1 - .5i)x_2$$

Setting $x_2 = 2$ produces the (complex) eigenvector $\mathbf{v} = \begin{bmatrix} 2 - i \\ 2 \end{bmatrix}$.

Since Re $\mathbf{v} = \begin{bmatrix} 2 \\ 2 \end{bmatrix}$ and Im $\mathbf{v} = \begin{bmatrix} -1 \\ 0 \end{bmatrix}$, take $P = \begin{bmatrix} 2 & -1 \\ 2 & 0 \end{bmatrix}$. Finally, compute
$$P^{-1}AP = \frac{1}{2}\begin{bmatrix} 0 & 1 \\ -2 & 2 \end{bmatrix}\begin{bmatrix} 1.52 & -.7 \\ .56 & .4 \end{bmatrix}\begin{bmatrix} 2 & -1 \\ 2 & 0 \end{bmatrix} = \begin{bmatrix} .96 & -.28 \\ .28 & .96 \end{bmatrix}$$

This final matrix, which has the proper form, is C.

23. See the paragraph before Example 5.

25. See the paragraph before Example 5 and the discussion in Section 5.2 after Example 4.

29. Write $\mathbf{x} = \text{Re } \mathbf{x} + i(\text{Im } \mathbf{x})$, so that $A\mathbf{x} = A(\text{Re } \mathbf{x}) + i\, A(\text{Im } \mathbf{x})$. Since A is real, so are $A(\text{Re } \mathbf{x})$ and $A(\text{Im } \mathbf{x})$. Thus $A(\text{Re } \mathbf{x})$ is the real part of $A\mathbf{x}$ and $A(\text{Im } \mathbf{x})$ is the imaginary part of $A\mathbf{x}$.

Copyright © 2021 Pearson Education, Inc.

MATLAB Complex Eigenvalues
The command **[P D] = eig(A)** (mentioned in Section 5.3) works for matrices with complex eigenvalues. In this case P and the diagonal matrix D have some complex entries. For a 2×2 real matrix with a complex eigenvalue, MATLAB tends to place the eigenvalue $a - bi$ (where $b > 0$) as the $(2, 2)$-entry of D. MATLAB does not produce matrices for a factorization $A = PCP^{-1}$ of the sort described in this section.

For any matrix P, the commands **real(P)** and **imag(P)** produce the real and imaginary parts of the entries in P, displayed as matrices the same size as P.

5.6 - Discrete Dynamical Systems

This section presents the climax to a crescendo of ideas that began in Section 1.10 and flowed through parts of Chapters 4 and 5.

KEY IDEAS

A solution of a first order homogeneous difference equation

$$\mathbf{x}_{k+1} = A\mathbf{x}_k \qquad (k = 0, 1, 2, \ldots) \tag{1}$$

is a sequence $\{\mathbf{x}_k\}$ that satisfies (1) and is described by a formula for each $\mathbf{x}_k$ that does not depend on the preceding terms in the sequence other than the initial term $\mathbf{x}_0$. In Section 5.1, you saw how a solution can be constructed when $\mathbf{x}_0$ is an eigenvector. When $\mathbf{x}_0$ is *not* an eigenvector, look for an eigenvector decomposition of $\mathbf{x}_0$:

$$\mathbf{x}_0 = c_1 \mathbf{v}_1 + \cdots + c_n \mathbf{v}_n \qquad \text{Each } \mathbf{v}_i \text{ is an eigenvector.} \tag{2}$$

To make (2) possible for any $\mathbf{x}_0$ in $\mathbb{R}^n$, the section assumes that the $n \times n$ matrix A has n linearly independent eigenvectors. If $\mathbf{x}_k = A^k \mathbf{x}_0$, then

$$\mathbf{x}_k = c_1 (\lambda_1)^k \mathbf{v}_1 + \cdots + c_n (\lambda_n)^k \mathbf{v}_n$$

When $\{\mathbf{x}_k\}$ describes the "state" of a system at discrete times (denoted by $k = 0, 1, 2, \ldots$), the *long-term behavior* of this dynamical system is a description of what happens to $\mathbf{x}_k$ as $k \to \infty$. The text focuses on the following important situation.

Copyright © 2021 Pearson Education, Inc.

Let A be an $n \times n$ matrix with n linearly independent eigenvectors, corresponding to eigenvalues such that $|\lambda_1| \geq 1 > |\lambda_j|$ for $j = 2, \ldots, n$. If $\mathbf{x}_0$ is given by (2) with $c_1 \neq 0$, then for all sufficiently large k,

$$\mathbf{x}_{k+1} \approx \lambda_1 \mathbf{x}_k \qquad \text{Each entry in } \mathbf{x}_k \text{ grows by a factor of } \lambda_1.$$

$$\mathbf{x}_k \approx c_1 (\lambda_1)^k \mathbf{v}_1 \qquad \mathbf{x}_k \text{ is approximately a multiple of } \mathbf{v}_1, \text{ and so the ratio between any two entries in } \mathbf{x}_k \text{ is nearly the same as the corresponding ratio for } \mathbf{v}_1.$$

STUDY NOTES

When the two approximations above are true in an application, the eigenvalue λ_1 and the eigenvector $\mathbf{v}_1$ have interesting physical interpretations. Make sure you can describe these on an exam. (See the last four sentences in the solution of Example 1, for instance.)

The predator-prey model is rather primitive and provides only a starting point for more refined models. Still, you might enjoy considering what the model in Example 1 predicts if $\mathbf{x}_0$ happens to be a multiple of $\mathbf{v}_2 = (5, 1)$, or if initially there are *more* than 5 owls for every 1 thousand rats, assuming $p = .104$.

The graphical descriptions of solutions to difference equations should help you understand what can happen to $\mathbf{x}_k$ as $k \to \infty$. I hope you enjoy studying the figures even if your class does not have time to cover this part of the section. Only the simplest cases are shown, but these cases form the foundation for studying *nonlinear* dynamical systems which are widely used (but require calculus techniques not covered here). Even for nonlinear systems, eigenvalues and eigenvectors of certain matrices play an important role.

SOLUTIONS TO EXERCISES

1. a. The eigenvectors $\mathbf{v}_1 = \begin{bmatrix} 1 \\ 1 \end{bmatrix}$ and $\mathbf{v}_2 = \begin{bmatrix} -1 \\ 1 \end{bmatrix}$ form a basis for $\mathbb{R}^2$. To find the action of A on

$\mathbf{x}_0 = \begin{bmatrix} 9 \\ 1 \end{bmatrix}$ express $\mathbf{x}_0$ in terms of $\mathbf{v}_1$ and $\mathbf{v}_2$. That is, find c_1, c_2 such that $\mathbf{x}_0 = c_1\mathbf{v}_1 + c_2\mathbf{v}_2$:

$$\begin{bmatrix} 1 & -1 & 9 \\ 1 & 1 & 1 \end{bmatrix} \sim \begin{bmatrix} 1 & 0 & 5 \\ 0 & 1 & -4 \end{bmatrix} \Rightarrow \mathbf{x}_0 = 5\mathbf{v}_1 - 4\mathbf{v}_2$$

Since $\mathbf{v}_1$, $\mathbf{v}_2$ are eigenvectors (for the eigenvalues 3 and 1/3):

$$\mathbf{x}_1 = A\mathbf{x}_0 = 5A\mathbf{v}_1 - 4A\mathbf{v}_2 = 5 \cdot 3\mathbf{v}_1 - 4 \cdot (1/3)\mathbf{v}_2$$

$$= \begin{bmatrix} 15 \\ 15 \end{bmatrix} - \begin{bmatrix} -4/3 \\ 4/3 \end{bmatrix} = \begin{bmatrix} 49/3 \\ 41/3 \end{bmatrix}$$

b. Each time A acts on a linear combination of $\mathbf{v}_1$ and $\mathbf{v}_2$, the $\mathbf{v}_1$ term is multiplied by the eigenvalue 3 and the $\mathbf{v}_2$ term is multiplied by the eigenvalue 1/3.

Copyright © 2021 Pearson Education, Inc.

$$\mathbf{x}_2 = A\mathbf{x}_1 = A[5(3)\mathbf{v}_1 - 4(1/3)\mathbf{v}_2] = 5(3)^2\mathbf{v}_1 - 4(1/3)^2\mathbf{v}_2$$

In general, $\mathbf{x}_k = 5(3)^k\mathbf{v}_1 - 4(1/3)^k\mathbf{v}_2$ for $k \geq 0$.

7. **a.** The matrix A in Exercise 1 has eigenvalues 3 and 1/3. Since $|3| > 1$ and $|1/3| < 1$, the origin is a saddle point.

b. The direction of greatest attraction is determined by $\mathbf{v}_2 = \begin{bmatrix} -1 \\ 1 \end{bmatrix}$, the eigenvector corresponding to the eigenvalue with absolute value less than 1. The direction of greatest repulsion is determined by $\mathbf{v}_1 = \begin{bmatrix} 1 \\ 1 \end{bmatrix}$, the eigenvector corresponding to the eigenvalue greater than 1.

c. The drawing below shows: (1) lines through the eigenvectors and the origin, (2) arrows toward the origin (showing attraction) on the line through $\mathbf{v}_2$ and arrows away from the origin (showing repulsion) on the line through $\mathbf{v}_1$, (3) several typical trajectories (with arrows) that show the general flow of points. No specific points other than $\mathbf{v}_1$ and $\mathbf{v}_2$ were computed. This type of drawing is about all that one can make without using a computer to plot points.

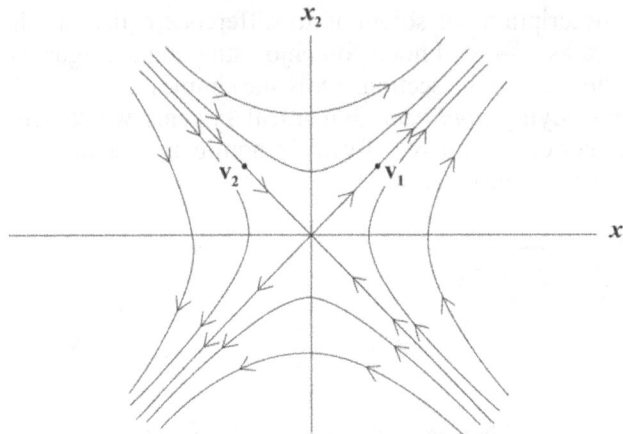

Remark: Sketching trajectories for a dynamical system in which the origin is an attractor or a repeller is more difficult than the sketch in Exercise 7. There has been no discussion of the direction in which the trajectories "bend" as they move toward or away from the origin. For instance, if you rotate Figure 1 of Section 5.6 through a quarter-turn and relabel the axes so that x_1 is on the horizontal axis, then the new figure corresponds to the matrix A with the diagonal entries .8 and .64 interchanged. In general, if A is a diagonal matrix, with positive diagonal entries a and d, unequal to 1, then the trajectories lie on the axes or on curves whose equations have the form $x_2 = r(x_1)^s$, where $s = (\ln d)/(\ln a)$ and r depends on the initial point $\mathbf{x}_0$. (See *Encounters with Chaos*, by Denny Gulick, New York: McGraw-Hill, 1992, pp. 147–150.)

Study Tip: If your instructor wants you to graph trajectories when the origin is an attractor or repeller, there will need to be some class discussion of exactly how to do this.

Copyright © 2021 Pearson Education, Inc.

13. $A = \begin{bmatrix} .8 & .3 \\ -.4 & 1.5 \end{bmatrix}$. First find the eigenvalues:

$$\det(A - \lambda I) = (.8 - \lambda)(1.5 - \lambda) - (.3)(-.4) = \lambda^2 - 2.3\lambda + 1.32$$

$$= (\lambda - 1.2)(\lambda - 1.1) \qquad \text{Use the quadratic formula, if needed.}$$

$$= 0$$

Since both eigenvalues, 1.2 and 1.1, are greater than 1, the origin is a repeller. For the direction of greatest repulsion, find the eigenvector for the larger eigenvalue, 1.2:

$$[(A - 1.2I) \quad \mathbf{0}] = \begin{bmatrix} -.4 & .3 & 0 \\ -.4 & .3 & 0 \end{bmatrix} \sim \begin{bmatrix} 1 & -3/4 & 0 \\ 0 & 0 & 0 \end{bmatrix}, \mathbf{x} = x_2 \begin{bmatrix} 3/4 \\ 1 \end{bmatrix}$$

Any multiple of $\begin{bmatrix} 3/4 \\ 1 \end{bmatrix}$, such as $\begin{bmatrix} 3 \\ 4 \end{bmatrix}$, determines the direction of greatest repulsion.

MATLAB Plotting Trajectories

Given a vector **x**, the command **x = A*x** will compute the "next" point on the trajectory. Use the up-arrow (↑) and <Enter> to repeat the command, over and over.

The following steps create a "trajectory" matrix T whose columns are the points **x**, A**x**, A^2**x**, ..., A^{15}**x**. (Change 15 to any integer you wish.)

```
T = x                          Put x in the first column of T.
for j=1:15                     This loop repeats the next two lines 15 times.
    x = A*x           ;        Compute the next point on the trajectory.
    T = [T   x]                Store the new point in T.
end                            End of the loop
```

After you type the line beginning with "for," MATLAB will suspend all calculations (while you type additional lines) until you type "end" and press <Enter>.

If you want MATLAB itself to plot the points in T, use the commands:

plot(T(l,:),T(2,:),'ob'), grid

The 'o' produces a small circle at each point on the trajectory. The 'b' makes the circle blue. If you have the data for another trajectory stored in a matrix S, you can plot both trajectories on the same graph:

plot(T(l,:),T(2,:),'ob',S(l,:),S(2,:),'*g'),grid   g is for green.

Each new **plot** command erases the previous graph. If you want the new graph added to the previous graph, issue the command **hold on** before the next **plot** command.

For Exercise 17, the following commands will produce a graph of the first entry in each of the first nine columns of the matrix T constructed above.

k = 0:8
plot(k,T(1,1:9),'-'), grid

To graph the sum of the two entries in the first nine columns of T, with k already defined, enter

Copyright © 2021 Pearson Education, Inc.

plot(k,T(1,1:9)+T(2,1:9),'-'), grid

The command ./ between two vectors of equal lengths divides each entry in the first vector by the corresponding entry in the second vector. Thus, to plot the quotient of the two entries in each column of T, use

plot(k,T(l,1:9)./ T(2,1:9),'-'), grid

5.7 - Applications to Differential Equations

If you plan to take a course in differential equations, the material in this section will be a valuable reference. If you have already studied differential equations, you may gain new understanding as you work through this section.

KEY IDEAS

A basic solution of the differential equation $\mathbf{x}' = A\mathbf{x}$ is an *eigenfunction* $\mathbf{x}(t) = \mathbf{v}e^{\lambda t}$, where λ is an eigenvalue of A and $\mathbf{v}$ is a corresponding eigenvector. In all examples and exercises in this section, every solution of $\mathbf{x}' = A\mathbf{x}$ is a linear combination of eigenfunctions. (This is because A is diagonalizable. The general case is usually handled in a full course in differential equations.) An initial condition, $\mathbf{x}(0) = \mathbf{x}_0$, determines the weights for the linear combination of eigenfunctions.

The eigenvalues of A determine the nature of the origin for the dynamical system described by $\mathbf{x}' = A\mathbf{x}$. Most of the discussion involves the case when A is 2×2. If one eigenvalue is negative and one is positive, the origin is a saddle point. If the real parts of the eigenvalues are negative (this includes the case when both eigenvalues are real and negative), the origin is an *attractor* of the dynamical system. If the real parts are positive, the origin is a *repeller*. If an eigenvalue is complex, then the trajectory of a corresponding eigenfunction forms a *spiral*—either toward the origin, away from the origin, or on an ellipse around the origin, depending on the real part of the eigenvalue.

Study Tip: Note that conditions on eigenvalues here differ from those in Section 5.6. For differential equations, the real parts of the eigenvalues determine the nature of the trajectories; for difference equations, the absolute values of the eigenvalues are important. You can remember this if you note that a basic solution $\mathbf{v}e^{\lambda t}$ of $\mathbf{x}' = A\mathbf{x}$ tends to $\mathbf{0}$ (as $t \to \infty$) only if the real part of λ is negative. In contrast, a basic solution $\lambda^k\mathbf{v}$ of $\mathbf{x}_{k+1} = A\mathbf{x}_k$ tends to $\mathbf{0}$ (as $k \to \infty$) only if the absolute value of λ is less than 1.

SOLUTIONS TO EXERCISES

1. The eigenfunctions for $\mathbf{x}' = A\mathbf{x}$ are $\mathbf{v}_1e^{4t}$ and $\mathbf{v}_2e^{7t}$. The general solution of $\mathbf{x}' = A\mathbf{x}$ has the form

Copyright © 2021 Pearson Education, Inc.

$$c_1\begin{bmatrix} -3 \\ 1 \end{bmatrix}e^{4t} + c_2\begin{bmatrix} -1 \\ 1 \end{bmatrix}e^{2t}$$

The initial condition $\mathbf{x}(0) = (-6, 1)$ determines c_1 and c_2:

$$c_1\begin{bmatrix} -3 \\ 1 \end{bmatrix}e^{4(0)} + c_2\begin{bmatrix} -1 \\ 1 \end{bmatrix}e^{2(0)} = \begin{bmatrix} -6 \\ 1 \end{bmatrix}$$

$$\begin{bmatrix} -3 & -1 & -6 \\ 1 & 1 & 1 \end{bmatrix} \sim \begin{bmatrix} 1 & 1 & 1 \\ -3 & -1 & -6 \end{bmatrix} \sim \cdots \sim \begin{bmatrix} 1 & 0 & 5/2 \\ 0 & 1 & -3/2 \end{bmatrix}$$

Thus $c_1 = 5/2$, $c_2 = -3/2$, and $\mathbf{x}(t) = \dfrac{5}{2}\begin{bmatrix} -3 \\ 1 \end{bmatrix}e^{4t} - \dfrac{3}{2}\begin{bmatrix} -1 \\ 1 \end{bmatrix}e^{2t}$.

Checkpoint: Let A be the matrix $\begin{bmatrix} 1 & 3/2 \\ -1/2 & -1 \end{bmatrix}$, obtained by dividing the matrix in Exercise 3 by 2. Now the eigenvalues of A are .5 and $-.5$. Is the origin an attractor or a saddle point for the equation $\mathbf{x}' = A\mathbf{x}$?

7. Use the eigenvectors $\mathbf{v}_1 = \begin{bmatrix} 1 \\ 3 \end{bmatrix}$ and $\mathbf{v}_2 = \begin{bmatrix} 1 \\ 1 \end{bmatrix}$ (found in Exercise 5) to create $P = [\mathbf{v}_1 \quad \mathbf{v}_2]$. Match the eigenvectors with the eigenvalues in $D = \begin{bmatrix} 4 & 0 \\ 0 & 6 \end{bmatrix}$. The details of the substitution of $\mathbf{x} = P\mathbf{y}$ into $\mathbf{x}' = A\mathbf{x}$ are given in the answer section of the text.

Helpful Hint: The idea of changing variables to uncouple a differential equation is fairly common in engineering texts. Exercises 7 and 8 test your understanding of the value of a diagonalization. (You might see such a question on an exam.)

13. An eigenvalue of A is $\lambda = 1 + 3i$, with eigenvector $\mathbf{v} = (1 + i, 2)$. The complex eigenfunctions $\mathbf{v}e^{\lambda t}$ and $\bar{\mathbf{v}}e^{\bar{\lambda}t}$ provide a basis for the solution space of all complex solutions of $\mathbf{x}' = A\mathbf{x}$. The general (complex) solution is

$$c_1\begin{bmatrix} 1+i \\ 2 \end{bmatrix}e^{(1+3i)t} + c_2\begin{bmatrix} 1-i \\ 2 \end{bmatrix}e^{(1-3i)t} \qquad (c_1 \text{ and } c_2 \text{ are complex})$$

Use the real and imaginary parts of $\mathbf{v}e^{(1+3i)t}$ to build the general real solution. Rewrite $\mathbf{v}e^{(1+3i)t}$ as:

$$\begin{bmatrix} 1+i \\ 2 \end{bmatrix}e^{(1+3i)t} = \begin{bmatrix} 1+i \\ 2 \end{bmatrix}(\cos 3t + i\sin 3t)e^{t}$$

$$= \begin{bmatrix} \cos 3t - \sin 3t \\ 2\cos 3t \end{bmatrix}e^{t} + i\begin{bmatrix} \sin 3t + \cos 3t \\ 2\sin 3t \end{bmatrix}e^{t}$$

Copyright © 2021 Pearson Education, Inc.

The general real solution has the form

$$c_1 \begin{bmatrix} \cos 3t - \sin 3t \\ 2\cos 3t \end{bmatrix} e^t + c_2 \begin{bmatrix} \sin 3t + \cos 3t \\ 2\sin 3t \end{bmatrix} e^t \qquad (c_1 \text{ and } c_2 \text{ are real})$$

The trajectories are spirals because the eigenvalues are complex. The spirals tend away from the origin because the real parts of the eigenvalues are positive.

19. Substitute $R_1 = 1/5$, $R_2 = 1/3$, $C_1 = 4$, and $C_2 = 3$ into the formula for A given in Example 1, and use a matrix program to find the eigenvalues and eigenvectors:

$$A = \begin{bmatrix} -2 & 3/4 \\ 1 & -1 \end{bmatrix}, \qquad \lambda_1 = -.5: \ \mathbf{v}_1 = \begin{bmatrix} 1 \\ 2 \end{bmatrix}, \qquad \lambda_2 = -2.5: \ \mathbf{v}_2 = \begin{bmatrix} -3 \\ 2 \end{bmatrix}$$

General solution: $\mathbf{x}(t) = c_1 \begin{bmatrix} 1 \\ 2 \end{bmatrix} e^{-.5t} + c_2 \begin{bmatrix} -3 \\ 2 \end{bmatrix} e^{-2.5t}$.

The condition $\mathbf{x}(0) = \begin{bmatrix} 4 \\ 4 \end{bmatrix}$ implies that $\begin{bmatrix} 1 & -3 \\ 2 & 2 \end{bmatrix} \begin{bmatrix} c_1 \\ c_2 \end{bmatrix} = \begin{bmatrix} 4 \\ 4 \end{bmatrix}$. By a matrix program, $c_1 = 5/2$ and $c_2 = -1/2$, so that

$$\begin{bmatrix} v_1(t) \\ v_2(t) \end{bmatrix} = \mathbf{x}(t) = \frac{5}{2}\begin{bmatrix} 1 \\ 2 \end{bmatrix} e^{-.5t} - \frac{1}{2}\begin{bmatrix} -3 \\ 2 \end{bmatrix} e^{-2.5t}$$

Helpful Hint: (for 21 and 22) Find the general real solution before you use the initial condition to find the constants c_1 and c_2. Otherwise, your c_1 and c_2 will probably be complex, and you will have to do unnecessary complex arithmetic to write the solution using only real scalars.

Answer to Checkpoint: One eigenvalue is positive and one is negative, so the origin is a saddle point. If you consider the *difference* equation $\mathbf{x}_{k+1} = A\mathbf{x}_k$ (with the same matrix A), then the origin is an attractor, because both eigenvalues are less than 1 in absolute value. Be careful if you have a test that covers both Sections 5.6 and 5.7.

MATLAB   Solutions of Differential Equations

If you use the command **[P   D] = eig(A)**, your eigenvectors should be multiples of those in the text's answer (when the eigenspaces are one-dimensional). To test whether a vector **v** is a multiple of a vector **w**, compute **v./w**. This divides each entry in **v** by the corresponding entry in **w**. If **v** is a multiple of **w**, the result of **v./w** should be a vector whose entries are all equal.

5.8 - Iterative Estimates for Eigenvalues

The algorithms in this section illustrate another use of the eigenvector decomposition described in Section 5.6. Other methods for eigenvalue estimation were mentioned in Section 5.2.

Copyright © 2021 Pearson Education, Inc.

STUDY NOTES

Throughout the section, we suppose that the initial vector $\mathbf{x}_0$ can be written as $\mathbf{x}_0 = c_1\mathbf{v}_1 + \ldots + c_n\mathbf{v}_n$, where $\mathbf{v}_1, \ldots, \mathbf{v}_n$ are eigenvectors of A and $c_1 \neq 0$. (In practice, you will not know $c_1\mathbf{v}_1, \ldots, c_n\mathbf{v}_n$. This eigenvector decomposition is used only to explain why the power method works.)

The Power Method: Assume the eigenvalue λ_1 for $\mathbf{v}_1$ is a strictly dominant eigenvalue (so that $|\lambda_1| > |\lambda_j|$ for $j = 2, \ldots, n$). Then, for large k, the line through $A^k\mathbf{x}_0$ and $\mathbf{0}$ nearly coincides with the line through $\mathbf{v}_1$ and $\mathbf{0}$. The vector $A^k\mathbf{x}_0$ itself may never approach a multiple of $\mathbf{v}_1$ (see Exercise 21), but if each $A^k\mathbf{x}_0$ is scaled so its largest entry is 1, then the scaled vectors approach an eigenvector (a multiple of $\mathbf{v}_1$) as $k \to \infty$.

The Inverse Power Method: You must start with an initial estimate α for a particular eigenvalue, say λ_2, and α must be closer to λ_2 than to any other eigenvalue of A. In this case, $1/(\lambda_2 - \alpha)$ is a strictly dominant eigenvalue of the matrix $B = (A - \alpha I)^{-1}$. The inverse power method avoids computing B. Instead of multiplying $\mathbf{x}_k$ by B to get $\mathbf{x}_{k+1}$ (suitably scaled), you solve the equation $(A - \alpha I)\mathbf{y}_k = \mathbf{x}_k$ for $\mathbf{y}_k$ and then scale $\mathbf{y}_k$ to produce $\mathbf{x}_{k+1}$.

SOLUTIONS TO EXERCISES

1. The vectors in the sequence $\begin{bmatrix} 1 \\ 0 \end{bmatrix}, \begin{bmatrix} 1 \\ .25 \end{bmatrix}, \begin{bmatrix} 1 \\ .3158 \end{bmatrix}, \begin{bmatrix} 1 \\ .3298 \end{bmatrix}, \begin{bmatrix} 1 \\ .3326 \end{bmatrix}$ approach an

   eigenvector $\mathbf{v}_1$. Of these vectors, the last one, $\mathbf{x}_4$, is probably the best estimate of $\mathbf{v}_1$. To compute an estimate of λ_1, multiply one of the vectors by A and examine its entries. Again, the best information probably comes from $A\mathbf{x}_4 = \begin{bmatrix} 4.9978 \\ 1.6652 \end{bmatrix}$ whose entries are approximately

   λ_1 times the entries in $\mathbf{x}_4$. From the first entry, the estimate of λ_1 is 4.9978.

   The computed value of $A\mathbf{x}_4$ can be used as an estimate of the direction of the eigenspace.

Copyright © 2021 Pearson Education, Inc.

Study Tip: Exercises 1–6 make good exam questions because they test your understanding of the power method without requiring extensive calculation.

7. The data in the table below and the tables in Exercise 19 were produced by MATLAB, which carried more decimal places than shown here.

k	0	1	2	3	4	5
$\mathbf{x}_k$	$\begin{bmatrix}1\\0\end{bmatrix}$	$\begin{bmatrix}.75\\1\end{bmatrix}$	$\begin{bmatrix}1\\.9565\end{bmatrix}$	$\begin{bmatrix}.9932\\1\end{bmatrix}$	$\begin{bmatrix}1\\.9990\end{bmatrix}$	$\begin{bmatrix}.9998\\1\end{bmatrix}$
$A\mathbf{x}_k$	$\begin{bmatrix}6\\8\end{bmatrix}$	$\begin{bmatrix}11.5\\11.0\end{bmatrix}$	$\begin{bmatrix}12.70\\12.78\end{bmatrix}$	$\begin{bmatrix}12.959\\12.946\end{bmatrix}$	$\begin{bmatrix}12.9927\\12.9948\end{bmatrix}$	$\begin{bmatrix}12.9990\\12.9987\end{bmatrix}$
μ_k	8	11.5	12.78	12.959	12.9948	12.9990

The exact eigenvalues are 13 and –2. The subspaces determined by $A^k\mathbf{x}$ are lines whose slopes alternate above and below the slope of the eigenspace. (The eigenspace is the line $x_2 = x_1.$)

13. If the eigenvalues close to 4 and –4 have different absolute values, then one of these eigenvalues is a strictly dominant eigenvalue, so the power method will work. But the power method depends on powers of the quotients λ_2/λ_1 and λ_3/λ_1 going to zero. If $|\lambda_2/\lambda_1|$ is close to 1, its powers will go to zero slowly.

15. Suppose $A\mathbf{x} = \lambda\mathbf{x}$, with $\mathbf{x} \neq 0$. For any α, $A\mathbf{x} - \alpha I\mathbf{x} = (\lambda - \alpha)\mathbf{x}$, and $(A - \alpha I)\mathbf{x} = (\lambda - \alpha)\mathbf{x}$. If α is *not* an eigenvalue of A, then $A - \alpha I$ is invertible and $\lambda - \alpha$ is not 0; hence

$$\mathbf{x} = (A - \alpha I)^{-1}(\lambda - \alpha)\mathbf{x} \text{ and } (\lambda - \alpha)^{-1}\mathbf{x} = (A - \alpha I)^{-1}\mathbf{x}$$

This last equation shows that $\mathbf{x}$ is an eigenvector of $(A - \alpha I)^{-1}$ corresponding to the eigenvalue $(\lambda - \alpha)^{-1}$.

19. **a**. The data in the table on the next page show that $\mu_6 = 30.2887 = \mu_7$ to four decimal places. Actually, to six places, the largest eigenvalue is 30.288685, with eigenvector (.957629, .688937, 1, .943782).

Copyright © 2021 Pearson Education, Inc.

k	0	1	2	3	4	5	6	7
$\mathbf{x}_k$	$\begin{bmatrix} 1 \\ 0 \\ 0 \\ 0 \end{bmatrix}$	$\begin{bmatrix} 1 \\ .7 \\ .8 \\ .7 \end{bmatrix}$	$\begin{bmatrix} .99 \\ .71 \\ 1 \\ .93 \end{bmatrix}$	$\begin{bmatrix} .961 \\ .691 \\ 1 \\ .942 \end{bmatrix}$	$\begin{bmatrix} .9581 \\ .6893 \\ 1 \\ .9436 \end{bmatrix}$	$\begin{bmatrix} .9577 \\ .6890 \\ 1 \\ .9438 \end{bmatrix}$	$\begin{bmatrix} .957637 \\ .688942 \\ 1 \\ .943778 \end{bmatrix}$	$\begin{bmatrix} .957630 \\ .688938 \\ 1 \\ .943781 \end{bmatrix}$
$A\mathbf{x}_k$	$\begin{bmatrix} 10 \\ 7 \\ 8 \\ 7 \end{bmatrix}$	$\begin{bmatrix} 26.2 \\ 18.8 \\ 26.5 \\ 24.7 \end{bmatrix}$	$\begin{bmatrix} 29.4 \\ 21.1 \\ 30.6 \\ 28.8 \end{bmatrix}$	$\begin{bmatrix} 29.05 \\ 20.90 \\ 30.32 \\ 28.61 \end{bmatrix}$	$\begin{bmatrix} 29.01 \\ 20.87 \\ 30.29 \\ 28.59 \end{bmatrix}$	$\begin{bmatrix} 29.006 \\ 20.868 \\ 30.289 \\ 28.586 \end{bmatrix}$	$\begin{bmatrix} 29.0054 \\ 20.8671 \\ 30.2887 \\ 28.5859 \end{bmatrix}$	$\begin{bmatrix} 29.0053 \\ 20.8670 \\ 30.2887 \\ 28.5859 \end{bmatrix}$
μ_k	10	26.5	30.6	30.32	30.29	30.2892	30.2887	30.2887

b. The inverse power method (with $\alpha = 0$) produces $v_1 = \mu_1^{-1} = .010141$, and $v_2 = .0101501$, which seems to be accurate to at least four places. Actually, v_2 is accurate to six places, v_3 is accurate to eight places, and v_4 is accurate to ten places. The convergence is so rapid because the next-to-smallest eigenvalue is near .85, which is much farther away from 0 than .0101501. The vector $\mathbf{x}_4$ gives an estimate for the eigenvector that is accurate to seven places in each entry.

k	0	1	2	3	4
$\mathbf{x}_k$	$\begin{bmatrix} 1 \\ 0 \\ 0 \\ 0 \end{bmatrix}$	$\begin{bmatrix} -.6098 \\ 1 \\ -.2439 \\ .1463 \end{bmatrix}$	$\begin{bmatrix} -.60401 \\ 1 \\ -.25105 \\ .14890 \end{bmatrix}$	$\begin{bmatrix} -.603973 \\ 1 \\ -.251134 \\ .148953 \end{bmatrix}$	$\begin{bmatrix} -.6039723 \\ 1 \\ -.2511351 \\ .1489534 \end{bmatrix}$
$A\mathbf{x}_k$	$\begin{bmatrix} 25 \\ -41 \\ 10 \\ -6 \end{bmatrix}$	$\begin{bmatrix} -59.56 \\ 98.61 \\ -24.76 \\ 14.68 \end{bmatrix}$	$\begin{bmatrix} -59.5041 \\ 98.5211 \\ -24.7420 \\ 14.6750 \end{bmatrix}$	$\begin{bmatrix} -59.5044 \\ 98.5217 \\ -24.7423 \\ 14.6751 \end{bmatrix}$	$\begin{bmatrix} -59.50438 \\ 98.52170 \\ -24.74226 \\ 14.67515 \end{bmatrix}$
μ_k	-.024	.010141	.0101501	.010150059	.0101500484

Copyright © 2021 Pearson Education, Inc.

MATLAB   Power Method and Inverse Power Method

Use **format long** to display 15 decimal digits in your data. The algorithms below assume that A has a strictly dominant eigenvalue, and the initial vector is **x**, with largest entry 1 (in magnitude). (If your initial vector is called **x0**, rename it by entering **x = x0**.)

The Power Method   When the following sequence of commands is performed over and over, the values of **x** approach (in many cases) an eigenvector for a strictly dominant eigenvalue:

y = A*x		(1)
[t r] = max(abs(y)); mu = y(r)	mu = estimate for eigenvalue	(2)
x = y/y(r)	Estimate for the eigenvector	(3)

In (2), t is the absolute value of the largest entry in **y** and r is the index of that entry. As these commands are repeated, the numbers that appear in **y**(r) are the μ_k that approach the dominant eigenvalue.

Recall that MATLAB commands can be recalled by the up-arrow key ($\uparrow$). After entering (1) – (3), your keystrokes can be

$\uparrow\uparrow\uparrow$ \<Enter\> $\uparrow\uparrow\uparrow$ \<Enter\> $\uparrow\uparrow\uparrow$ \<Enter\>

and so on. Alternatively, you could enclose lines (1) – (3) in a loop (See the MATLAB box after section 5.6.)

The Inverse Power Method   Store the initial estimate of the eigenvalue in the variable **a**, and enter the command **C = A - a*eye(n)**, where n is the number of columns of A. Then enter the commands

y = C\X	Solves $(A - aI)y = x$	(1)
[t r] = max(abs(y)); nu = a + 1/y(r)	nu = estimated eigenvalue	(2)
x = y/y(r)	Estimate for the eigenvector	(3)

As these commands are repeated (using $\uparrow\uparrow\uparrow$ \<Enter\> each time), lines (2) and (3) produce the sequences $\{v_k\}$ and $\{\mathbf{x}_k\}$ described in the text.

Displaying Data   If your computer screen displays only 24 or 25 lines, vectors in the sequence $\{\mathbf{x}_k\}$ tend to scroll off the screen soon after you compute them. To see more vectors at once, and to compare their entries more easily, you can *display* them as row vectors. Change (1) to **y = A*x; y'** (power method) or **y = C\x; y'** (inverse power), and for both methods, change (3) to **x = y/y(r); x'**.

For even more data on your screen, use the command **format compact**, which removes extra lines between data displays. The simple command **format** returns everything to normal.

5.9 - Applications to Markov Chains

This section builds on the population movement example in Section 1.10. You should review that example now. Markov chains are widely used in applications and there is a rich theory connected with them. The simple examples and exercises in this section provide a basic foundation on which

Copyright © 2021 Pearson Education, Inc.

you can build later as needed. Two of the examples here were analyzed from a different point of view in Section 5.2.

KEY IDEAS

A probability vector is a list of nonnegative numbers that sum to one. A Markov chain is a sequence of probability vectors $\{\mathbf{x}_k\}$ that satisfy a difference equation $\mathbf{x}_{k+1} = P\mathbf{x}_k$ ($k = 0, 1, \ldots$) for some stochastic matrix P (whose columns are themselves probability vectors).

Theorem 10 shows that P always has 1 as an eigenvalue. Advanced texts show that eigenspace associated with the eigenvalue 1 always includes at least one probability vector, which then is a steady-state vector for P, because $P\mathbf{q} = \mathbf{q}$.

Our main interest is in a *regular* stochastic matrix P. In this case the steady-state vector is unique, according to Theorem 11. The key to predicting the distant future for a Markov chain associated with such a P is to find the steady-state vector $\mathbf{q}$, since the sequence $\{\mathbf{x}_k\}$ converges to $\mathbf{q}$ no matter what the initial state.

SOLUTIONS TO EXERCISES

1. a. To set up the stochastic matrix P, label the columns N (for news) and M (for music) in some order; use the *same* order for the rows. (Failure to keep the same order is a common source of error in this type of problem.) The data should be arranged so you read *down* a column and then to the right along a row.

$$\text{From:}$$
$$\begin{array}{cc} N & M \end{array} \quad \underline{\text{To:}}$$
$$\begin{bmatrix} .7 & .6 \\ .3 & .4 \end{bmatrix} \begin{array}{l} \text{News} \\ \text{Music} \end{array}$$

b. You are told that 100% of the listeners are listening to the news at 8:15 a.m., so start the Markov chain then, with $\mathbf{x}_0 = \begin{bmatrix} 1 \\ 0 \end{bmatrix}$.

c. There are two breaks between 8:15 and 9:25, so you need $\mathbf{x}_2$

$$\mathbf{x}_1 = P\mathbf{x}_0 = \begin{bmatrix} .7 & .6 \\ .3 & .4 \end{bmatrix} \begin{bmatrix} 1 \\ 0 \end{bmatrix} = \begin{bmatrix} .7 \\ .3 \end{bmatrix}$$

$$\mathbf{x}_2 = P\mathbf{x}_1 = \begin{bmatrix} .7 & .6 \\ .3 & .4 \end{bmatrix} \begin{bmatrix} .7 \\ .3 \end{bmatrix} = \begin{bmatrix} .67 \\ .33 \end{bmatrix}$$

The entries in $\mathbf{x}_2$ show that after two station breaks, 67% of the audience is listening to the news and 33% is listening to music.

Study Tip: When you compute a typical probability vector $P\mathbf{x}$, be sure to *compute all* of the entries in the product $P\mathbf{x}$. Then check your work by verifying that the entries sum to 1.

Copyright © 2021 Pearson Education, Inc.

7. To find the steady state vector for a regular stochastic matrix P:

(i)   set up the matrix $P - I$;

(ii) find the general solution of $(P - I)\mathbf{x} = \mathbf{0}$;

(iii) choose a basis vector for $\text{Nul}(P - I)$ whose entries sum to 1.

$$P = \begin{bmatrix} .7 & .1 & .1 \\ .2 & .8 & .2 \\ .1 & .1 & .7 \end{bmatrix}, P - I = \begin{bmatrix} .7 & .1 & .1 \\ .2 & .8 & .2 \\ .1 & .1 & .7 \end{bmatrix} - \begin{bmatrix} 1 & 0 & 0 \\ 0 & 1 & 0 \\ 0 & 0 & 1 \end{bmatrix} = \begin{bmatrix} -.3 & .1 & .1 \\ .2 & -.2 & .2 \\ .1 & .1 & -.3 \end{bmatrix}$$

Solve $(P - I)\mathbf{x} = \mathbf{0}$:

$$\begin{bmatrix} -.3 & .1 & .1 & 0 \\ .2 & -.2 & .2 & 0 \\ .1 & .1 & -.3 & 0 \end{bmatrix} \sim \begin{bmatrix} .1 & .1 & -.3 & 0 \\ .2 & -.2 & .2 & 0 \\ -.3 & .1 & .1 & 0 \end{bmatrix}$$

Interchange rows 1 and 3
Scale every row by 10

$$\sim \cdots \sim \begin{bmatrix} 1 & 0 & -1 & 0 \\ 0 & 1 & -2 & 0 \\ 0 & 0 & 0 & 0 \end{bmatrix} \begin{array}{l} x_1 = x_3 \\ ; x_2 = 2x_3; \\ x_3 \text{ is free} \end{array} \begin{bmatrix} x_1 \\ x_2 \\ x_3 \end{bmatrix} = \begin{bmatrix} x_3 \\ 2x_3 \\ x_3 \end{bmatrix} = x_3 \begin{bmatrix} 1 \\ 2 \\ 1 \end{bmatrix}$$

The entries in $\begin{bmatrix} 1 \\ 2 \\ 1 \end{bmatrix}$ sum to 4, so $\mathbf{q} = \dfrac{1}{4}\begin{bmatrix} 1 \\ 2 \\ 1 \end{bmatrix} = \begin{bmatrix} 1/4 \\ 1/2 \\ 1/4 \end{bmatrix}$ or $\begin{bmatrix} .25 \\ .50 \\ .25 \end{bmatrix}$.

Study Tip: Notice that the column sums are all zero for the matrix $P - I$ of Exercise 7. This always happens (see Exercise 27), and so you have a fast way to check your arithmetic for the entries in $I - P$.

Warning: You may have noticed that in Exercise 7, the rows were scaled by 10 to avoid decimals. A common mistake is to do this only to P, before forming $I - P$. That changes P drastically. The scaling was permissible because it was applied to all the coefficients in an *equation*.

13. **a.** From Exercise 3, $P = \begin{bmatrix} .95 & .45 \\ .05 & .55 \end{bmatrix}$. So $P - I = \begin{bmatrix} -.05 & .45 \\ .05 & -.45 \end{bmatrix}$. Solve $(P - I)\mathbf{x} = \mathbf{0}$:

$$\begin{bmatrix} -.05 & .45 & 0 \\ .05 & -.45 & 0 \end{bmatrix} \sim \begin{bmatrix} -.05 & .45 & 0 \\ 0 & 0 & 0 \end{bmatrix} \sim \begin{bmatrix} 1 & -9 & 0 \\ 0 & 0 & 0 \end{bmatrix} \begin{array}{l} x_1 = 9x_2 \\ x_2 \text{ is free} \end{array}$$

A basis for $\text{Nul}(P - I)$ is $\left\{ \begin{bmatrix} 9 \\ 1 \end{bmatrix} \right\}$; the steady-state vector is $\mathbf{q} = \begin{bmatrix} .9 \\ .1 \end{bmatrix}$.

b. The description in Exercise 3 may be interpreted as saying that the "state" of any specified person (in some group of students) on day k is predicted by a probability vector, say, $\mathbf{x}_k$. The second entry in $\mathbf{x}_k$ is the probability that the person is ill on day k. The starting vector for a specified person is (1,0) if the person is well today, and (0,1) if the

Copyright © 2021 Pearson Education, Inc.

person is ill. This situation applies to each person, because the exercise says, for example, that *every* healthy student has a 95% probability of being healthy the next day. That is, the stochastic matrix P applies to each person in the group.

The question in part (b) is about $\mathbf{x}_k$ for a large value of k. By Theorem 11, $\mathbf{x}_k$ approaches $\mathbf{q}$, so it is reasonable to assume that $\mathbf{q}$ may be used to answer a question about $\mathbf{x}_k$. Thus, the probability is .10 that after many days a specific student is ill. The second question essentially asks, "If $\mathbf{x}_0 = (0,1)$, does this have any affect on $\mathbf{x}_k$ for large k?" No, by Theorem 11, because the sequence $\{\mathbf{x}_k\}$ approaches $\mathbf{q}$ no matter what $\mathbf{x}_0$ is.

19. **a.** The product $S\mathbf{x}$ equals the sum of the entries in $\mathbf{x}$. Thus, by definition, $\mathbf{x}$ is a probability vector if and only if its entries are nonnegative and $S\mathbf{x} = 1$.

   b. Let $P = [\mathbf{p}_1 \ \mathbf{p}_2 \ \cdots \ \mathbf{p}_n]$, where the $\mathbf{p}_i$ are probability vectors. By matrix multiplication and part (a),
   $$SP = [S\mathbf{p}_1 \ \ S\mathbf{p}_2 \ \ \cdots \ \ S\mathbf{p}_n] = [1 \ \ 1 \ \ \cdots \ \ 1] = S$$

   c. By part (b), $S(P\mathbf{x}) = (SP)\mathbf{x} = S\mathbf{x} = 1$. The entries in $P\mathbf{x}$ are obviously nonnegative, because P and $\mathbf{x}$ have only nonnegative entries. By (a), the condition $S(P\mathbf{x}) = 1$ shows that $P\mathbf{x}$ is a probability vector.

MATLAB random stochastic matrices

Once you specify a value for n, the command **A=rand(n)** produces a random $n \times n$ matrix whose entries are in the interval $[0,1]$. To create a stochastic matrix, we need to divide each column by the column sum:
for j=1:n
P(:,j)=A(:,j)/sum(A(:,j))
end

Chapter 5 - Supplementary Exercises

In this chapter, Exercises 1-23 consist of true/false questions, whose level of difficulty varies. Some are similar to the ones that appear in many sections of the text, in which a word or phrase is sometimes missing or slightly misstated. Some follow fairly easily from a theorem: others may need careful reasoning. A few may require an argument that uses several ideas. In each case, think carefully about the statement and attempt to write a solution. The text provides the true/false answer, but you must supply the justification or counterexample.

25. **a.** Suppose $A\mathbf{x} = \lambda\mathbf{x}$, with $\mathbf{x} \neq \mathbf{0}$. Then $(5I - A)\mathbf{x} = 5\mathbf{x} - A\mathbf{x} = 5\mathbf{x} - \lambda\mathbf{x} = (5 - \lambda)\mathbf{x}$. The eigenvalue is $5 - \lambda$.

Copyright © 2021 Pearson Education, Inc.

b. $(5I - 3A + A^2)\mathbf{x} = 5\mathbf{x} - 3A\mathbf{x} + A(A\mathbf{x}) = 5\mathbf{x} - 3(\lambda\mathbf{x}) + \lambda^2\mathbf{x} = (5 - 3\lambda + \lambda^2)\mathbf{x}$. The eigenvalue is $5 - 3\lambda + \lambda^2$.

31. If $I - A$ were not invertible, then the equation $(I - A)\mathbf{x} = \mathbf{0}$. would have a nontrivial solution $\mathbf{x}$. Then $\mathbf{x} - A\mathbf{x} = \mathbf{0}$ and $A\mathbf{x} = 1 \cdot \mathbf{x}$, which shows that A would have 1 as an eigenvalue. This cannot happen if all the eigenvalues are less than 1 in magnitude. So $I - A$ must be invertible.

37. Replace A by $A - \lambda$ in the determinant formula from Exercise 30 in Chapter 3 Supplementary Exercises.

$$\det(A - \lambda I) = (a - b - \lambda)^{n-1}[a - \lambda + (n-1)b]$$

This determinant is zero only if $a - b - \lambda = 0$ or $a - \lambda + (n-1)b = 0$. Thus λ is an eigenvalue of A if and only if $\lambda = a - b$ or $\lambda = a + (n-1)b$. From the formula for $\det(A - \lambda I)$ above, the algebraic multiplicity is $n-1$ for $a-b$ and 1 for $a + (n-1)b$.

43. If p is a polynomial of order 2, then a calculation such as in Exercise 41 shows that the characteristic polynomial of C_p is $p(\lambda) = (-1)^2 p(\lambda)$, so the result is true for $n = 2$. Suppose the result is true for $n = k$ for some $k \geq 2$, and consider a polynomial p of degree $k + 1$. Then expanding $\det(C_p - \lambda I)$

by cofactors down the first column, the determinant of $C_p - \lambda I$ equals

$$(-\lambda)\det\begin{bmatrix} -\lambda & 1 & \cdots & 0 \\ \vdots & & & \vdots \\ 0 & & & 1 \\ -a_1 & -a_2 & \cdots & -a_k - \lambda \end{bmatrix} + (-1)^{k+1} a_0$$

The $k \times k$ matrix shown is $C_q - \lambda I$, where $q(t) = a_1 + a_2 t + \cdots + a_k t^{k-1} + t^k$. By the induction assumption, the determinant of $C_q - \lambda I$ is $(-1)^k q(\lambda)$. Thus

$$\det(C_p - \lambda I) = (-1)^{k+1} a_0 + (-\lambda)(-1)^k q(\lambda)$$
$$= (-1)^{k+1}[a_0 + \lambda(a_1 + \cdots + a_k \lambda^{k-1} + \lambda^k)]$$
$$= (-1)^{k+1} p(\lambda)$$

So the formula holds for $n = k + 1$ when it holds for $n = k$. By the principle of induction, the formula for $\det(C_p - \lambda I)$ is true for all $n \geq 2$.

Copyright © 2021 Pearson Education, Inc.

Chapter 5 - Glossary Checklist

Check your knowledge by attempting to write definitions of the terms below. Then compare your work with the definitions given in the text's Glossary. Ask your instructor which definitions, if any, might appear on a test.

algebraic multiplicity: The multiplicity of an eigenvalue as . . . .

attractor (of a dynamical system in $\mathbb{R}^2$):   The origin of $\mathbb{R}^2$ when all trajectories tend . . . .

$\mathcal{B}$-matrix (for T):   A matrix $[T]_\mathcal{B}$ for a linear transformation $T : V \rightarrow V$ relative to a basis $\mathcal{B}$ for V, with the property that . . . .

characteristic equation (of A):   . . . .

characteristic polynomial (of A):   . . . .

companion matrix: A special form of a matrix whose characteristic . . . is . . . .

complex eigenvalue: A nonreal root of the characteristic equation of an $n \times n$ matrix A, when . . . .

complex eigenvector: A nonzero vector $\mathbf{x}$ in $\mathbb{C}^n$ such that . . . , where . . . .

decoupled system: A difference equation $\mathbf{y}_{k+1} = A\mathbf{y}_k$, or a differential equation, in which A is a . . . .

determinant (of a square matrix A):   A number det A computed from A; equal to . . . .

diagonalizable (matrix):   A matrix that may be written in factored form as . . . .

difference equation (or **linear recurrence relation**):   An equation of the form . . . whose solution is . . . .

discrete linear dynamical system (or briefly, a **dynamical system**):   A difference equation of the form . . . that describes . . . .

eigenfunction (of the equation $\mathbf{x}'(t) = A\mathbf{x}(t)$):   A function of the form . . . .

eigenspace (of A corresponding to λ):   The set of . . . solutions of . . . .

eigenvalue (of A):   A scalar λ such that . . . .

eigenvector (of A):   A . . . vector $\mathbf{x}$ such that . . . .

eigenvector basis: A basis consisting entirely of . . . .

eigenvector decomposition (of $\mathbf{x}$):   An equation $\mathbf{x} = $ . . . .

fundamental set of solutions (for $\mathbf{x}' = A\mathbf{x}$):   A basis for . . . .

Im x: The vector in $\mathbb{R}^n$ formed from . . . .

invariant subspace (for A):   A subspace H such that . . . .

inverse power method: An algorithm for estimating . . .

Copyright © 2021 Pearson Education, Inc.

matrix for *T* relative to bases $\mathcal{B}$ and $\mathcal{C}$:   A matrix M for a linear transformation $T : V \to V$ with the property that . . . .

power method: An algorithm for estimating . . . .

repeller (of a dynamical system in $\mathbb{R}^2$):   The origin in $\mathbb{R}^2$ when all trajectories . . . tend . . . .

Rayleigh quotient: $R(\mathbf{x}) = $ . . . . An estimate of . . . .

Re x: The vector in $\mathbb{R}^n$ formed from . . . .

saddle point (of a dynamical system in $\mathbb{R}^2$):   The origin in $\mathbb{R}^2$ when . . . .

similar (matrices):   Matrices A and B such that . . . .

spiral point (of a dynamical system in $\mathbb{R}^2$):   The origin in $\mathbb{R}^2$ when . . . .

stage-matrix model: A difference equation $\mathbf{x}_{k+1} = A\mathbf{x}_k$ where $\mathbf{x}_k$ lists . . . .

strictly dominant eigenvalue: An eigenvalue λ of a matrix A with the property that . . . .

trace (of a square matrix A):   The . . . , denoted by tr A.

trajectory: The graph of a solution $\{\mathbf{x}_0, \mathbf{x}_1, \mathbf{x}_2, \dots\}$ of a . . . .

        Also, the graph of $\mathbf{x}(t)$ for $t \geq 0$, when . . . .

Copyright © 2021 Pearson Education, Inc.

6 Orthogonality and Least-Squares

6.1 - Inner Product, Length, and Orthogonality

The concepts of length, distance, and orthogonality introduced in this section are essential for many geometric descriptions in the rest of the text.

STUDY NOTES

The first half of the section is computational and easily learned. The second half, however, requires more attention. Read it carefully. The concepts of orthogonality and orthogonal complements are the foundation for the rest of the chapter. In fact, Theorem 3 is sometimes called the Fundamental Theorem of Linear Algebra.

SOLUTIONS TO EXERCISES

1. $\mathbf{u} = \begin{bmatrix} -1 \\ 2 \end{bmatrix}$, $\mathbf{v} = \begin{bmatrix} 2 \\ 3 \end{bmatrix}$, $\mathbf{u} \cdot \mathbf{u} = (-1)^2 + 2^2 = 5$,

$$\mathbf{v} \cdot \mathbf{u} = 2(-1) + 3(2) = 4, \frac{\mathbf{v} \cdot \mathbf{u}}{\mathbf{u} \cdot \mathbf{u}} = \frac{4}{5}$$

7. $\mathbf{w} = \begin{bmatrix} 3 \\ -1 \\ -5 \end{bmatrix}$, $\quad \|\mathbf{w}\|^2 = \mathbf{w}^T\mathbf{w} = 3^2 + (-1)^2 + (-5)^2 = 35.$ So $\|\mathbf{w}\| = \sqrt{35}$.

13. $\mathbf{x} = \begin{bmatrix} 10 \\ -3 \end{bmatrix}$, $\mathbf{y} = \begin{bmatrix} -1 \\ -5 \end{bmatrix}$

$\|\mathbf{x} - \mathbf{y}\|^2 = [10 - (-1)]^2 + [-3 - (-5)]^2 = 121 + 4 = 125$

$\text{dist}(\mathbf{x}, \mathbf{y}) = \|\mathbf{x} - \mathbf{y}\| = \sqrt{125}$, or $5\sqrt{5}$

19. See the definition of $\|\mathbf{v}\|$.

21. See the discussion of Figure 5.

23. See the box following Example 6.

Copyright © 2021 Pearson Education, Inc.

25. The absolute value sign is missing. See the box before Example 2.

27. See Theorem 3.

29. Theorem l(b): $(\mathbf{u} + \mathbf{v}) \cdot \mathbf{w} = (\mathbf{u} + \mathbf{v})^T\mathbf{w} = (\mathbf{u}^T + \mathbf{v}^T)\mathbf{w} = \mathbf{u}^T\mathbf{w} + \mathbf{v}^T\mathbf{w} = \mathbf{u} \cdot \mathbf{w} + \mathbf{v} \cdot \mathbf{w}$. The second and third equalities used Theorems 3(b) and 2(c), respectively, from Section 2.1. Theorem l(c): $(c\mathbf{u}) \cdot \mathbf{v} = (c\mathbf{u})^T\mathbf{v} = (c\mathbf{u}^T)\mathbf{v} = c(\mathbf{u}^T\mathbf{v}) = c(\mathbf{u} \cdot \mathbf{v})$, by Theorems 3(c) and 2(d) in Section 2.1. Also, $\mathbf{u} \cdot (c\mathbf{v}) = \mathbf{u}^T(c\mathbf{v}) = c\mathbf{u}^T\mathbf{v} = c(\mathbf{u} \cdot \mathbf{v})$.

33. When $\mathbf{v} = \begin{bmatrix} a \\ b \end{bmatrix}$, the set H of all vectors $\begin{bmatrix} x \\ y \end{bmatrix}$ that are orthogonal to $\mathbf{v}$ is the subspace of vectors whose entries satisfy $ax + by = 0$. If $a \neq 0$, then $x = -(b/a)y$, with y a free variable. Then H is a line through the origin. A natural choice for a basis for this subspace is $\begin{bmatrix} -b \\ a \end{bmatrix}$. If $a = 0$ and $b \neq 0$, then the vectors in H satisfy $by = 0$. Since b is nonzero, $y = 0$ and x is free. A basis for H is $\begin{bmatrix} 1 \\ 0 \end{bmatrix}$. Note, however, that $\begin{bmatrix} -b \\ a \end{bmatrix}$ is also a basis, since $a = 0$ and $b \neq 0$.

Finally, if a and b are both zero, then H is $\mathbb{R}^2$ itself, because the equation $0x + 0y = 0$ places no restrictions on x or y.

35. If $\mathbf{y}$ is orthogonal to $\mathbf{u}$ and $\mathbf{v}$, then $\mathbf{y} \cdot \mathbf{u} = 0$ and $\mathbf{y} \cdot \mathbf{v} = 0$, and hence by a property of the inner product, $\mathbf{y} \cdot (\mathbf{u} + \mathbf{v}) = \mathbf{y} \cdot \mathbf{u} + \mathbf{y} \cdot \mathbf{v} = 0 + 0 = 0$. So $\mathbf{y}$ is orthogonal to $\mathbf{u} + \mathbf{v}$.

37. Take a typical vector $\mathbf{w} = c_1\mathbf{v}_1 + \cdots + c_p\mathbf{v}_p$ in W. If $\mathbf{x}$ is orthogonal to each $\mathbf{v}_j$, then using the linearity of the inner product (Theorem l(b) and l(c)), $\mathbf{w} \cdot \mathbf{x} = (c_1\mathbf{v}_1 + \cdots + c_p\mathbf{v}_p)'\mathbf{x} = c_1\mathbf{v}_1 \cdot \mathbf{x} + \cdots + c_p\mathbf{v}_p \cdot \mathbf{x} = 0$. So $\mathbf{x}$ is orthogonal to each $\mathbf{w}$ in W.

39. Suppose $\mathbf{x}$ is in W and $W^\perp$. Then, since $\mathbf{x}$ is in $W^\perp$, $\mathbf{x}$ is orthogonal to every vector in W, including $\mathbf{x}$ itself. So $\mathbf{x} \cdot \mathbf{x} = 0$. This is true only if $\mathbf{x} = 0$. This problem shows that $W \cap W^\perp$ is the zero subspace.

MATLAB   The inner product of real column vectors $\mathbf{u}$ and $\mathbf{v}$ is **u'*v** (and **v'*u**); the length of $\mathbf{v}$ is **norm(v)**. See the MATLAB note for Section 2.1.

6.2 - Orthogonal Sets

Orthogonal sets and orthogonal bases are used throughout the chapter. The "orthogonal projection" discussed in this section is an important special case of the orthogonal projections studied in Section 6.3.

Copyright © 2021 Pearson Education, Inc.

_____

STUDY NOTES

The proofs of Theorems 4 and 5 are worth studying because they involve a calculation you will see and use several times.

   The subsection entitled *An Orthogonal Projection* is simple but extremely important. Also, the geometric interpretation of Theorem 5 will be helpful when you study Theorem 8 in the next section.

   The attention paid to Theorems 6 and 7 will depend on what your instructor plans to do later in the chapter. In some cases, an instructor may discuss Theorems 6 and 7 only for square matrices. The $m \times n$ case is needed later, for Theorems 10, 12 and 15. Remember: the term *orthogonal matrix* applies *only to a square matrix*. Also, the columns of an orthogonal matrix must be *orthonormal*, not simply orthogonal.

_____

SOLUTIONS TO EXERCISES

1. $\mathbf{u} = \begin{bmatrix} -1 \\ 4 \\ -3 \end{bmatrix}$, $\mathbf{v} = \begin{bmatrix} 5 \\ 2 \\ 1 \end{bmatrix}$, $\mathbf{w} = \begin{bmatrix} 3 \\ -4 \\ -7 \end{bmatrix}$, $\mathbf{u} \cdot \mathbf{v} = -5 + 8 - 3 = 0$, $\mathbf{u} \cdot \mathbf{w} = -3 - 16 + 21 = 2 \neq 0$. The

   set $\{\mathbf{u}, \mathbf{v}, \mathbf{w}\}$ is not orthogonal. There is no need to check $\mathbf{v} \cdot \mathbf{w}$.

7. $\mathbf{u}_1 = \begin{bmatrix} 2 \\ -3 \end{bmatrix}$, $\mathbf{u}_2 = \begin{bmatrix} 6 \\ 4 \end{bmatrix}$, $\mathbf{x} = \begin{bmatrix} 9 \\ -7 \end{bmatrix}$. $\mathbf{u}_1 \cdot \mathbf{u}_2 = 12 - 12 = 0$, so $\{\mathbf{u}_1, \mathbf{u}_2\}$ is an orthogonal set.

   Since the vectors are nonzero, $\mathbf{u}_1$ and $\mathbf{u}_2$ are linearly independent, by Theorem 4. But two such vectors in $\mathbb{R}^2$ automatically form a basis for $\mathbb{R}^2$. So $\{\mathbf{u}_1, \mathbf{u}_2\}$ is an orthogonal basis for $\mathbb{R}^2$. By Theorem 5,

$$\mathbf{x} = \frac{\mathbf{x} \cdot \mathbf{u}_1}{\mathbf{u}_1 \cdot \mathbf{u}_1}\mathbf{u}_1 + \frac{\mathbf{x} \cdot \mathbf{u}_2}{\mathbf{u}_2 \cdot \mathbf{u}_2}\mathbf{u}_2 = \frac{18 + 21}{4 + 9}\mathbf{u}_1 + \frac{54 - 28}{36 + 16}\mathbf{u}_2 = 3\begin{bmatrix} 2 \\ -3 \end{bmatrix} + \frac{1}{2}\begin{bmatrix} 6 \\ 4 \end{bmatrix}$$

13. $\mathbf{y} = \begin{bmatrix} 2 \\ 3 \end{bmatrix}$, $\mathbf{u} = \begin{bmatrix} 4 \\ -7 \end{bmatrix}$. The orthogonal projection of $\mathbf{y}$ onto $\mathbf{u}$ is

$$\hat{\mathbf{y}} = \frac{\mathbf{y} \cdot \mathbf{u}}{\mathbf{u} \cdot \mathbf{u}}\mathbf{u} - \frac{8 - 21}{16 + 49}\mathbf{u} = \frac{-13}{65}\mathbf{u} = \frac{-1}{5}\begin{bmatrix} 4 \\ -7 \end{bmatrix} = \begin{bmatrix} -4/5 \\ 7/5 \end{bmatrix}$$

   The component of $\mathbf{y}$ orthogonal to $\mathbf{u}$ is $\mathbf{y} - \hat{\mathbf{y}} = \begin{bmatrix} 2 \\ 3 \end{bmatrix} - \begin{bmatrix} -4/5 \\ 7/5 \end{bmatrix} = \begin{bmatrix} 14/5 \\ 8/5 \end{bmatrix}$.   Thus,

$$\mathbf{y} = \hat{\mathbf{y}} + (\mathbf{y} - \hat{\mathbf{y}}) = \begin{bmatrix} -4/5 \\ 7/5 \end{bmatrix} + \begin{bmatrix} 14/5 \\ 8/5 \end{bmatrix}.$$

Copyright © 2021 Pearson Education, Inc.

19. $\mathbf{u} = \begin{bmatrix} -.6 \\ .8 \end{bmatrix}$, $\mathbf{v} = \begin{bmatrix} .8 \\ .6 \end{bmatrix}$, $\mathbf{u} \cdot \mathbf{v} = -.48 + .48 = 0$, so $\{\mathbf{u}, \mathbf{v}\}$ is an orthogonal set. Also,

   $\|\mathbf{u}\|^2 = \mathbf{u} \cdot \mathbf{u} = (-.6)^2 + (.8)^2 = .36 + .64 = 1$. Similarly, $\|\mathbf{v}\|^2 = \mathbf{v} \cdot \mathbf{v} = 1$. Thus $\{\mathbf{u}, \mathbf{v}\}$ is an orthonormal set.

23. See Example 3, for instance.

25. See the paragraph following Theorem 5.

27. See the paragraph following Example 5.

29. The matrix must also be square. See the paragraph before Example 7.

33. $(U\mathbf{x}) \cdot (U\mathbf{y}) = (U\mathbf{x})^T(U\mathbf{y}) = \mathbf{x}^T U^T U\mathbf{y} = \mathbf{x}^T\mathbf{y} = \mathbf{x} \cdot \mathbf{y}$ (because $U^TU = I$). If $\mathbf{y} = \mathbf{x}$, Theorem 7(b) says that $\|U\mathbf{x}\|^2 = \|\mathbf{x}\|^2$, which implies part (a).
    Part (c) of Theorem 7 follows immediately from part (b).

Study Tip: If your instructor emphasizes orthogonal matrices, work Exercises 35–37. (They make good test questions.) In each case, mention explicitly how you use the fact that the matrices are square. Don't read the solutions below until you have first *written* your own solution.

35. If U has orthonormal columns, then $U^TU = I$, by Theorem 6. If U is also *square*, then the equation $U^TU = I$ implies that U is invertible, by the Invertible Matrix Theorem.

37. Since U and V are orthogonal, each is invertible. By Theorem 6 in Section 2.2, UV is invertible and $(UV)^{-1} = V^{-1}U^{-1} = V^TU^T = (UV)^T$ (by Theorem 3 in Section 2.1). Thus UV is an orthogonal matrix.

Mastering Linear Algebra Concepts: Orthogonal Basis

To the review sheet(s) you have on "basis," add the concepts of an orthogonal basis and an orthonormal basis for a subspace. You need to know what special properties they possess.

- basic definitions

- equivalent descriptions   Theorems 4 and 6

- geometric interpretation   Figs. 1 and 6

- special cases   Matrix with orthonormal columns

- examples and counterexamples   Examples 2 and 5

- algorithms and computations   Example 2

- connections with other concepts   Orthogonal matrix

Copyright © 2021 Pearson Education, Inc.

MATLAB   Orthogonality

In Exercises 1–10 and 17–22, the fastest way (counting the keystrokes) in MATLAB to test a set such as {$\mathbf{u}_1$, $\mathbf{u}_2$, $\mathbf{u}_3$} for orthogonality is to use a matrix **U = [u1   u2   u3]** whose columns are the vectors from the set, and test whether **U′*U** is a diagonal matrix. See the proof of Theorem 6.

For column vectors **y** and **u**, the orthogonal projection of **y** onto **u** is

(y′*u)/(u′*u)*u

The parentheses (and the final *) are essential. MATLAB computes the scalar quotient $(\mathbf{y}^T\mathbf{u})/(\mathbf{u}^T\mathbf{u})$ and then multiplies **u** by this scalar.

6.3 - Orthogonal Projections

A familiar idea in Euclidean geometry is to construct a line segment perpendicular to a line or plane. This section treats an analogous situation in $\mathbb{R}^n$, namely, the orthogonal projection of a vector (a point in $\mathbb{R}^n$) onto a subspace. The case when the subspace is a line through the origin was already examined in Section 6.2.

KEY IDEAS

If **y** is in $\mathbb{R}^n$ and if W is a subspace of $\mathbb{R}^n$, then the orthogonal projection of **y** onto W, denoted by $\hat{\mathbf{y}}$ or proj$_W$ **y**, has two important properties:

(i) $\mathbf{y} - \hat{\mathbf{y}}$ is orthogonal to W (so **y** is the sum of a vector $\hat{\mathbf{y}}$ in W and a vector $\mathbf{y} - \hat{\mathbf{y}}$ in $W^\perp$), and

(ii) $\hat{\mathbf{y}}$ is the closest point in W to **y**.

Properties (i) and (ii) are described in the Orthogonal Decomposition Theorem and the Best Approximation Theorem. You should learn the statements of both theorems. (By now you probably know that whenever a theorem has an official name, an instructor has an easy time asking test questions about it.) When you need one of these theorems in a discussion (homework or test question), you should mention the theorem by name.

If your class covers Theorem 10, then the paragraph following the theorem will help you understand the difference between an *orthogonal matrix* (which must be square) and a rectangular matrix with orthonormal columns.

SOLUTIONS TO EXERCISES

1. $\mathbf{u}_1 = \begin{bmatrix} 0 \\ 1 \\ -4 \\ -1 \end{bmatrix}$, $\mathbf{u}_2 = \begin{bmatrix} 3 \\ 5 \\ 1 \\ 1 \end{bmatrix}$, $\mathbf{u}_3 = \begin{bmatrix} 1 \\ 0 \\ 1 \\ -4 \end{bmatrix}$, $\mathbf{u}_4 = \begin{bmatrix} 5 \\ -3 \\ -1 \\ 1 \end{bmatrix}$, $\mathbf{x} = \begin{bmatrix} 10 \\ -8 \\ 2 \\ 0 \end{bmatrix}$. You could calculate all the inner

products in the decomposition:

Copyright © 2021 Pearson Education, Inc.

$$\mathbf{x} = \frac{\mathbf{x} \cdot \mathbf{u}_1}{\mathbf{u}_1 \cdot \mathbf{u}_1}\mathbf{u}_1 + \frac{\mathbf{x} \cdot \mathbf{u}_2}{\mathbf{u}_2 \cdot \mathbf{u}_2}\mathbf{u}_2 + \frac{\mathbf{x} \cdot \mathbf{u}_3}{\mathbf{u}_3 \cdot \mathbf{u}_3}\mathbf{u}_3 + \frac{\mathbf{x} \cdot \mathbf{u}_4}{\mathbf{u}_4 \cdot \mathbf{u}_4}\mathbf{u}_4 \qquad (1)$$

$$\underbrace{\qquad\qquad\qquad\qquad\qquad\qquad}_{\text{in Span}\{\mathbf{u}_1, \mathbf{u}_2, \mathbf{u}_3\}} \quad \underbrace{\qquad\qquad}_{\text{in Span}\{\mathbf{u}_4\}}$$

However, once you know the vector in Span$\{\mathbf{u}_4\}$, the vector in Span$\{\mathbf{u}_1, \mathbf{u}_2, \mathbf{u}_3\}$ is determined completely by (1). So all you need is

$$\frac{\mathbf{x} \cdot \mathbf{u}_4}{\mathbf{u}_4 \cdot \mathbf{u}_4}\mathbf{u}_4 = \frac{50+24-2+0}{25+9+1+1}\mathbf{u}_4 = 2\mathbf{u}_4 = \begin{bmatrix} 10 \\ -6 \\ -2 \\ 2 \end{bmatrix}$$

The vector in Span$\{\mathbf{u}_1, \mathbf{u}_2, \mathbf{u}_3\}$ is $\mathbf{x} - 2\mathbf{u}_4 = \begin{bmatrix} 10 \\ -8 \\ 2 \\ 0 \end{bmatrix} - \begin{bmatrix} 10 \\ -6 \\ -2 \\ 2 \end{bmatrix} = \begin{bmatrix} 0 \\ -2 \\ 4 \\ -2 \end{bmatrix}$.

Notice $\mathbf{x} = \begin{bmatrix} 10 \\ -8 \\ 2 \\ 0 \end{bmatrix} = \begin{bmatrix} 0 \\ -2 \\ 4 \\ -2 \end{bmatrix} + \begin{bmatrix} 10 \\ -6 \\ -2 \\ 2 \end{bmatrix}$.

Study Tip: One way to check whether $\text{proj}_W\,\mathbf{y}$ is computed correctly is to verify that $\mathbf{y} - \text{proj}_W\,\mathbf{y}$ is orthogonal to each vector in the orthogonal basis $\{\mathbf{u}_1, \ldots, \mathbf{u}_p\}$ for W. A faster check that will catch most errors (but not all) is to verify that $\mathbf{y} - \text{proj}_W\,\mathbf{y}$ is orthogonal to $\text{proj}_W\,\mathbf{y}$.

7. $\mathbf{y} = \begin{bmatrix} 1 \\ 3 \\ 5 \end{bmatrix}$, $\mathbf{u}_1 = \begin{bmatrix} 1 \\ 3 \\ -2 \end{bmatrix}$, $\mathbf{u}_2 = \begin{bmatrix} 5 \\ 1 \\ 4 \end{bmatrix}$. First, make sure that $\{\mathbf{u}_1, \mathbf{u}_2\}$ is an orthogonal basis for

Span$\{\mathbf{u}_1, \mathbf{u}_2\}$. This is easy, since $\mathbf{u}_1$ and $\mathbf{u}_2$ are nonzero and $\mathbf{u}_1 \cdot \mathbf{u}_2 = 0$. Next, by the Orthogonal Decomposition Theorem, $\mathbf{y}$ is the sum of $\text{proj}_W\,\mathbf{y}$ and $\mathbf{y} - \text{proj}_W\,\mathbf{y}$, where $W = \text{Span}\{\mathbf{u}_1, \mathbf{u}_2\}$.

$$\text{proj}_W\,\mathbf{y} = \frac{\mathbf{y} \cdot \mathbf{u}_1}{\mathbf{u}_1 \cdot \mathbf{u}_1}\mathbf{u}_1 + \frac{\mathbf{y} \cdot \mathbf{u}_2}{\mathbf{u}_2 \cdot \mathbf{u}_2}\mathbf{u}_2 = \frac{1+9-10}{1+9+4}\mathbf{u}_1 + \frac{5+3+20}{25+1+16}\mathbf{u}_2$$

$$= 0\mathbf{u}_1 + \frac{2}{3}\mathbf{u}_2 = \begin{bmatrix} 10/3 \\ 2/3 \\ 8/3 \end{bmatrix}$$

and

Copyright © 2021 Pearson Education, Inc.

$$\mathbf{y} - \text{proj}_W \mathbf{y} = \begin{bmatrix} 1 \\ 3 \\ 5 \end{bmatrix} - \begin{bmatrix} 10/3 \\ 2/3 \\ 8/3 \end{bmatrix} = \begin{bmatrix} -7/3 \\ 7/3 \\ 7/3 \end{bmatrix}$$

As a check, scale $\mathbf{y} - \text{proj}_W \mathbf{y} = \begin{bmatrix} -1 \\ 1 \\ 1 \end{bmatrix}$, and observe that the scaled vector is obviously orthogonal to $\mathbf{u}_1$ and $\mathbf{u}_2$. Thus $\mathbf{y} - \text{proj}_W \mathbf{y}$ is in $W^\perp$, as it should be.

Warning: The formula for $\text{proj}_W \mathbf{y}$ applies only if $\{\mathbf{u}_1, \dots, \mathbf{u}_p\}$ is an *orthogonal* basis for W. That's why you should check orthogonality, as in Exercise 7, if you are not sure that the basis is orthogonal. If an orthogonal basis is not available, then other methods can be used to compute $\widehat{\mathbf{y}}$.

13. $\mathbf{z} = \begin{bmatrix} 3 \\ -7 \\ 2 \\ 3 \end{bmatrix}$, $\mathbf{v}_1 = \begin{bmatrix} 2 \\ -1 \\ -3 \\ 1 \end{bmatrix}$, $\mathbf{v}_2 = \begin{bmatrix} 1 \\ 1 \\ 0 \\ -1 \end{bmatrix}$. Note that $\mathbf{v}_1$ and $\mathbf{v}_2$ are orthogonal. By the Best

Approximation Theorem, the closest point in $\text{Span}\{\mathbf{v}_1, \mathbf{v}_2\}$ to $\mathbf{z}$ is the orthogonal projection $\widehat{\mathbf{z}}$, where

$$\widehat{\mathbf{z}} = \frac{\mathbf{z} \cdot \mathbf{v}_1}{\mathbf{v}_1 \cdot \mathbf{v}_1} \mathbf{v}_1 + \frac{\mathbf{z} \cdot \mathbf{v}_2}{\mathbf{v}_2 \cdot \mathbf{v}_2} \mathbf{v}_2 = \frac{10}{15} \mathbf{v}_1 + \frac{-7}{3} \mathbf{v}_2 = \frac{2}{3} \begin{bmatrix} 2 \\ -1 \\ -3 \\ 1 \end{bmatrix} - \frac{7}{3} \begin{bmatrix} 1 \\ 1 \\ 0 \\ -1 \end{bmatrix} = \begin{bmatrix} -1 \\ -3 \\ -2 \\ 3 \end{bmatrix}$$

Check: $\mathbf{z} - \widehat{\mathbf{z}} = \begin{bmatrix} 4 \\ -4 \\ 4 \\ 0 \end{bmatrix}$. The vector $\begin{bmatrix} 1 \\ -1 \\ 1 \\ 0 \end{bmatrix}$ is orthogonal to both $\mathbf{v}_1$ and $\mathbf{v}_2$.

19. By the Orthogonal Decomposition Theorem, $\mathbf{u}_3$ is the sum of a vector in $W - \text{Span}\{\mathbf{u}_1, \mathbf{u}_2\}$ and a vector $\mathbf{v}$ orthogonal to W. First,

$$\text{proj}_W \mathbf{u}_3 = \frac{-2}{6} \mathbf{u}_1 + \frac{2}{30} \mathbf{u}_2 = \begin{bmatrix} -2/6 \\ -2/6 \\ 4/6 \end{bmatrix} + \begin{bmatrix} 10/30 \\ -2/30 \\ 4/30 \end{bmatrix} = \begin{bmatrix} 0 \\ -2/5 \\ 4/5 \end{bmatrix}$$

Then

$$\mathbf{v} = \mathbf{u}_3 - \text{proj}_W \mathbf{u}_3 = \begin{bmatrix} 0 \\ 0 \\ 1 \end{bmatrix} - \begin{bmatrix} 0 \\ -2/5 \\ 4/5 \end{bmatrix} = \begin{bmatrix} 0 \\ 2/5 \\ 1/5 \end{bmatrix}$$

Copyright © 2021 Pearson Education, Inc.

Not only is $\mathbf{v}$ orthogonal to W, but also any multiple of $\mathbf{v}$ is in $W^{\perp}$.

Study Tip: It would be a good idea to try Exercise 20 and compare the result with Exercise 19. Then think about the following problem: Suppose that $\{\mathbf{u}_1, \mathbf{u}_2\}$ is an orthogonal set of nonzero vectors in $\mathbb{R}^3$. How would you find an orthogonal basis of $\mathbb{R}^3$ that contains $\mathbf{u}_1$ and $\mathbf{u}_2$? You might discuss this with your instructor.

21. See the calculations for $\mathbf{z}_2$ in Example 1 or the box after Example 6 in Section 6.1.

23. See the last paragraph in the proof of Theorem 8, or see the second paragraph after the statement of Theorem 9.

25. The Best Approximation Theorem says that the best approximation to $\mathbf{y}$ is $\text{proj}_W \mathbf{y}$.

27. See the subsection "A Geometric Interpretation of the Orthogonal Projection."

29. Theorem 10 applies to the column space W of U because the columns of U are linearly independent and hence form a basis for W.

31. By the Orthogonal Decomposition Theorem, each $\mathbf{x}$ in $\mathbb{R}^n$ can be written uniquely as $\mathbf{x} = \mathbf{p} + \mathbf{u}$, with $\mathbf{p}$ in Row A and $\mathbf{u}$ in (Row A)$^{\perp}$. By Theorem 3 in Section 6.1, $\mathbf{u}$ is in Nul A.

Next, suppose that $A\mathbf{x} = \mathbf{b}$ is consistent. Let $\mathbf{x}$ be a solution, and write $\mathbf{x} = \mathbf{p} + \mathbf{u}$, as above. Then $A\mathbf{p} = A(\mathbf{x} - \mathbf{u}) = A\mathbf{x} - A\mathbf{u} = \mathbf{b} - \mathbf{0} = \mathbf{b}$. So the equation $A\mathbf{x} = \mathbf{b}$ has at least one solution $\mathbf{p}$ in Row A.

Finally, suppose that $\mathbf{p}$ and $\mathbf{p}_1$ are both in Row A and satisfy $A\mathbf{x} = \mathbf{b}$. Then $\mathbf{p} - \mathbf{p}_1$ is in Nul A because

$$A\,(\mathbf{p} - \mathbf{p}_1) = A\mathbf{p} - A\mathbf{p}_1 = \mathbf{b} - \mathbf{b} = \mathbf{0}$$

The equations $\mathbf{p} = \mathbf{p}_1 + (\mathbf{p} - \mathbf{p}_1)$ and $\mathbf{p} = \mathbf{p} + \mathbf{0}$ both decompose $\mathbf{p}$ as the sum of a vector in Row A and a vector in (Row A)$^{\perp}$. By the uniqueness of the orthogonal decomposition (Theorem 8), $\mathbf{p}_1 = \mathbf{p}$, so $\mathbf{p}$ is unique.

37. From Exercise 42 of Section 6.2, U should have orthonormal columns, because U is formed by normalizing the columns of the matrix A in Exercise 42 whose columns are orthogonal. Verify this by computing $U^T U$. The result should be the 4×4 identity matrix.

The closest point to $\mathbf{y}$ in Col U is the orthogonal projection of $\mathbf{y}$ onto Col U. By Theorem 10, this closest point is $UU^T\mathbf{y}$. The MATLAB command is **U*U'*y**. The result of this computation should be the (column) vector $(1.2, .4, 1.2, 1.2, .4, 1.2, .4, .4)$.

Warning: It was hard work to make the arithmetic simple in the exercises for this section, to avoid distractions for you and to save you time. You might not be so lucky on an exam. Even if a problem is designed to be numerically simple, there is always a chance that a minor error will make the calculations messy. In such a case, don't despair. Carry out the arithmetic as best you can, showing the details of your work (patterned after the solutions in this *Study Guide*). Chances are that you will get substantial credit for showing that you understand the concepts.

Copyright © 2021 Pearson Education, Inc.

MATLAB   Orthogonal Projections

The orthogonal projection of **y** onto a single vector was described in the MATLAB note for Section 6.2. The orthogonal projection onto the set spanned by an orthogonal set of vectors is the sum of the one-dimensional projections. Another way to construct this projection is to normalize the orthogonal vectors, place them in the columns of a matrix U, and use Theorem 10. For instance, if $\{\mathbf{y}_1, \mathbf{y}_2, \mathbf{y}_3\}$ is an orthogonal set of nonzero vectors, then the matrix

U = [y1/norm(y1) y2/norm(y2) y3/norm(y3)]

has orthonormal columns, and **U*(U′*y)** produces the orthogonal projection of **y** onto the subspace spanned by $\{\mathbf{y}_1, \mathbf{y}_2, \mathbf{y}_3\}$. (The parentheses around **U′*y** speeds up the computation of **U*U′*y** by avoiding a matrix-matrix product.)

6.4 - The Gram-Schmidt Process

This section has a nice geometric appeal. The Gram-Schmidt process is well-liked by students and faculty because it is easily learned. Although the process is seldom used in practical computations, it has important generalizations to spaces other than $\mathbb{R}^n$ (to be discussed briefly in Section 6.7).

KEY IDEAS

When the Gram-Schmidt process is applied to $\{\mathbf{x}_1, \ldots, \mathbf{x}_p\}$, the first step is to set $\mathbf{v}_1 = \mathbf{x}_1$. For $k = 2, \ldots, n$, the kth step consists of subtracting from $\mathbf{x}_k$ its projection onto the subspace spanned by the previous **x**'s. At each step the projection is easy to compute because an orthogonal basis for the appropriate subspace has already been constructed.

The QR factorization of a matrix A encapsulates the result of applying the Gram-Schmidt process to the columns of A, just as the LU factorization of a matrix encodes the row operations that reduce a matrix to echelon form. Also, just as the LU factorization can be implemented via multiplication by elementary matrices, so can the QR factorization be constructed via multiplication by orthogonal matrices.

SOLUTIONS TO EXERCISES

1. $\mathbf{x}_1 = \begin{bmatrix} 3 \\ 0 \\ -1 \end{bmatrix}, \mathbf{x}_2 = \begin{bmatrix} 8 \\ 5 \\ -6 \end{bmatrix}$. Set $\mathbf{v}_1 = \mathbf{x}_1$ and compute

$$\mathbf{v}_2 = \mathbf{x}_2 - \frac{\mathbf{x}_2 \cdot \mathbf{v}_1}{\mathbf{v}_1 \cdot \mathbf{v}_1} \mathbf{v}_1 = \begin{bmatrix} 8 \\ 5 \\ -6 \end{bmatrix} - \frac{30}{10} \begin{bmatrix} 3 \\ 0 \\ -1 \end{bmatrix} = \begin{bmatrix} -1 \\ 5 \\ -3 \end{bmatrix}$$

Copyright © 2021 Pearson Education, Inc.

Check: $\mathbf{v}_2 \cdot \mathbf{v}_1 = -3 + 0 + 3 = 0$. So an orthogonal basis is $\left\{ \begin{bmatrix} 3 \\ 0 \\ -1 \end{bmatrix}, \begin{bmatrix} -1 \\ 5 \\ -3 \end{bmatrix} \right\}$.

7. $\mathbf{x}_1 = \begin{bmatrix} 2 \\ -5 \\ 1 \end{bmatrix}$, $\mathbf{x}_2 = \begin{bmatrix} 4 \\ -1 \\ 2 \end{bmatrix}$. From Exercise 3, use $\mathbf{v}_1 = \begin{bmatrix} 2 \\ -5 \\ 1 \end{bmatrix}$ and $\mathbf{v}_2 = \begin{bmatrix} 3 \\ 3/2 \\ 3/2 \end{bmatrix}$ as an orthogonal

basis for $W = \text{Span}\{\mathbf{x}_1, \mathbf{x}_2\}$. Scale $\mathbf{v}_2$ to $(2, 1, 1)$ before normalizing, and then obtain

$$\mathbf{u}_1 = \frac{1}{\sqrt{30}} \begin{bmatrix} 2 \\ -5 \\ 1 \end{bmatrix} = \begin{bmatrix} 2/\sqrt{30} \\ -5/\sqrt{30} \\ 1/\sqrt{30} \end{bmatrix}, \quad \mathbf{u}_2 = \frac{1}{\sqrt{6}} \begin{bmatrix} 2 \\ 1 \\ 1 \end{bmatrix} = \begin{bmatrix} 2/\sqrt{6} \\ 1/\sqrt{6} \\ 1/\sqrt{6} \end{bmatrix}$$

Study Tip: If you need to normalize a vector by hand, first consider scaling the entries in the vector to make them small integers, if possible.

13. $A = \begin{bmatrix} 5 & 9 \\ 1 & 7 \\ -3 & -5 \\ 1 & 5 \end{bmatrix}$, $Q = \begin{bmatrix} 5/6 & -1/6 \\ 1/6 & 5/6 \\ -3/6 & 1/6 \\ 1/6 & 3/6 \end{bmatrix}$. Let

$$R = Q^{\mathrm{T}}A = \begin{bmatrix} 5/6 & 1/6 & -3/6 & 1/6 \\ -1/6 & 5/6 & 1/6 & 3/6 \end{bmatrix} \begin{bmatrix} 5 & 9 \\ 1 & 7 \\ -3 & -5 \\ 1 & 5 \end{bmatrix} = \begin{bmatrix} 36/6 & 72/6 \\ 0 & 36/6 \end{bmatrix} = \begin{bmatrix} 6 & 12 \\ 0 & 6 \end{bmatrix}$$

As a check, compute $QR = \begin{bmatrix} 5/6 & -1/6 \\ 1/6 & 5/6 \\ -3/6 & 1/6 \\ 1/6 & 3/6 \end{bmatrix} \begin{bmatrix} 6 & 12 \\ 0 & 6 \end{bmatrix} = \begin{bmatrix} 5 & 54/6 \\ 1 & 42/6 \\ -3 & -30/6 \\ 1 & 30/6 \end{bmatrix} = A$.

Remark: The reason the R in Exercise 13 works is that the columns of Q form an orthonormal basis for Col A (since they were obtained by the Gram-Schmidt process). Thus $QQ^{\mathrm{T}}\mathbf{y} = \mathbf{y}$ for all $\mathbf{y}$ in Col A, by Theorem 10 in Section 6.3. In particular, $QQ^{\mathrm{T}}A = A$. So if R is $Q^{\mathrm{T}}A$, then $QR = Q(Q^{\mathrm{T}}A) = A$.

17. See the remark after Example 5 in Section 6.2, and the reference there to Exercise 40.
19. See (1) in the statement of Theorem 11.
21. See the solution of Example 4.

Copyright © 2021 Pearson Education, Inc.

27. Use the definition of matrix multiplication. When $A = QR$, the first p columns of A are determined by the action of Q on the first p columns of R. So, if $A = [A_1 \quad A_2]$, make the same column-partition of R as $[R_1 \quad R_2]$, where R_1 has p columns. Then

$$A = Q[R_1 \quad R_2] = [QR_1 \quad QR_2] = [A_1 \quad A_2]$$

Is QR_1 a QR factorization of A_1? Unfortunately, no. The second factor in a QR factorization should be square and upper triangular with positive entries on the diagonal. Since R has those properties, its first p columns have zeros in rows $p + 1$ to n. (This is a key

observation.) So, partition R_1 into two blocks, $R_1 = \begin{bmatrix} R_{11} \\ 0 \end{bmatrix}$, where R_{11} is square and upper

triangular. The entries on the diagonal of R_{11} are positive because they come from R. Then

$A_1 = QR_1 = Q\begin{bmatrix} R_{11} \\ 0 \end{bmatrix}$. You might consider left-multiplying R_{11} by Q, but partitioned matrix

multiplication does not work that way. QR_{11} is not defined, because Q has more columns than R_{11} has rows. (Why?)

The final idea needed is to view the product QR_1 as a product of block matrices, by partitioning the *columns* of Q to match the row partition of R_1. Write $Q = [Q_1 \quad Q_2]$, where Q_1 consists of the first p columns of Q. The matrix Q_1 has orthonormal columns, because the columns come from Q and so the columns are unit vectors and are pairwise orthogonal. Finally,

$$A_1 = QR_1 = [Q_1 \quad Q_2]\begin{bmatrix} R_{11} \\ 0 \end{bmatrix} = Q_1 R_{11} + Q_2 0 = Q_1 R_{11}$$

This concludes the construction, because the properties of Q_1 and R_{11} have already been discussed. This solution has followed a path that a good student might find. Now that you see how things fit together, you should be able to write a short proof that begins with appropriate partitions of Q and R. I encourage you to try it. The shorter proof is hidden at the end of the solutions for Section 6.5.

MATLAB   The Gram-Schmidt Process

If A has only two columns, then the Gram-Schmidt process is

    vl = A(:,1)
    v2 = A(:,2) - (A(:,2)'*vl)/(vl' * vl)*vl

If A has three columns, add the command

    v3 = A(:,3) - (A(:,3)'*vl)/(v1'*vl)*vl - (A(:,3)'*v2)/(v2'*v2)*v2

Copyright © 2021 Pearson Education, Inc.

6.5 - Least-Squares Problems

The basic geometric principles in this section provide the foundation for all the applications in Sections 6.6–6.8.

KEY IDEAS

A least-squares solution of $A\mathbf{x} = \mathbf{b}$ is any vector $\mathbf{x}$ that makes $A\mathbf{x}$ as close as possible to $\mathbf{b}$. (Learn the formal definition, too.) If the columns of A are linearly dependent, then there are many least-squares solutions of $A\mathbf{x} = \mathbf{b}$. They can all be found by row reducing the augmented matrix for the normal equations $A^T A\mathbf{x} = A^T\mathbf{b}$ (Theorem 13).

If the columns of A are linearly independent, then there is only one least-squares solution. To find it, solve the normal equations or compute $(A^T A)^{-1}A^T\mathbf{b}$ (Theorem 14). If a QR factorization of A is available, say $A = QR$, solve the equation $R\mathbf{x} = Q^T\mathbf{b}$ (Theorem 15 and the Numerical Note).

STUDY NOTES

The material up to and including Figure 2 needs to be read carefully several times, so you understand what the term "least-squares solution" means. Be careful to distinguish between $\hat{\mathbf{x}}$ and $\hat{\mathbf{b}}$. A common mistake is to think that $\hat{\mathbf{x}}$ itself somehow has the least-squares norm or is the closest point to $\mathbf{b}$. Look at Fig. 2 again. The vector closest to $\mathbf{b}$ is $A\hat{\mathbf{x}}$, not $\hat{\mathbf{x}}$.

Theorem 13 provides a common way to find least-squares solutions. One way to remember the normal equations is to observe that they look the same as $A\mathbf{x} = \mathbf{b}$ with A^T left-multiplied on each side of the equation. It is *completely wrong*, however, to try to *derive* the normal equations from $A\mathbf{x} = \mathbf{b}$ via left-multiplication by A^T. If the equation $A\mathbf{x} = \mathbf{b}$ has no solution, then the equation itself is a false statement about every vector $\mathbf{x}$. Matrix algebra on such a false statement is meaningless.

SOLUTIONS TO EXERCISES

1. $A = \begin{bmatrix} -1 & 2 \\ 2 & -3 \\ -1 & 3 \end{bmatrix}$, $\mathbf{b} = \begin{bmatrix} 4 \\ 1 \\ 2 \end{bmatrix}$

$A^T A = \begin{bmatrix} -1 & 2 & -1 \\ 2 & -3 & 3 \end{bmatrix}\begin{bmatrix} -1 & 2 \\ 2 & -3 \\ -1 & 3 \end{bmatrix} = \begin{bmatrix} 6 & -11 \\ -11 & 22 \end{bmatrix}$, $A^T\mathbf{b} = \begin{bmatrix} -1 & 2 & -1 \\ 2 & -3 & 3 \end{bmatrix}\begin{bmatrix} 4 \\ 1 \\ 2 \end{bmatrix} = \begin{bmatrix} -4 \\ 11 \end{bmatrix}$

   a. The normal equations: $\begin{bmatrix} 6 & -11 \\ -11 & 22 \end{bmatrix}\begin{bmatrix} x_1 \\ x_2 \end{bmatrix} = \begin{bmatrix} -4 \\ 11 \end{bmatrix}$

   b. Since $A^T A$ is only 2×2, $(A^T A)^{-1}$ is easy to compute, and

   $\hat{\mathbf{x}} = \begin{bmatrix} 6 & -11 \\ -11 & 22 \end{bmatrix}^{-1}\begin{bmatrix} -4 \\ 11 \end{bmatrix} = \frac{1}{11}\begin{bmatrix} 22 & 11 \\ 11 & 6 \end{bmatrix}\begin{bmatrix} -4 \\ 11 \end{bmatrix} = \frac{1}{11}\begin{bmatrix} 33 \\ 22 \end{bmatrix} = \begin{bmatrix} 3 \\ 2 \end{bmatrix}$

Copyright © 2021 Pearson Education, Inc.

Warning: It is important to distinguish between the normal equations $A^T A \hat{\mathbf{x}} = A^T \mathbf{b}$ and the formula $\hat{\mathbf{x}} = (A^T A)^{-1} A^T \mathbf{b}$. Both equations describe $\hat{\mathbf{x}}$ (implicitly or explicitly), but the formula for $\hat{\mathbf{x}}$ holds only when A has linearly independent columns. Note that the expression $(A^T A)^{-1} A^T$ cannot be simplified when A is not invertible.

7. $A = \begin{bmatrix} 1 & -2 \\ -1 & 2 \\ 0 & 3 \\ 2 & 5 \end{bmatrix}$, $\mathbf{b} = \begin{bmatrix} 3 \\ 1 \\ -4 \\ 2 \end{bmatrix}$, $A^T A = \begin{bmatrix} 1 & -1 & 0 & 2 \\ -2 & 2 & 3 & 5 \end{bmatrix} \begin{bmatrix} 1 & -2 \\ -1 & 2 \\ 0 & 3 \\ 2 & 5 \end{bmatrix} = \begin{bmatrix} 6 & 6 \\ 6 & 42 \end{bmatrix}$

$A^T \mathbf{b} = \begin{bmatrix} 1 & -1 & 0 & 2 \\ -2 & 2 & 3 & 5 \end{bmatrix} \begin{bmatrix} 3 \\ 1 \\ -4 \\ 2 \end{bmatrix} = \begin{bmatrix} 6 \\ -6 \end{bmatrix}$

The normal equations: $\begin{bmatrix} 6 & 6 \\ 6 & 42 \end{bmatrix} \begin{bmatrix} x_1 \\ x_2 \end{bmatrix} = \begin{bmatrix} 6 \\ -6 \end{bmatrix}$

The particular numbers in $A^T A$ suggest that the normal equations might be solved easily via row operations:

$\begin{bmatrix} 6 & 6 & 6 \\ 6 & 42 & -6 \end{bmatrix} \sim \begin{bmatrix} 6 & 6 & 6 \\ 0 & 36 & -12 \end{bmatrix} \sim \begin{bmatrix} 1 & 1 & 1 \\ 0 & 1 & -1/3 \end{bmatrix} \sim \begin{bmatrix} 1 & 0 & 4/3 \\ 0 & 1 & -1/3 \end{bmatrix}$

Thus $\hat{\mathbf{x}} = \begin{bmatrix} 4/3 \\ -1/3 \end{bmatrix}$. The least-squares error is $\| A\hat{\mathbf{x}} - \mathbf{b} \|$, so compute

$A\hat{\mathbf{x}} - \mathbf{b} = \begin{bmatrix} 1 & -2 \\ -1 & 2 \\ 0 & 3 \\ 2 & 5 \end{bmatrix} \begin{bmatrix} 4/3 \\ -1/3 \end{bmatrix} - \begin{bmatrix} 3 \\ 1 \\ -4 \\ 2 \end{bmatrix} = \begin{bmatrix} 2 \\ -2 \\ -1 \\ 1 \end{bmatrix} - \begin{bmatrix} 3 \\ 1 \\ -4 \\ 2 \end{bmatrix} = \begin{bmatrix} -1 \\ -3 \\ 3 \\ -1 \end{bmatrix}$

$\| A\hat{\mathbf{x}} - \mathbf{b} \|^2 = 1 + 9 + 9 + 1 = 20$, and $\| A\hat{\mathbf{x}} - \mathbf{b} \| = \sqrt{20} = 2\sqrt{5}$

Study Tip: A good way to check your work in Exercises 1–8 is to verify, that $A\hat{\mathbf{x}} - \mathbf{b}$ is orthogonal to each column of A.

Warning: The matrices in Exercises 9–12 are special—their columns are orthogonal. That is why these exercises are not difficult. See Example 4. In general, if the columns of A are not orthogonal, finding the orthogonal projection of $\mathbf{b}$ onto Col(A) takes more work. See Exercise 31.

Copyright © 2021 Pearson Education, Inc.

13. $A\mathbf{u} = \begin{bmatrix} 3 & 4 \\ -2 & 1 \\ 3 & 4 \end{bmatrix} \begin{bmatrix} 5 \\ -1 \end{bmatrix} = \begin{bmatrix} 11 \\ -11 \\ 11 \end{bmatrix},\ \mathbf{b} - A\mathbf{u} = \begin{bmatrix} 11 \\ -9 \\ 5 \end{bmatrix} - \begin{bmatrix} 11 \\ -11 \\ 11 \end{bmatrix} = \begin{bmatrix} 0 \\ 2 \\ -6 \end{bmatrix},\ \|\mathbf{b} - A\mathbf{u}\| = \sqrt{40}$

$A\mathbf{v} = \begin{bmatrix} 3 & 4 \\ -2 & 1 \\ 3 & 4 \end{bmatrix} \begin{bmatrix} 5 \\ -2 \end{bmatrix} = \begin{bmatrix} 7 \\ -12 \\ 7 \end{bmatrix},\ \mathbf{b} - A\mathbf{v} = \begin{bmatrix} 11 \\ -9 \\ 5 \end{bmatrix} - \begin{bmatrix} 7 \\ -12 \\ 7 \end{bmatrix} = \begin{bmatrix} 4 \\ 3 \\ -2 \end{bmatrix},\ \|\mathbf{b} - A\mathbf{u}\| = \sqrt{29}$

Obviously, $A\mathbf{u}$ is not the closest point of Col A to $\mathbf{b}$, because $A\mathbf{v}$ is closer. Hence $\mathbf{u}$ is *not* the least-squares solution of $A\mathbf{x} = \mathbf{b}$.

17. See the beginning of the section. The distance from $A\mathbf{x}$ to $\mathbf{b}$ is $\| A\mathbf{x} - \mathbf{b} \|$.

19. See the comments about equation (1).

21. See Theorem 13.

23. If $\hat{\mathbf{x}}$ is the least-squares solution, then $A\,\hat{\mathbf{x}}$ is the point in the column space of A closest to $\mathbf{b}$. See Figure 1 and the paragraph preceding it.

25. See the comments after Example 4.

29. **a.** If A has linearly independent columns, then the equation $A\mathbf{x} = \mathbf{0}$ has only the trivial solution. By Exercise 27, $A^T A \mathbf{x} = \mathbf{0}$ also has only the trivial solution. Since $A^T A$ is *square*, it must be invertible, by the Invertible Matrix Theorem.

b. Since the n linearly independent columns of A belong to $\mathbb{R}^m$, m could not be less than n.

c. The n linearly independent columns of A form a basis for Col A, so the rank of A is n.

MATLAB   The Backslash Command

When A has linearly dependent columns, you can write the general description of all least–squares solutions on paper after you row reduce the augmented matrix for the normal equations: **ref([A'*A A'*b]).**   When A has linearly independent columns, enter the MATLAB "backslash" command

  x = (A'*A)\(A'*b)

to solve the normal equations. You can also enter **inv(A'*A)*(A'*b)** ; or use **ref** , as above. For Exercises 15 and 16, see the Numerical Note in Section 6.5 in the text and use the backslash command **R\(Q'*b)** to solve $R\mathbf{x} = Q^T\mathbf{b}$.

  When is A is not square but has linearly independent columns, this procedure of first forming Q and R and then solving $R\mathbf{x} = Q^T\mathbf{b}$ is exactly what MATLAB does (internally) when the MATLAB command **A\b** is used to solve $A\mathbf{x} = \mathbf{b}$. You should use the normal equations or QR for computations here, instead of just using the backslash. This will give you a solid conceptual background for applying least-squares techniques later in your career. For more about the backslash, see the MATLAB box for Section 2.5.

Copyright © 2021 Pearson Education, Inc.

6.6 - Machine Learning and Linear Models

This section of the text will be a valuable reference for any person who is interested in data analytics or data that requires statistical analysis. Many graduate fields require such work, often in connection with doctoral research. Even most undergraduates will take a course where least-squares lines are used.

KEY IDEA

Linear algebra unifies the study of many problems in statistics and data analysis. All the examples in this section, from ordinary linear regression (using a least-squares line) to multiple regression, concern just one idea: find a least-squares solution of $X\boldsymbol{\beta} = \mathbf{y}$. Only the design matrix X varies. The exercises help you practice choosing X. The least-squares solution $\hat{\boldsymbol{\beta}}$ always satisfies the normal equations $X^T X \hat{\boldsymbol{\beta}} = X^T \mathbf{y}$.

STUDY NOTES

Don't confuse the least-squares line in Fig.1 with the lines and planes in Section 6.5 onto which we projected various vectors $\mathbf{b}$. The line is nothing more than a special case of the curves in Figures 4–5. In each case, the "linearity" of the model lies not in the curve, but rather in the fact that the unknown parameters (or *weights*) β_0, β_1, ... occur linearly in the formula for the curve, just as the variables x_1, x_2, ... occur in an ordinary linear equation.

Any 4×4 submatrix of the design matrix in Example 4 is called a Vandermonde matrix. One can show that if at least four of the values x_1, ..., x_n are distinct, then the least-squares solution $\hat{\boldsymbol{\beta}}$ will be unique, by Theorem 14 in Section 6.5.

FURTHER READING

An important generalization of the discussion here is to *multivariate* analysis, which involves several $\mathbf{y}$ vectors rather than just one. In this case the basic equation is $XB = Y$, where each column of Y is a data set for one dependent variable, and each column of B is a set of parameters to be determined. That is, $X[\boldsymbol{\beta}_1 \ ) \ \boldsymbol{\beta}_p] = [\mathbf{y}_1 \ ) \ \mathbf{y}_p]$. For more information, see the classic text by T. W. Anderson, *An Introduction to Multivariate Statistical Analysis*, John Wiley & Sons, New York, 1984. The preface of the text says, "A knowledge of matrix algebra is a prerequisite [for understanding the text]." Most modern multivariate statistics texts rely heavily on matrix notation and matrix algebra, as do many techniques in data analysis and machine learning.

Copyright © 2021 Pearson Education, Inc.

SOLUTIONS TO EXERCISES

1. Place the x-coordinates of the data in the second column of X and the y-coordinates in the vector **y**. So $X = \begin{bmatrix} 1 & 0 \\ 1 & 1 \\ 1 & 2 \\ 1 & 3 \end{bmatrix}$ and $\mathbf{y} = \begin{bmatrix} 1 \\ 1 \\ 2 \\ 2 \end{bmatrix}$. Compute

$$\underbrace{\begin{bmatrix} 1 & 1 & 1 & 1 \\ 0 & 1 & 2 & 3 \end{bmatrix}}_{X^T} \underbrace{\begin{bmatrix} 1 & 0 \\ 1 & 1 \\ 1 & 2 \\ 1 & 3 \end{bmatrix}}_{X} = \begin{bmatrix} 4 & 6 \\ 6 & 14 \end{bmatrix}, \quad \underbrace{\begin{bmatrix} 1 & 1 & 1 & 1 \\ 0 & 1 & 2 & 3 \end{bmatrix}}_{X^T} \underbrace{\begin{bmatrix} 1 \\ 1 \\ 2 \\ 2 \end{bmatrix}}_{\mathbf{y}} = \begin{bmatrix} 6 \\ 11 \end{bmatrix}$$

The matrix normal equation and its solution are:

$$\begin{bmatrix} 4 & 6 \\ 6 & 14 \end{bmatrix} \begin{bmatrix} \beta_0 \\ \beta_1 \end{bmatrix} = \begin{bmatrix} 6 \\ 11 \end{bmatrix}$$

$$\begin{bmatrix} \beta_0 \\ \beta_1 \end{bmatrix} = \begin{bmatrix} 4 & 6 \\ 6 & 14 \end{bmatrix}^{-1} \begin{bmatrix} 6 \\ 11 \end{bmatrix} = \frac{1}{20} \begin{bmatrix} 14 & -6 \\ -6 & 4 \end{bmatrix} \begin{bmatrix} 6 \\ 11 \end{bmatrix} = \frac{1}{20} \begin{bmatrix} 18 \\ 8 \end{bmatrix} = \begin{bmatrix} .9 \\ .4 \end{bmatrix}$$

The least-squares line, $y = \beta_0 + \beta_1 x$, is $y = .9 + .4x$.

7. Using the line "learned" in Exercise 1, $y = .9 + .4x$, plug in for x: $y = .9 + .4(3) = 2.1$. The machine returns the point $(3, 2.1)$. The original data for Exercise 1 includes the point $(3, 2)$. Since least squares line is a "best fit", rather than an "exact fit" line, we should not be surprised that the estimated point is slightly different from the actual point.

13. a. $\mathbf{y} = X\boldsymbol{\beta} + \boldsymbol{\varepsilon}$, where $\mathbf{y} = \begin{bmatrix} 1.8 \\ 2.7 \\ 3.4 \\ 3.8 \\ 3.9 \end{bmatrix}$, $X = \begin{bmatrix} 1 & 1 \\ 2 & 4 \\ 3 & 9 \\ 4 & 16 \\ 5 & 25 \end{bmatrix}$, $\boldsymbol{\beta} = \begin{bmatrix} \beta_1 \\ \beta_2 \end{bmatrix}$, $\boldsymbol{\varepsilon} = \begin{bmatrix} \varepsilon_1 \\ \varepsilon_2 \\ \varepsilon_3 \\ \varepsilon_4 \\ \varepsilon_5 \end{bmatrix}$.

b. In this problem, $X^T X$ is invertible. You can use your matrix program to solve the normal equations without explicitly computing the entries in $X^T X$. For details, see the MATLAB box below or the corresponding box in the appendix for your matrix program. In any case,

$$\boldsymbol{\beta} = \begin{bmatrix} \beta_1 \\ \beta_2 \end{bmatrix} = \begin{bmatrix} 1.76 \\ -.20 \end{bmatrix} \quad \text{(to two decimal places)}$$

The desired least-squares equation is $y = 1.76x - .20x^2$.

Copyright © 2021 Pearson Education, Inc.

c. The machine would return the value $y = 1.76(6) - .20(6)^2 = 3.36$.

19. Let **1** be the vector in $\mathbb{R}^3$ with 1 in each entry, let $\mathbf{t} = (0, ..., 12)$, and for $k = 2$ and 3, let $\mathbf{t}.\char`^k$ denote the vector whose entries are the kth powers of the entries in **t**. (See the MATLAB box below.) Then the design matrix is:

$$X = [\mathbf{1} \quad \mathbf{t} \quad \mathbf{t}.\char`^2 \quad \mathbf{t}.\char`^3]$$

The observation vector **y** lists the measured positions of the plane.

a. Numerical solution of the normal equations yields

$$\boldsymbol{\beta} = (-.8558, 4.7025, 5.5554, -.0274)$$

The least-squares polynomial (position of the plane at time t) is

$$y = -.8558 + 4.7025t + 5.5554t^2 - .0274t^3$$

b. The velocity is the derivative of the position function:

$$v(t) = 4.7025 + 11.1108t - .0822t^2$$

When $t = 4.5$ seconds, the machine would estimate $v(4.5) = 53.0$ ft/sec.

21. From equation (1),

$$X^T X = \begin{bmatrix} 1 & \cdots & 1 \\ x_1 & \cdots & x_n \end{bmatrix} \begin{bmatrix} 1 & x_1 \\ \vdots & \vdots \\ 1 & x_n \end{bmatrix} = \begin{bmatrix} n & \Sigma x \\ \Sigma x & \Sigma x^2 \end{bmatrix}$$

$$X^T \mathbf{y} = \begin{bmatrix} 1 & \cdots & 1 \\ x_1 & \cdots & x_n \end{bmatrix} \begin{bmatrix} y_1 \\ \vdots \\ y_n \end{bmatrix} = \begin{bmatrix} \Sigma y \\ \Sigma xy \end{bmatrix}$$

The equations (7) in the text follow immediately from the usual matrix normal equation $X^T X \boldsymbol{\beta} = X^T \mathbf{y}$.

22. *Note:* The formulas you should derive are

$$\boldsymbol{\beta}_0 = \frac{(\Sigma x^2)(\Sigma y) - (\Sigma x)(\Sigma xy)}{n \Sigma x^2 - (\Sigma x)^2}, \quad \boldsymbol{\beta}_1 = \frac{n \Sigma xy - (\Sigma x)(\Sigma y)}{n \Sigma x^2 - (\Sigma x)^2}$$

Some statistics texts present other equivalent formulas for $\boldsymbol{\beta}_0$ and $\boldsymbol{\beta}_1$.

25. The equation to be proved is $\| \mathbf{y} \|^2 = \| X\boldsymbol{\beta} \|^2 + \| \mathbf{y} - X\boldsymbol{\beta} \|^2$. This follows from the Pythagorean Theorem (in Section 6.1) and the figure below.

Copyright © 2021 Pearson Education, Inc.

Appendix: The Geometry of a Linear Model

The column space of the design matrix X is sometimes called the **design subspace**. If $\hat{\boldsymbol{\beta}}$ is the least-squares solution of $\mathbf{y} = X\boldsymbol{\beta}$, then the residual vector $\boldsymbol{\varepsilon} = \mathbf{y} - X\hat{\boldsymbol{\beta}}$, is orthogonal to the design subspace, and the equation $\mathbf{y} = X\hat{\boldsymbol{\beta}} + \boldsymbol{\varepsilon}$ is an orthogonal decomposition of the observed $\mathbf{y}$ into the sum of the least-squares predicted $\hat{\mathbf{y}}$ and the residual vector $\boldsymbol{\varepsilon}$.

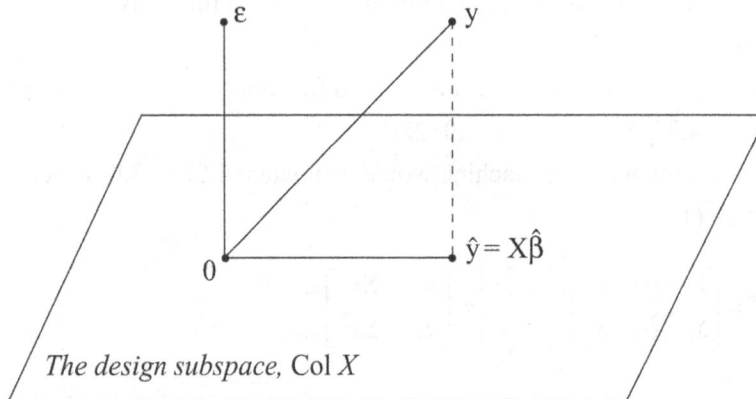

The design subspace, Col X

MATLAB   Least-squares Solutions and Functions of Vectors

Once you create the design matrix X and the observation vector $\mathbf{y}$, your computations for least-squares solutions here are the same as those described in the MATLAB box for Section 6.5. Here, A and $\mathbf{b}$ are replaced by X and $\mathbf{y}$, respectively. The MATLAB command

   ref([X'*X X'*y])

leads to the general description of all least-squares solutions. When X has linearly independent columns, the command **(X'*X)\(X'*y)** creates the least-squares solution. In subsequent courses, you may choose simply to use **X\y**, which also produces a least- squares solution, except when X is square and singular (or nearly singular).

   To construct the design matrix for an exercise in this section, you may need MATLAB's ability to compute functions of vectors. If $\mathbf{x}$ is a vector and k is a positive integer, then **x.^k** is a vector the same size as $\mathbf{x}$ whose entries are the kth powers of the entries in $\mathbf{x}$. The function **cos(x).^k** was mentioned in the MATLAB box for Section 4.3.   The exponential function, **exp(x)**, and natural logarithm function, **log(x)**, also act on each entry in $\mathbf{x}$. The entries in the vector **exp(-.02*x)**, for example, are computed by applying the function $e^{-.02x}$ to the corresponding entries in $\mathbf{x}$.

Copyright © 2021 Pearson Education, Inc.

6.7 - Inner Product Spaces

Three examples of inner product spaces are described here, in Examples 1, 2, and 7. Corresponding applications appear in the next section. Material in Sections 6.7 and 6.8 will be useful for many careers, particularly science, engineering, and mathematics. If your course does not cover this now, the text and *Study Guide* can help you learn it later on your own.

KEY IDEAS

The concepts of length and orthogonality in R^n have analogues in a number of other vector spaces. The definition of an inner product identifies the basic properties needed for a theory that parallels the familiar theory for R^n. Two useful facts, the Cauchy-Schwarz inequality and the triangle inequality, were not developed earlier, but they are important for applications both in R^n and in other inner product spaces. Every mathematics major will need to know these facts in other undergraduate courses.

The inner product in Example 1 is used in Section 6.8 to describe weighted least-squares problems. The inner product in Examples 2–6 provides a more sophisticated approach to the least-squares curve fitting discussed in Section 6.6. See the "trend analysis" in Section 6.8.

Be sure to read the paragraph preceding Example 6. The idea of "best approximation" to a function is of fundamental importance in mathematics. The most common applications of best approximation (such as Fourier series, introduced in Section 6.8) involve the inner product in Example 7.

SOLUTIONS TO EXERCISES

1. The inner product is $<\mathbf{x}, \mathbf{y}> = 4x_1y_1 + 5x_2y_2$. Let $\mathbf{x} = (1, 1)$, $\mathbf{y} = (5, -1)$.

    a. $\|\mathbf{x}\|^2 = 4 \cdot 1 \cdot 1 + 5 \cdot 1 \cdot 1 = 9$, $\|\mathbf{x}\| = 3$

$$\|\mathbf{y}\|^2 = 4 \cdot 5 \cdot 5 + 5(-1)(-1) = 105, \ \|\mathbf{y}\| = \sqrt{105}$$

$$|<\mathbf{x}, \mathbf{y}>|^2 = |4 \cdot 1 \cdot 5 + 5 \cdot 1(-1)|^2 = |15|^2 = 225$$

    b. A vector $\mathbf{z} = (z_1, z_2)$ is orthogonal to $\mathbf{y}$ if and only if $<\mathbf{z}, \mathbf{y}> = 0$, that is,

$$4 \cdot z_1 \cdot 5 + 5 \cdot z_2 \cdot (-1) = 0, \quad 20z_1 - 5z_2 = 0, \quad \text{and} \quad z_2 = 4z_1$$

Thus (z_1, z_2) is orthogonal to $\mathbf{y}$ if and only if $z_2 = 4z_1$.

7. Given $p(t) = 4 + t$ and $q(t) = 5 - 4t^2$. The orthogonal projection $\hat{q}$ of q onto the subspace spanned by p is $\dfrac{<q, p>}{<p, p>} \, p$. The notation of Example 5 organizes the calculations nicely:

Copyright © 2021 Pearson Education, Inc.

Polynomial p q

Vector of values: $\begin{bmatrix} 3 \\ 4 \\ 5 \end{bmatrix}$, $\begin{bmatrix} 1 \\ 5 \\ 1 \end{bmatrix}$ ← value at -1
← value at 0
← value at 1

The inner product $<q, p>$ equals the (standard) inner product of the two corresponding vectors in $\mathbb{R}^3$: $<q, p> = 1(3)+5(4)+1(5) = 28$. Similarly, $<p, p> = 3^2 + 4^2 + 5^2 = 50$. Thus

$$\hat{q}(t) = \frac{28}{50}(4+t) = \frac{56}{25} + \frac{14}{25}t$$

13. Suppose A is invertible and $<\mathbf{u}, \mathbf{v}> = (A\mathbf{u})^T(A\mathbf{v})$, for $\mathbf{u}, \mathbf{v}$ in $\mathbb{R}^n$. Note that $<\mathbf{u}, \mathbf{v}>$ is in $\mathbb{R}$, and check each axiom in the definition of an inner product space:

  i. $\langle \mathbf{u}, \mathbf{v} \rangle = (A\mathbf{u}) \bullet (A\mathbf{v}) = (A\mathbf{v}) \bullet (A\mathbf{u})$ Property of dot product
      $= \langle \mathbf{v}, \mathbf{u} \rangle$

  ii. $\langle \mathbf{u} + \mathbf{v}, \mathbf{w} \rangle = [A(\mathbf{u} + \mathbf{v})] \bullet (A\mathbf{w}) = [A\mathbf{u} + A\mathbf{v}] \bullet (A\mathbf{w})$ Matrix multiplication
      $= (A\mathbf{u}) \bullet (A\mathbf{w}) + (A\mathbf{v}) \bullet (A\mathbf{w})$ Property of dot product
      $= \langle \mathbf{u}, \mathbf{w} \rangle + \langle \mathbf{v}, \mathbf{w} \rangle$

  iii. $\langle c\mathbf{u} \, \mathbf{v} \rangle = [A(c\mathbf{u})] \bullet (A\mathbf{v}) = [c(A\mathbf{u})] \bullet (A\mathbf{v})$ Matrix multiplication
      $= c(A\mathbf{u}) \bullet (A\mathbf{v})$ Property of dot product
      $= c\langle \mathbf{u}, \mathbf{v} \rangle$

  iv. $<\mathbf{u}, \mathbf{u}> = (A\mathbf{u}) \bullet (A\mathbf{u}) = \|A\mathbf{u}\|^2 \geq 0$, and this quantity is zero if and only if the vector $A\mathbf{u}$ is **0**. But $A\mathbf{u} = \mathbf{0}$ if and only if $\mathbf{u} = \mathbf{0}$, because A is invertible.

Another method for verifying the axioms is to use properties of the transpose operation. The calculations are similar. However, for (i), you need to use the fact that the transpose of a scalar (which is a 1×1 matrix) is the scalar itself: $<\mathbf{u}, \mathbf{v}> = <\mathbf{u}, \mathbf{v}>^T = [(A\mathbf{u})^T(A\mathbf{v})]^T = (A\mathbf{v})^T(A\mathbf{u})^{TT} = (A\mathbf{v})^T(A\mathbf{u}) = <\mathbf{v}, \mathbf{u}>$.

19. See the definition of inner product.

21. Use two properties of the definition of an inner product space.

23. See the definition of inner product.

31. In the space $C[-1, 1]$ with the integral inner product, the polynomials t and 1 are orthogonal, because

$$\langle t, 1 \rangle = \int_{-1}^{1} t \cdot 1 \, dt = \frac{1}{2}t^2 \Big|_{-1}^{1} = \frac{1}{2}(1)^2 - \frac{1}{2}(-1)^2 = 0$$

So 1 and t can be in an orthogonal basis for Span$\{1, t, t^2\}$. Next, compute proj$_W$ t^2, the orthogonal projection of the vector t^2 onto the subspace W spanned by 1 and t.

$$\langle t^2, 1 \rangle = \int_{-1}^{1} t^2 \cdot 1 \, dt = \frac{1}{3}t^3 \Big|_{-1}^{1} = \frac{1}{3}(1)^3 - \frac{1}{3}(-1)^3 = \frac{2}{3}$$

Copyright © 2021 Pearson Education, Inc.

$$\langle 1, 1 \rangle = \int_{-1}^{1} 1 \cdot 1 \, dt = t \Big|_{-1}^{1} = 1 - (-1) = 2$$

$$\langle t^2, t \rangle = \int_{-1}^{1} t^2 \cdot t \, dt = \frac{1}{4} t^4 \Big|_{-1}^{1} = \frac{1}{4}(1)^4 - \frac{1}{4}(-1)^4 = 0$$

There is no need to compute $\langle t, t \rangle$, because t^2 is orthogonal to t. Thus

$$\text{proj}_W t^2 = \frac{\langle t^2, 1 \rangle}{\langle 1, 1 \rangle} 1 + \frac{\langle t^2, t \rangle}{\langle t, t \rangle} t = \frac{2/3}{2} 1 + 0 = \frac{1}{3}$$

A polynomial orthogonal to W is $t^2 - \text{proj}_W t^2 = t^2 - \frac{1}{3}$. Another choice is this polynomial scaled by 3, namely, $3t^2 - 1$. Thus, the polynomials, 1, t, and $3t^2 - 1$ form an orthogonal basis for Span$\{1, t, t^2\}$.

Can you find the next Legendre polynomial, a cubic polynomial that is orthogonal to each of the first three Legendre polynomials?

6.8 - Applications of Inner Product Spaces

Of the three applications in this section, the discussion of Fourier series is by far the most important. Such series have great practical value, particularly in mathematics, engineering and the physical sciences. Calculations with Fourier series are simple because sine and cosine functions are orthogonal. This fact is often overlooked in undergraduate courses that do not assume a linear algebra background.

KEY IDEAS

The text gives the normal equations for the weighted least-squares solution of $A\mathbf{x} = \mathbf{y}$. When applied to a least-squares line problem, the most common situation, the normal equations are usually written as $(WX)^T WX \boldsymbol{\beta} = (WX)^T \mathbf{y}$ where W is the (diagonal) weighting matrix, X is the design matrix, $\boldsymbol{\beta}$ is the least-squares parameter vector, and $\mathbf{y}$ is the observation vector.

Trend analysis is really a least-squares regression problem of the type described in Section 6.6, with data points $(x_1, y_1), \ldots, (x_n, y_n)$ fitted by a curve of the form

$$y = \beta_0 f_0(x) + \beta_1 f_1(x) + \cdots + \beta_k f_k(x)$$

where the functions $f_0, \ldots, f_k$ are polynomials that are orthogonal with respect to an inner product on P_{n-1} defined by $\langle p, q \rangle = p(x_1)q(x_1) + \cdots + p(x_n)q(x_n)$. Usually, $x_1, \ldots, x_n$ are arranged to be evenly spaced and sum to zero, and the functions $f_1, \ldots, f_k$ are of degree 3 or 4 or less.

In $C[0, 2\pi]$ with the integral inner product, the set

$$\{1, \cos t, \cos 2t, \ldots, \cos nt, \sin t, \sin 2t, \ldots, \sin nt\} \qquad (*)$$

is orthogonal. The nth order Fourier approximation to some f in $C[0, 2\pi]$ is simply the orthogonal projection of f onto the subspace W of trigonometric polynomials spanned by the functions in (*). The Fourier coefficients of f are the weights in the usual formula for the orthogonal projection of f onto W.

Copyright © 2021 Pearson Education, Inc.

If an application involves an interval $[0, T]$ instead of $[0, 2\pi]$, then the inner product requires an integral over $[0, T]$, and the appropriate orthogonal set is obtained by replacing t in each function in (*) with $2\pi t/T$.

SOLUTIONS TO EXERCISES

1. For the data $(-2, 0)$, $(-1, 0)$, $(0, 2)$, $(1, 4)$, $(2, 4)$, construct

$$X = \begin{bmatrix} 1 & -2 \\ 1 & -1 \\ 1 & 0 \\ 1 & 1 \\ 1 & 2 \end{bmatrix} \text{Design matrix,} \quad \boldsymbol{\beta} = \begin{bmatrix} \beta_0 \\ \beta_1 \end{bmatrix} \text{Parameter vector,} \quad \mathbf{y} = \begin{bmatrix} 0 \\ 0 \\ 2 \\ 4 \\ 4 \end{bmatrix} \text{Observation vector}$$

Since the first and last data points are about half as reliable as the other points, a suitable weighting matrix is

$$W = \begin{bmatrix} 1 & 0 & 0 & 0 & 0 \\ 0 & 2 & 0 & 0 & 0 \\ 0 & 0 & 2 & 0 & 0 \\ 0 & 0 & 0 & 2 & 0 \\ 0 & 0 & 0 & 0 & 1 \end{bmatrix}, \quad \text{so } WX = \begin{bmatrix} 1 & -2 \\ 2 & -2 \\ 2 & 0 \\ 2 & 2 \\ 1 & 2 \end{bmatrix}, \quad \text{and } W\mathbf{y} = \begin{bmatrix} 0 \\ 0 \\ 4 \\ 8 \\ 4 \end{bmatrix}$$

The remaining calculations are the same as in ordinary least-squares, except that the *weighted* design matrix WX and the *weighted* observation vector $W\mathbf{y}$ appear in place of X and $\mathbf{y}$, respectively.

$$(WX)^T (WX) = \begin{bmatrix} 1 & 2 & 2 & 2 & 1 \\ -2 & -2 & 0 & 2 & 2 \end{bmatrix} \begin{bmatrix} 1 & -2 \\ 2 & -2 \\ 2 & 0 \\ 2 & 2 \\ 1 & 2 \end{bmatrix} = \begin{bmatrix} 14 & 0 \\ 0 & 16 \end{bmatrix}$$

$$(WX)^T (W\mathbf{y}) = \begin{bmatrix} 1 & 2 & 2 & 2 & 1 \\ -2 & -2 & 0 & 2 & 2 \end{bmatrix} \begin{bmatrix} 0 \\ 0 \\ 4 \\ 8 \\ 4 \end{bmatrix} = \begin{bmatrix} 28 \\ 24 \end{bmatrix}$$

The normal equations and solution are

$$\begin{bmatrix} 14 & 0 \\ 0 & 16 \end{bmatrix} \begin{bmatrix} \beta_0 \\ \beta_1 \end{bmatrix} = \begin{bmatrix} 28 \\ 24 \end{bmatrix}, \quad \begin{bmatrix} \beta_0 \\ \beta_1 \end{bmatrix} = \begin{bmatrix} 1/14 & 0 \\ 0 & 1/16 \end{bmatrix} \begin{bmatrix} 28 \\ 24 \end{bmatrix} = \begin{bmatrix} 2 \\ 3/2 \end{bmatrix}$$

Copyright © 2021 Pearson Education, Inc.

The equation of the least-squares line is $y = 2 + (3/2)x$.

7. $\|\cos kt\|^2 = \int_0^{2\pi} \cos kt \cdot \cos kt \, dt = \int_0^{2\pi} \frac{1 + \cos 2kt}{2} \, dt$

$$= \left(\frac{1}{2}t + \frac{\sin 2kt}{4k} \right) \Big|_0^{2\pi} = (\frac{1}{2} \cdot 2\pi + 0) - 0 = \pi \quad \text{(if } k \neq 0)$$

$\|\sin kt\|^2 = \int_0^{2\pi} \sin kt \cdot \sin kt \, dt = \int_0^{2\pi} \frac{1 - \cos 2kt}{2} \, dt$

$$= \left(\frac{1}{2}t + \frac{\sin 2kt}{4k} \right) \Big|_0^{2\pi} = \left(\frac{1}{2} \cdot 2\pi + 0 \right) - 0 = \pi \quad \text{(if } k \neq 0)$$

9. $f(t) = 2\pi - t$. The definite integrals of $t \cos kt$ and $t \sin kt$, shown below, were computed in Example 4. The Fourier coefficients of f are:

$$\frac{a_0}{2} = \frac{1}{2} \cdot \frac{1}{\pi} \int_0^{2\pi} (2\pi - t) \, dt = \frac{1}{2\pi}(-\frac{1}{2})(2\pi - t)^2 \Big|_0^{2\pi} = 0 + \frac{1}{4\pi}(2\pi)^2 = \pi$$

and for $k > 0$,

$$a_k = \frac{1}{\pi} \int_0^{2\pi} (2\pi - t) \cos kt \, dt = \frac{1}{\pi} \int_0^{2\pi} 2\pi \cos kt \, dt - \frac{1}{\pi} \int_0^{2\pi} t \cos kt \, dt$$

$$= 0 - 0 = 0$$

$$b_k = \frac{1}{\pi} \int_0^{2\pi} (2\pi - t) \sin kt \, dt = \frac{1}{\pi} \int_0^{2\pi} 2\pi \sin kt \, dt - \frac{1}{\pi} \int_0^{2\pi} t \sin kt \, dt$$

$$= 0 - \left(-\frac{2}{k} \right) = \frac{2}{k}$$

The third-order Fourier approximation to f is

$$\pi + 2\sin t + \sin 2t + \frac{2}{3}\sin 3t$$

13. Let f and g be in $C[0, 2\pi]$ and let m be a nonnegative integer. Then the linearity of the inner product shows that $\langle (f + g), \cos mt \rangle = \langle f, \cos mt \rangle + \langle g, \cos mt \rangle$ and $\langle (f + g), \sin mt \rangle = \langle f, \sin mt \rangle + \langle g, \sin mt \rangle$.

Dividing these identities respectively by $\langle \cos mt, \cos mt \rangle$ and $\langle \sin mt, \sin mt \rangle$ shows that the Fourier coefficients a_m and b_m for $f + g$ are the sums of the corresponding Fourier coefficients of f and of g.

Copyright © 2021 Pearson Education, Inc.

The argument for Exercise 13 is a special case of a general principle. In any inner product space, the mapping $\mathbf{y} \mapsto \dfrac{\langle \mathbf{y}, \mathbf{u} \rangle}{\langle \mathbf{u}, \mathbf{u} \rangle} \mathbf{u}$ is linear, for any nonzero $\mathbf{u}$. To verify this, take any $\mathbf{x}$ and $\mathbf{y}$ in the space and any scalar c. Then

$$\frac{\langle \mathbf{x} + \mathbf{y}, \mathbf{u} \rangle}{\langle \mathbf{u}, \mathbf{u} \rangle} \mathbf{u} = \frac{\langle \mathbf{x}, \mathbf{u} \rangle}{\langle \mathbf{u}, \mathbf{u} \rangle} \mathbf{u} + \frac{\langle \mathbf{y}, \mathbf{u} \rangle}{\langle \mathbf{u}, \mathbf{u} \rangle} \mathbf{u} \text{, and } \frac{\langle c\mathbf{x}, \mathbf{u} \rangle}{\langle \mathbf{u}, \mathbf{u} \rangle} \mathbf{u} = \frac{c\langle \mathbf{x}, \mathbf{u} \rangle}{\langle \mathbf{u}, \mathbf{u} \rangle} \mathbf{u} = c\frac{\langle \mathbf{x}, \mathbf{u} \rangle}{\langle \mathbf{u}, \mathbf{u} \rangle} \mathbf{u}$$

Similarly, if $\mathbf{u}_1, \ldots, \mathbf{u}_p$ are any nonzero vectors, then the mapping

$$\mathbf{y} \mapsto \frac{\langle \mathbf{y}, \mathbf{u}_1 \rangle}{\langle \mathbf{u}_1 \ \mathbf{u}_1 \rangle} \mathbf{u}_1 + \cdots + \frac{\langle \mathbf{y}, \mathbf{u}_p \rangle}{\langle \mathbf{u}_p \ \mathbf{u}_p \rangle} \mathbf{u}_p$$

is a linear transformation. Thus, if $\{\mathbf{u}_1, \ldots, \mathbf{u}_p\}$ is an orthogonal basis for a subspace W, then the mapping $\mathbf{y} \mapsto \text{proj}_W \mathbf{y}$ is a linear transformation.

In particular, if W is the vector space of trigonometric polynomials of order at most n, and if f and g are in $C[0, 2\pi]$, then

$$\text{proj}_W (f + g) = \text{proj}_W f + \text{proj}_W g$$

That is, the nth order Fourier approximation to $f + g$ is the sum of the nth order Fourier approximations to f and to g. Can you use the linearity of the mapping $f \mapsto \text{proj}_W f$ and the final result of Example 4, to produce (with practically no work) the answer to Exercise 9? [*Hint:* The nth order Fourier approximation to a constant function is the function itself.]

MATLAB   Graphing Functions

After you find f_4 and f_5 by hand computations, you can use **plot** to graph them. For instance, to plot $f(t) = \sin t + \sin 3t$, you can write

```
t = linspace(0,2*pi);
y = sin(t) + sin(3*t);
plot(t,y)
```

See the MATLAB box for Section 4.1 for more details.

Chapter 6 - Supplementary Exercises

In this chapter, Exercises 1-19 consist of true/false questions, whose level of difficulty varies. Some are similar to the ones that appear in many sections of the text, in which a word or phrase is sometimes missing or slightly misstated. Some follow fairly easily from a theorem; others may need careful reasoning. A few may require an argument that uses several ideas. In each case, think

Copyright © 2021 Pearson Education, Inc.

carefully about the statement and attempt to write a solution. The text provides the true/false answer, but you must supply the justification or counterexample.

25. Let **u** be a unit vector, and let $Q = I - 2\mathbf{u}\mathbf{u}^T$. Since $(\mathbf{u}\mathbf{u}^T)^T = \mathbf{u}^{TT}\mathbf{u}^T = \mathbf{u}\mathbf{u}^T$,

$$Q^T = (I - 2\mathbf{u}\mathbf{u}^T)^T = I - 2(\mathbf{u}\mathbf{u}^T)^T = I - 2\mathbf{u}\mathbf{u}^T = Q$$

Then

$$QQ^T = Q^2 = (I - 2\mathbf{u}\mathbf{u}^T)^2 = I - 2\mathbf{u}\mathbf{u}^T - 2\mathbf{u}\mathbf{u}^T + 4(\mathbf{u}\mathbf{u}^T)(\mathbf{u}\mathbf{u}^T)$$

Since **u** is a unit vector, $\mathbf{u}^T\mathbf{u} = \mathbf{u} \cdot \mathbf{u} = 1$, so $(\mathbf{u}\mathbf{u}^T)(\mathbf{u}\mathbf{u}^T) = \mathbf{u}(\mathbf{u}^T)(\mathbf{u})\mathbf{u}^T = \mathbf{u}\mathbf{u}^T$, and

$$QQ^T = I - 2\mathbf{u}\mathbf{u}^T - 2\mathbf{u}\mathbf{u}^T + 4\mathbf{u}\mathbf{u}^T = I$$

Thus Q is an orthogonal matrix.

31. **a.** The row-column calculation of $A\mathbf{u}$ shows that each row of A is orthogonal to every **u** in Nul A. So each row of A is in $(\text{Nul } A)^\perp$. Since $(\text{Nul } A)^\perp$ is a subspace, it must contain all linear combinations of the rows of A; hence $(\text{Nul } A)^\perp$ contains Row A.

b. If rank $A = r$, then dim Nul $A = n - r$ by the Rank Theorem. By Exercsie 24(c) in Section 6.3, dim Nul $A + \dim(\text{Nul } A)^\perp = n$, so $\dim(\text{Nul } A)^\perp$ must be r. But Row A is an r-dimensional subspace of $(\text{Nul } A)^\perp$ by the Rank Theorem and part (a). Therefore, Row $A = (\text{Nul } A)^\perp$.

c. Replace A by A^T in part (b) and conclude that Row $A^T = (\text{Nul } A^T)^\perp$. Since Row $A^T = \text{Col } A$, Col $A = (\text{Nul } A^T)^\perp$.

37. Compute that $\|\Delta\mathbf{x}\| / \|\mathbf{x}\| = 7.178 \times 10^{-8}$ and

cond$(A) \times (\|\Delta\mathbf{b}\| / \|\mathbf{b}\|) = 23683 \times (2.832 \times 10^{-4}) = 6.707$. Observe that the relative change in **x** is *much* smaller than the relative change in **b**. In fact the theoretical bound on the relative change in **x** is 6.707 (to four significant figures). This exercise shows that even when a condition number is large, the relative error in the solution need not be as large as you suspect.

Chapter 6 - Glossary Checklist

Check your knowledge by attempting to write definitions of the terms below. Then compare your work with the definitions given in the text's Glossary. Ask your instructor which definitions, if any, might appear on a test.

Copyright © 2021 Pearson Education, Inc.

angle (between nonzero vectors **u** and **v** in $\mathbb{R}^2$ or $\mathbb{R}^3$):   The angle φ between the . . . . Related to the scalar product by: $\mathbf{u} \cdot \mathbf{v} = $ . . . .

best approximation: The closest point . . . .

Cauchy-Schwarz inequality: . . . for all **u**, **v**.

component of y orthogonal to u (for $\mathbf{u} \neq \mathbf{0}$):   The vector . . . .

design matrix: The matrix X in the linear model . . . , where the columns of X are determined in some way by . . . .

distance between u and v: . . . , denoted by dist(**u**, **v**).

Fourier approximation (of order n):   The closest point in . . . to . . . .

Fourier coefficients: The weights used to make . . . .

Fourier series: An infinite series that . . . in $C[0, 2\pi]$, with the inner product given by a definite integral.

fundamental subspaces (of A):   The . . . space and . . . space of A, and the . . . space and . . . space of A^T, with Col A^T commonly called the . . . .

general least-squares problem: Given an $m \times n$ matrix A and a vector **b** in $\mathbb{R}^m$, find . . . such that . . . for all . . . .

Gram-Schmidt process: An algorithm for producing . . . .

inner product: The scalar . . . , usually written as $\mathbf{u} \cdot \mathbf{v}$, where **u** and **v** are vectors in R^n viewed as . . . . Also called the . . . of **u** and **v**. In general, a function on a vector space that assigns to each pair of vectors **u** and **v** a number . . . , subject to certain axioms.

inner product space: A vector space on which is defined . . . .

least-squares line: The line . . . that minimizes . . . in the equation . . . .

least-squares solution (of $A\mathbf{x} = \mathbf{b}$):   A vector . . . such that . . . .

length (of **v**):   The scalar $\|\mathbf{x}\| = $ . . . ; also called the . . . of **v**.

linear model (in statistics):   Any equation of the form . . . , where X and **y** are known and β is to be chosen to minimize . . . .

mean-deviation form (of a vector):   A vector whose entries . . . .

multiple regression: A linear model involving . . . variables and . . . .

normal equations: The system of equations represented by . . . , whose solution yields all . . . solutions of $A\mathbf{x} = \mathbf{b}$. In statistics, a common notation is . . . .

normalizing (a vector **v**):   The process of creating a . . . vector **u** that . . . .

observation vector: The vector . . . in the linear model. . . . are the observed values of . . . .

orthogonal basis: A basis that . . . .

orthogonal complement (of W):   The set $W^\perp$ of . . . .

orthogonal matrix: A . . . matrix U such that . . . .

Copyright © 2021 Pearson Education, Inc.

orthogonal projection of y onto u (or onto the line through **u** and the origin, for $\mathbf{u} \neq \mathbf{0}$):   The vector $\hat{\mathbf{y}}$ defined by . . . .

orthogonal projection of y onto W: The unique vector $\hat{\mathbf{y}}$ such that . . . .
    Notation: $\hat{\mathbf{y}} = \text{proj}_W \mathbf{y}$.

orthogonal set: A set S of vectors such that . . . for . . . .

orthogonal to W: Orthogonal to every . . . .

orthonormal basis: A basis that is . . . .

orthonormal set: An . . . set of . . . .

parameter vector: The unknown vector . . . in the linear model . . . .

regression coefficients: The coefficients . . . in the . . . .

residual vector: The quantity . . . that appears in the general linear model: . . . , the difference between . . . and the . . . values (of y).

QR factorization: A factorization of an $m \times n$ matrix A with linearly independent columns, $A = QR$, where Q is an . . . matrix whose . . . , and R is an . . . matrix.

same direction (as a vector **v**):   A vector that is . . . .

scale (a vector):   Multiply a vector by . . . .

trend analysis: The use of . . . to fit data, with the inner product . . . .

triangle inequality: . . . .

trigonometric polynomial: A linear combination of . . . and . . . functions such as . . . .

unit vector: A vector **v** such that . . . .

weighted least squares: Least-squares problems with a . . . inner product such as $\langle \mathbf{x}, \mathbf{y} \rangle = $ . . . .

Copyright © 2021 Pearson Education, Inc.

7 Symmetric Matrices and Quadratic Forms

7.1 - Diagonalization of Symmetric Matrices

To prepare for this section, review Section 5.3. Focus on the Diagonalization Theorem, Example 3, and Theorem 6. Also, review Example 3 in Section 6.2.

STUDY NOTES

If a symmetric matrix has distinct eigenvalues, as in Example 2, then the ordinary diagonalization process produces a matrix P with *orthogonal* columns, because eigenvectors from different eigenspaces are automatically orthogonal. However, the P you need here must have ortho**normal** columns. Forgetting to normalize the columns of P is the main error students make in this section.

The statements in Theorem 3 (The Spectral Theorem), together with the general approach used in Example 3, lead to the following outline for orthogonally diagonalizing any symmetric matrix.

Procedure for Orthogonally Diagonalizing a Symmetric Matrix A

1. Find the eigenvalues of A.

2. For each eigenvalue of A, construct an orthonormal basis for the eigenspace.

   a. If the eigenspace has a basis $\{\mathbf{v}\}$, normalize $\mathbf{v}$ to produce a unit vector $\mathbf{u}$.

   b. If the eigenspace has a basis $\{\mathbf{v}_1, \mathbf{v}_2\}$, first produce an orthogonal basis $\{\mathbf{v}_1, \mathbf{z}_2\}$, where

   $$\mathbf{z}_2 = \mathbf{v}_2 - \frac{\mathbf{v}_2 \cdot \mathbf{v}_1}{\mathbf{v}_1 \cdot \mathbf{v}_1} \mathbf{v}_1$$

   Then normalize $\mathbf{v}_1$ and $\mathbf{z}_2$ to produce an orthonormal basis $\{\mathbf{u}_1, \mathbf{u}_2\}$ for the eigenspace.

   c. If the eigenspace has a basis $\{\mathbf{v}_1, ..., \mathbf{v}_k\}$, use the Gram-Schmidt process to construct an orthonormal basis $\{\mathbf{u}_1, ..., \mathbf{u}_k\}$ for the eigenspace.

3. The union of the bases for all the eigenspaces is an orthonormal basis for R^n (when A is $n \times n$). Use these vectors as the columns of an orthogonal matrix P.

4. Construct D from the eigenvalues, in an order corresponding to the columns of P. Repeat each eigenvalue according to the dimension of the eigenspace.

5. Finally, $A = PDP^{-1} = PDP^{T}$.

Copyright © 2021 Pearson Education, Inc.

The exercises in this section have been constructed so that mastery of the Gram-Schmidt process is not needed, because some courses may omit Section 6.4. However, you do need to understand the calculations in Step 2(b) above.

_____

SOLUTIONS TO EXERCISES

1. $A = \begin{bmatrix} 3 & 5 \\ 5 & -7 \end{bmatrix} = A^T$, because the (1,2) and (2,1) entries match. The entries on the main diagonal of A can have any values.

7. $P = \begin{bmatrix} .6 & .8 \\ .8 & -.6 \end{bmatrix} = [\mathbf{p}_1 \quad \mathbf{p}_2]$. To show that P is orthogonal by hand calculations, show that its columns are orthonormal: $\mathbf{p}_1 \cdot \mathbf{p}_2 = .48 - .48 = 0$, $\|\mathbf{p}_1\|^2 = (.6)^2 + (.8)^2 = 1$, and similarly, $\|\mathbf{p}_2\|^2 = 1$. Since P is square, P is an orthogonal matrix.

13. $A = \begin{bmatrix} 3 & 1 \\ 1 & 3 \end{bmatrix}$. Characteristic polynomial: $(3 - \lambda)^2 - 1 = \lambda^2 - 6\lambda + 8 = (\lambda - 4)(\lambda - 2)$. So the eigenvalues are 4 and 2.

For $\lambda = 4$: $[A - 4I \quad \mathbf{0}] = \begin{bmatrix} -1 & 1 & 0 \\ 1 & -1 & 0 \end{bmatrix} \sim \begin{bmatrix} 1 & -1 & 0 \\ 0 & 0 & 0 \end{bmatrix}$, $\begin{matrix} x_1 = x_2 \\ x_2 \text{ is free} \end{matrix}$

Take $x_2 = 1$ to get a basis for the eigenspace: $\begin{bmatrix} 1 \\ 1 \end{bmatrix}$. Then normalize to get a unit vector:

$\mathbf{u}_1 = \begin{bmatrix} 1/\sqrt{2} \\ 1/\sqrt{2} \end{bmatrix}$. (Don't forget this step.)

For $\lambda = 2$: $[A - 2I \quad \mathbf{0}] = \begin{bmatrix} 1 & 1 & 0 \\ 1 & 1 & 0 \end{bmatrix} \sim \begin{bmatrix} 1 & 1 & 0 \\ 0 & 0 & 0 \end{bmatrix}$, $\begin{matrix} x_1 = -x_2 \\ x_2 \text{ is free} \end{matrix}$

Take $x_2 = 1$ to get a basis for the eigenspace: $\begin{bmatrix} -1 \\ 1 \end{bmatrix}$. Then normalize to get a unit vector:

$\mathbf{u}_2 = \begin{bmatrix} -1/\sqrt{2} \\ 1/\sqrt{2} \end{bmatrix}$.

Set $P = [\mathbf{u}_1 \quad \mathbf{u}_2] = \begin{bmatrix} 1/\sqrt{2} & -1/\sqrt{2} \\ 1/\sqrt{2} & 1/\sqrt{2} \end{bmatrix}$. The corresponding D is $\begin{bmatrix} 4 & 0 \\ 0 & 2 \end{bmatrix}$.

Study Tip: The fact that eigenvectors for distinct eigenvalues are orthogonal gives you a check on your work. After you find $\mathbf{u}_2$ in Exercise 13, verify that $\mathbf{u}_2 \cdot \mathbf{u}_1 = 0$. Actually, when $\mathbf{u}_1$ and $\mathbf{u}_2$ are in $\mathbb{R}^2$, you can easily guess what $\mathbf{u}_2$ must be, once you know $\mathbf{u}_1$. If you do this, you should compute $A\mathbf{u}_2$ to make sure that $\mathbf{u}_2$ is indeed an eigenvector.

Copyright © 2021 Pearson Education, Inc.

19. Be sure to *work* this problem before reading the solution. Use Exercises 13–24 to sharpen your skills. They are critical for the rest of the chapter. Here, $A = \begin{bmatrix} 3 & -2 & 4 \\ -2 & 6 & 2 \\ 4 & 2 & 3 \end{bmatrix}$, and the eigenvalues are given: 7 and –2.

For $\lambda = 7$: $[A - 7I \quad \mathbf{0}] = \begin{bmatrix} -4 & -2 & 4 & 0 \\ -2 & -1 & 2 & 0 \\ 4 & 2 & -4 & 0 \end{bmatrix} \sim \begin{bmatrix} 1 & .5 & -1 & 0 \\ 0 & 0 & 0 & 0 \\ 0 & 0 & 0 & 0 \end{bmatrix}$

Thus, $x_1 = -.5x_2 + x_3$, with x_2 and x_3 free. Instead of describing all vectors in the eigenspace, you can produce a basis quickly by choosing two linearly independent solutions of $(A - 7I)\mathbf{x} = \mathbf{0}$. The natural choices are the vector corresponding to $x_2 = 1$ and $x_3 = 0$ and the vector for $x_2 = 0$ and $x_3 = 1$. However, in this particular problem, the coefficient $-.5$ of x_2 suggests that a better choice for the first vector is to take $x_2 = 2$ and $x_3 = 0$. In this case, the two vectors are $\mathbf{v}_1 = \begin{bmatrix} -1 \\ 2 \\ 0 \end{bmatrix}$ and $\mathbf{v}_2 = \begin{bmatrix} 1 \\ 0 \\ 1 \end{bmatrix}$.

This basis for the eigenspace is not orthogonal. Keep $\mathbf{v}_1$ and subtract from $\mathbf{v}_2$ its orthogonal projection onto $\mathbf{v}_1$. The new vector, $\mathbf{z}_2$, is an eigenvector for the eigenvalue 7 because it is a linear combination of the vectors $\mathbf{v}_2$ and $\mathbf{v}_1$ in the eigenspace for $\lambda = 7$:

$$\mathbf{z}_2 = \mathbf{v}_2 - \frac{\mathbf{v}_2 \cdot \mathbf{v}_1}{\mathbf{v}_1 \cdot \mathbf{v}_1} \mathbf{v}_1 = \begin{bmatrix} 1 \\ 0 \\ 1 \end{bmatrix} - \frac{-1}{5} \begin{bmatrix} -1 \\ 2 \\ 0 \end{bmatrix} = \begin{bmatrix} 4/5 \\ 2/5 \\ 1 \end{bmatrix} \qquad \text{Instead of } \mathbf{z}_2, \text{ use } \mathbf{z}_2' = \begin{bmatrix} 4 \\ 2 \\ 5 \end{bmatrix},$$

which is easier to normalize. Check that $\mathbf{z}_2' \cdot \mathbf{v}_1 = 0$. An orthonormal basis for the eigenspace is $\mathbf{u}_1 = \begin{bmatrix} -1/\sqrt{5} \\ 2/\sqrt{5} \\ 0 \end{bmatrix}$, $\mathbf{u}_2 = \begin{bmatrix} 4/\sqrt{45} \\ 2/\sqrt{45} \\ 5/\sqrt{45} \end{bmatrix}$.

For $\lambda = -2$: $[A + 2I \quad \mathbf{0}] = \begin{bmatrix} 5 & -2 & 4 & 0 \\ -2 & 8 & 2 & 0 \\ 4 & 2 & 5 & 0 \end{bmatrix} \sim \begin{bmatrix} 1 & -4 & -1 & 0 \\ 5 & -2 & 4 & 0 \\ 4 & 2 & 5 & 0 \end{bmatrix} \sim \begin{bmatrix} 1 & -4 & -1 & 0 \\ 0 & 18 & 9 & 0 \\ 0 & 18 & 9 & 0 \end{bmatrix}$

$$= \begin{bmatrix} 1 & -4 & -1 & 0 \\ 0 & 1 & 1/2 & 0 \\ 0 & 0 & 0 & 0 \end{bmatrix} \sim \begin{bmatrix} ① & 0 & 1 & 0 \\ 0 & ① & 1/2 & 0 \\ 0 & 0 & 0 & 0 \end{bmatrix}, \qquad \begin{array}{l} x_1 = -x_3 \\ x_2 = -(1/2)x_3 \\ x_3 \text{ is free} \end{array}$$

Copyright © 2021 Pearson Education, Inc.

Take $x_3 = 2$ to get a basis for the eigenspace, $\begin{bmatrix} -2 \\ -1 \\ 2 \end{bmatrix}$, and normalize to obtain $\mathbf{u}_3 = \begin{bmatrix} -2/3 \\ -1/3 \\ 2/3 \end{bmatrix}$.

Finally, set $P = [\mathbf{u}_1 \ \ \mathbf{u}_2 \ \ \mathbf{u}_3] = \begin{bmatrix} -1/\sqrt{5} & 4/\sqrt{45} & -2/3 \\ 2/\sqrt{5} & 2/\sqrt{45} & -1/3 \\ 0 & 5/\sqrt{45} & 2/3 \end{bmatrix}$ and $D = \begin{bmatrix} 7 & 0 & 0 \\ 0 & 7 & 0 \\ 0 & 0 & -2 \end{bmatrix}$.

What other answers might someone produce? If the vectors $\mathbf{v}_1$ and $\mathbf{v}_2$ are interchanged, the

first two columns of P probably will be $\begin{bmatrix} 1/\sqrt{2} \\ 0 \\ 1/\sqrt{2} \end{bmatrix}$ and $\begin{bmatrix} -1/\sqrt{18} \\ 4/\sqrt{18} \\ 1/\sqrt{18} \end{bmatrix}$. If the entries in D are

rearranged, the columns of P must be rearranged to correspond to the new entries in D.

Study Tip: The matrix in Exercise 20 has only two distinct eigenvalues (according to the text's information), so one or both eigenspaces will be at least two-dimensional. You will have to construct an orthonormal basis for such an eigenspace. Exercises 23 and 24 are good models for exam questions because they give you information from which you can orthogonally diagonalize A without extensive computations.

25. See Theorem 2 and the paragraph preceding the theorem.

27. See the matrix P in Example 2.

29. See the paragraph following formula (2), in which each $\mathbf{u}$ is a unit vector.

31. There are n real eigenvalues (Theorem 3), but they need not be distinct (Example 3).

35. By hypothesis, $A = PDP^{-1}$, where P is orthogonal and D is diagonal. Since A is invertible, 0 is not an eigenvalue and D is invertible. Then $A^{-1} = (PDP^{-1})^{-1} = (P^{-1})^{-1}D^{-1}P^{-1} = PD^{-1}P^{-1}$. Since D^{-1} is diagonal, A^{-1} is orthogonally diagonalizable.

    A second argument: By Theorem 2, A is symmetric. Since A is invertible, a property of transposes shows that $(A^{-1})^T = (A^T)^{-1} = A^{-1}$, so A^{-1} is symmetric. By Theorem 2, again, A^{-1} is orthogonally diagonalizable.

41. **a.** The matrix $B = \mathbf{u}\mathbf{u}^T$ is an outer product, or a rank 1 matrix. Given $\mathbf{x}$ in $\mathbb{R}^n$, $B\mathbf{x} = (\mathbf{u}\mathbf{u}^T)\mathbf{x}$ $= \mathbf{u}(\mathbf{u}^T\mathbf{x}) = (\mathbf{u}^T\mathbf{x})\mathbf{u}$, because $\mathbf{u}^T\mathbf{x}$ is a scalar. Using dot products, $B\mathbf{x} = (\mathbf{x} \cdot \mathbf{u})\mathbf{u}$. Since $\mathbf{u}$ is a unit vector, this is the orthogonal projection of $\mathbf{x}$ onto $\mathbf{u}$. See Section 6.2.

    b. B is symmetric, because $B^T = (\mathbf{u}\mathbf{u}^T)^T = \mathbf{u}^{TT}\mathbf{u}^T = \mathbf{u}\mathbf{u}^T = B$. Also, $B^2 = (\mathbf{u}\mathbf{u}^T)(\mathbf{u}\mathbf{u}^T) = \mathbf{u}(\mathbf{u}^T\mathbf{u})\mathbf{u}^T$ $= \mathbf{u}\mathbf{u}^T = B$, because $\mathbf{u}^T\mathbf{u} = 1$.

Copyright © 2021 Pearson Education, Inc.

c. Notice $B\mathbf{u} = (\mathbf{u}\mathbf{u}^T)\mathbf{u} = \mathbf{u}(\mathbf{u}^T\mathbf{u}) = (\mathbf{u}^T\mathbf{u})\mathbf{u}$. Since $\mathbf{u}^T\mathbf{u}$ is a scalar, it is the eigenvalue corresponding to the eigenvector $\mathbf{u}$.

MATLAB   Orthogonal Diagonalization

Use **[X,Y]=eig(A)** for eigenvalues and eigenvectors, as in Section 5.3. If you encounter a two-dimensional eigenspace with a basis $\{\mathbf{v}_1, \mathbf{v}_2\}$, use the command

    v2 = v2 – (v2′*v1)/(v1′*v1)*v1

or

    v2 = v2 – proj(v2,v1)

to make the *new* eigenvector $\mathbf{v}_2$ orthogonal to $\mathbf{v}_1$. See the MATLAB note for Section 6.4. After you normalize the eigenvectors and create P, check that $P^TP = I$, to verify that P is indeed an orthogonal matrix. For practice, you might use MATLAB to work Exercise 19.

7.2 - Quadratic Forms

Sections 7.1 and 7.2 provide the foundation for the rest of the chapter.

KEY IDEAS

The main point here is to learn how a change of variable, $\mathbf{x} = P\mathbf{y}$, with P an orthogonal matrix, can transform a quadratic form into a new quadratic form with no cross-product terms.

The equation $\mathbf{x} = P\mathbf{y}$ expresses $\mathbf{x}$ as a linear combination of the columns of P, using the entries in $\mathbf{y}$ as weights. If the columns of P are used as a basis $\mathcal{B}$ for $\mathbb{R}^n$, then the entries in $\mathbf{y}$ are the coordinates of $\mathbf{x}$ relative to the basis $\mathcal{B}$. See Section 4.4 (or Section 2.9). As Section 7.2 shows, the columns of P are eigenvectors of the matrix A of the quadratic form $\mathbf{x}^TA\mathbf{x}$.

If you were fortunate enough to study Section 5.6 or 5.7, you saw the *same* equation $\mathbf{x} = P\mathbf{y}$, with P constructed from eigenvectors of A! The key difference here is that P must be an orthogonal matrix as well as a matrix that diagonalizes A. Nothing less will do. The change of variable $\mathbf{x} = P\mathbf{y}$ will work only if $P^T = P^{-1}$ (and $A = PDP^{-1}$).

If your course covers the various classes of quadratic forms (or, equivalently, classes of symmetric matrices), you should learn both the definitions and the characterizations (in Theorem 5) of these classes. Exercise 32 describes another useful way to characterize quadratic forms, often used in multivariable calculus courses. (The 2×2 case can be generalized to $n\times n$ matrices.)

Copyright © 2021 Pearson Education, Inc.

SOLUTIONS TO EXERCISES

1. a. $\mathbf{x}^T A \mathbf{x} = [x_1 \ \ x_2] \begin{bmatrix} 5 & 1/3 \\ 1/3 & 1 \end{bmatrix} \begin{bmatrix} x_1 \\ x_2 \end{bmatrix} = [x_1 \ \ x_2] \begin{bmatrix} 5x_1 + (1/3)x_2 \\ (1/3)x_1 + x_2 \end{bmatrix} = 5x_1^2 + (2/3)x_1 x_2 + x_2^2$

b. When $\mathbf{x} = (6, 1)$, $\mathbf{x}^T A \mathbf{x} = 5(6)^2 + (2/3)(6)(1) + (1)^2 = 185$.

c. When $\mathbf{x} = (1, 3)$, $\mathbf{x}^T A \mathbf{x} = 5(1)^2 + (2/3)(1)(3) + (3)^2 = 16$.

7. The matrix of the quadratic form is $A = \begin{bmatrix} 1 & 5 \\ 5 & 1 \end{bmatrix}$. The characteristic polynomial is

$\lambda^2 - 2\lambda - 24 = (\lambda - 6)(\lambda + 4)$; eigenvalues are 6 and –4.

<u>For $\lambda = 6$</u>:   an eigenvector is $\begin{bmatrix} 1 \\ 1 \end{bmatrix}$, normalized: $\mathbf{u}_1 = \dfrac{1}{\sqrt{2}} \begin{bmatrix} 1 \\ 1 \end{bmatrix}$

<u>For $\lambda = -4$</u>:   an eigenvector is $\begin{bmatrix} -1 \\ 1 \end{bmatrix}$, normalized: $\mathbf{u}_2 = \dfrac{1}{\sqrt{2}} \begin{bmatrix} -1 \\ 1 \end{bmatrix}$

Thus $A = PDP^{-1}$ and $D = P^{-1}AP = P^T AP$, when $P = \dfrac{1}{\sqrt{2}} \begin{bmatrix} 1 & -1 \\ 1 & 1 \end{bmatrix}$ and $D = \begin{bmatrix} 6 & 0 \\ 0 & -4 \end{bmatrix}$.

The desired change of variable is $\mathbf{x} = P\mathbf{y}$, so that

$$\mathbf{x}^T A \mathbf{x} = (P\mathbf{y})^T A (P\mathbf{y}) = \mathbf{y}^T P^T A P \mathbf{y} = \mathbf{y}^T D \mathbf{y} = 6y_1^2 - 4y_2^2 \qquad (*)$$

Study Tip: To make the "change of variable" requested in Exercise 7, you should: (1) write the equation $\mathbf{x} = P\mathbf{y}$ and specify P; (2) show the matrix algebra in (*) that produces the new quadratic form; and (3) include the new quadratic form. Find out how much of this information you should supply if a problem like Exercise 7 were to appear on an exam.

13.   The matrix of the quadratic form is $A = \begin{bmatrix} 1 & -3 \\ -3 & 9 \end{bmatrix}$. The characteristic polynomial is

$\lambda^2 - 10\lambda = \lambda(\lambda - 10)$; the eigenvalues are 10 and 0. Thus the quadratic form is positive semidefinite. To find the change of variable, proceed as in Exercise 7:

<u>For $\lambda = 10$</u>:   an eigenvector is $\begin{bmatrix} 1 \\ -3 \end{bmatrix}$, normalized: $\mathbf{u}_1 = \dfrac{1}{\sqrt{10}} \begin{bmatrix} 1 \\ -3 \end{bmatrix}$

<u>For $\lambda = 0$</u>:   an eigenvector is $\begin{bmatrix} 3 \\ 1 \end{bmatrix}$, normalized: $\mathbf{u}_2 = \dfrac{1}{\sqrt{10}} \begin{bmatrix} 3 \\ 1 \end{bmatrix}$

Take $P = \dfrac{1}{\sqrt{10}} \begin{bmatrix} 1 & 3 \\ -3 & 1 \end{bmatrix}$ and $D = \begin{bmatrix} 10 & 0 \\ 0 & 0 \end{bmatrix}$. Since P orthogonally diagonalizes A, the desired change of variable is $\mathbf{x} = P\mathbf{y}$, and

$$\mathbf{x}^T A \mathbf{x} = (P\mathbf{y})^T A (P\mathbf{y}) = \mathbf{y}^T P^T A P \mathbf{y} = \mathbf{y}^T D \mathbf{y} = 10 y_1^2$$

Copyright © 2021 Pearson Education, Inc.

The new quadratic form is $10y_1^2$.

19. Because 8 is larger than 5, you should make the x_2^2 term as large as possible. The constraint $x_1^2 + x_2^2 = 1$ keeps x_2 from exceeding 1. When $x_1 = 0$ and $x_2 = 1$, the value of the quadratic form is $5(0) + 8(1) = 8$.

21. See the definition before Example 1.

23. See the paragraph following Example 3.

25. See the Diagonalization Theorem in Section 5.3.

27. $Q(\mathbf{x}) = 0$ when $\mathbf{x} = \mathbf{0}$.

29. See the Numerical Note after Example 6.

33. The text's answer showed that $\mathbf{x}^T B^T B \mathbf{x} \geq 0$ for all $\mathbf{x}$. To show $B^T B$ is positive definite, suppose that $\mathbf{x}^T B^T B \mathbf{x} = 0$. Then $(B\mathbf{x})^T B\mathbf{x} = 0$, so that $\|B\mathbf{x}\|^2 = 0$ and $B\mathbf{x} = \mathbf{0}$. If B is invertible, then $\mathbf{x} = \mathbf{0}$, which shows that, in this case, the form $\mathbf{x}^T B^T B \mathbf{x}$ is positive definite.

35. The quadratic forms $\mathbf{x}^T A \mathbf{x}$ and $\mathbf{x}^T B \mathbf{x}$ are both positive definite, by Theorem 5, because all eigenvalues of A and B are positive. Then for any nonzero $\mathbf{x}$, $\mathbf{x}^T(A + B)\mathbf{x} = \mathbf{x}^T(A\mathbf{x} + B\mathbf{x}) = \mathbf{x}^T A \mathbf{x} + \mathbf{x}^T B \mathbf{x} > 0$, so the quadratic form $\mathbf{x}^T(A + B)\mathbf{x}$ is positive definite. Also, the matrix $A + B$ is symmetric, because $(A + B)^T = A^T + B^T = A + B$. By Theorem 5, $A + B$ has positive eigenvalues.

Mastering Linear Algebra Concepts: Diagonalization and Quadratic Forms

Since the end of the course draws near, I recommend that you prepare a review sheet that contrasts the Diagonalization Theorem in Section 5.3 with Theorem 2 in Section 7.1. You might begin by copying the statements of these theorems, and then use the following questions to guide your review:

- What special properties does an orthogonal diagonalization have that are not present in all diagonalizations? Consider the eigenvalues, the eigenspaces, the eigenvectors, and the matrix P in PDP^{-1}.

- Suppose A is symmetric, and all eigenspaces are one-dimensional. What differences are there between a general diagonalization and an orthogonal diagonalization of A? What differences are there when A is symmetric and one eigenspace is two-dimensional?

- Why is an orthogonal diagonalization needed to simplify a quadratic form? Make the change of variable $\mathbf{x} = P\mathbf{y}$ in $\mathbf{x}^T A \mathbf{x}$ and show the algebra involved.

- If you studied Section 5.6 or 5.7, compare Figure 4 in Section 5.6 with Figure 3 in Section 5.7. How do the general shapes of the trajectories differ? Why do they differ? Could Figure 5 in Section 5.6 or Figure 5 in Section 5.7 be associated with a symmetric matrix?

Copyright © 2021 Pearson Education, Inc.

7.3 - Constrained Optimization

This section is important in its own right, since constrained optimization problems arise in many mathematical problems and applications. The main results of the section are also used in the following two sections.

KEY IDEAS

Theorem 6 gives the main idea. The maximum value of a quadratic form $\mathbf{x}^T A\mathbf{x}$ over the set of all unit vectors can be computed by finding the greatest eigenvalue of A; this maximum value is attained at a corresponding eigenvector. The key step in the proof is to diagonalize A by P, substitute $\mathbf{x} = P\mathbf{y}$, and use the fact that in this case, $\mathbf{x}$ and $\mathbf{y}$ have the same norm.

Example 6 presents a topic that is widely discussed in elementary economics texts.

SOLUTIONS TO EXERCISES

1. We are given an equality of two quadratic forms:
$$5x_1^2 + 6x_2^2 + 7x_3^2 + 4x_1x_2 - 4x_2x_3 = 9y_1^2 + 6y_2^2 + 3y_3^2$$

The matrix of the left quadratic form is $A = \begin{bmatrix} 5 & 2 & 0 \\ 2 & 6 & -2 \\ 0 & -2 & 7 \end{bmatrix}$.

The equality between the two quadratic forms indicates that the eigenvalues of A are 9, 6, 3. (Proof: The diagonal matrix D of the quadratic form $9y_1^2 + 6y_2^2 + 3y_3^2$ obviously has eigenvalues 9, 6, 3. Since A is similar to D, A has the same eigenvalues as D.) The standard calculations produce a unit eigenvector for each eigenvalue. Don't forget to normalize each eigenvector.

$$\lambda = 9: \mathbf{u}_1 = \begin{bmatrix} 1/3 \\ 2/3 \\ -2/3 \end{bmatrix}; \quad \lambda = 6: \mathbf{u}_2 = \begin{bmatrix} 2/3 \\ 1/3 \\ 2/3 \end{bmatrix}; \quad \lambda = 3: \mathbf{u}_3 = \begin{bmatrix} -2/3 \\ 2/3 \\ 1/3 \end{bmatrix}$$

These eigenvectors are mutually orthogonal because they correspond to distinct eigenvalues. So the desired change of variable is

$$\mathbf{x} = P\mathbf{y}, \text{ where } P = \begin{bmatrix} 1/3 & 2/3 & -2/3 \\ 2/3 & 1/3 & 2/3 \\ -2/3 & 2/3 & 1/3 \end{bmatrix}$$

Study Tip: Review the matrix algebra that leads from $\mathbf{x}^T A\mathbf{x}$ to $\mathbf{y}^T D\mathbf{y}$ when $\mathbf{x} = P\mathbf{y}$. Also, be sure you can show that $\|\mathbf{x}\| = \|\mathbf{y}\|$ when P is an orthogonal matrix.

Copyright © 2021 Pearson Education, Inc.

7. The matrix of $Q(\mathbf{x}) = -2x_1^2 - x_2^2 + 4x_1x_2 + 4x_2x_3$ is $A = \begin{bmatrix} -2 & 2 & 0 \\ 2 & -1 & 2 \\ 0 & 2 & 0 \end{bmatrix}$.

The hint in the exercise lists 2, –1, and –4 as the eigenvalues. The greatest eigenvalue is 2, not –4, because "greatest" here refers to the eigenvalue that is farthest to the right on the real line. The maximum value of $Q(\mathbf{x})$ (for $\mathbf{x}$ a unit vector) is attained at a unit eigenvector for $\lambda = 2$. Standard calculations produce the eigenvector:

$$\mathbf{v}_1 = \begin{bmatrix} 1/2 \\ 1 \\ 1 \end{bmatrix}, \text{ scaled to } \begin{bmatrix} 1 \\ 2 \\ 2 \end{bmatrix}, \text{ and normalized to } \mathbf{u}_1 = \begin{bmatrix} 1/3 \\ 2/3 \\ 2/3 \end{bmatrix}.$$

Warning: Exercise 7 illustrates the potential error of selecting –4 instead of 2 as the greatest eigenvalue.

12. This exercise can be done by using a theorem, but try to do it by direct computation, using the hint in the text.

13. If $m = M$ and $\mathbf{x}$ is the unit eigenvector $\mathbf{u}_1$, then $\mathbf{x}^T A\mathbf{x} = \mathbf{u}_1^T A\mathbf{u}_1 = \mathbf{u}_1^T(m\mathbf{u}_1) = m$. Otherwise, set $\alpha = (t - m)/(M - m)$. The hint in the text shows that $0 \leq \alpha \leq 1$ when $m \leq t \leq M$. For such α, let $\mathbf{x} = \sqrt{1 - \alpha}\,\mathbf{u}_n + \sqrt{\alpha}\,\mathbf{u}_1$. Then the vectors $\sqrt{1 - \alpha}\,\mathbf{u}_n$ and $\sqrt{\alpha}\,\mathbf{u}_1$ are orthogonal because they are eigenvectors for different eigenvalues (or one of the vectors is $\mathbf{0}$). By the Pythagorean Theorem,

$$\mathbf{x}^T\mathbf{x} = \|\mathbf{x}\|^2 = \left\|\sqrt{1-\alpha}\,\mathbf{u}_n\right\|^2 + \left\|\sqrt{\alpha}\,\mathbf{u}_1\right\|^2$$

$$= |1 - \alpha|\,\|\mathbf{u}_n\|^2 + |\alpha|\,\|\mathbf{u}_1\|^2 \qquad \text{because } \mathbf{u}_n \text{ and } \mathbf{u}_1 \text{ are unit vectors and } 0$$

$$= (1 - \alpha) + \alpha = 1$$

$\leq \alpha \leq 1$. Also, using the fact that $\mathbf{u}_n$ and $\mathbf{u}_1$ are orthogonal, compute

$$\mathbf{x}^T A\mathbf{x} = (\sqrt{1 - \alpha}\,\mathbf{u}_n + \sqrt{\alpha}\,\mathbf{u}_1)^T A(\sqrt{1 - \alpha}\,\mathbf{u}_n + \sqrt{\alpha}\,\mathbf{u}_1)$$

$$= (\sqrt{1 - \alpha}\,\mathbf{u}_n + \sqrt{\alpha}\,\mathbf{u}_1)^T(m\sqrt{1 - \alpha}\,\mathbf{u}_n + M\sqrt{\alpha}\,\mathbf{u}_1)$$

$$= |1 - \alpha|\,m\mathbf{u}_n^T\mathbf{u}_n + |\alpha|\,M\mathbf{u}_1^T\mathbf{u}_1 = (1 - \alpha)m + \alpha M = t$$

Thus the quadratic form $\mathbf{x}^T A\mathbf{x}$ assumes every value between m and M for a suitable unit vector $\mathbf{x}$.

Copyright © 2021 Pearson Education, Inc.

7.4 - The Singular Value Decomposition

This section is the capstone of the text. It completes the story of the linear transformation $\mathbf{x} \mapsto A\mathbf{x}$ for a general $m \times n$ matrix A and, in so doing, gives you an opportunity to review many basic concepts from Chapters 4–7. In addition, this section opens the door into the modern world of applied linear algebra. An understanding of the singular value decomposition is essential for advanced work in science and engineering that requires matrix computations.

KEY IDEAS

The first singular value σ_1 of an $m \times n$ matrix A is the maximum of $\|A\mathbf{x}\|$ over all unit vectors. This maximum value is attained at a unit eigenvector $\mathbf{v}_1$ of $A^T A$ corresponding to the greatest eigenvalue λ_1 of $A^T A$. The second singular value is the maximum of $\|A\mathbf{x}\|$ over all unit vectors orthogonal to $\mathbf{v}_1$. The following algorithm produces the singular value decomposition for A. (As mentioned in the text, other more reliable methods are used in professional software.)

Procedure for Computing a Singular Value Decomposition

1. Find an orthonormal basis $\{\mathbf{v}_1, \ldots, \mathbf{v}_n\}$ for $\mathbb{R}^n$ consisting of eigenvectors of $A^T A$, arranged so that the associated eigenvalues satisfy $\lambda_1 \geq \cdots \geq \lambda_r > 0$ and $\lambda_{r+1} = \cdots = 0$, where $r = \operatorname{rank} A$.

2. Construct the $n \times n$ orthogonal matrix $V = [\mathbf{v}_1 \ \cdots \ \mathbf{v}_n]$.

3. Let $\sigma_j = \sqrt{\lambda_j}$ $(1 \leq j \leq n)$, and construct the $m \times n$ diagonal matrix Σ whose (j, j)-entry is σ_j $(1 \leq j \leq n)$ and has zeros elsewhere.

4. The set $\{A\mathbf{v}_1, \ldots, A\mathbf{v}_r\}$ is orthogonal and $\sigma_j = \|A\mathbf{v}_j\|$. Compute $\mathbf{u}_j = (1/\sigma_j)A\mathbf{v}_j$ for $1 \leq j \leq r$.

5. Extend $\{\mathbf{u}_1, \ldots, \mathbf{u}_r\}$ to an orthonormal basis $\{\mathbf{u}_1, \ldots, \mathbf{u}_m\}$ for R^m. Write the $m \times m$ orthogonal matrix $U = [\mathbf{u}_1 \ \cdots \ \mathbf{u}_m]$.

6. $A = U\Sigma V^T$.

The diagram below illustrates how the SVD splits the action of A into first multiplication by V^T (which amounts to an orthogonal change of basis in $\mathbb{R}^n$), then a scaling by Σ in the directions of the standard basis vectors $\mathbf{e}_1, \ldots, \mathbf{e}_n$, and finally multiplication by U (an orthogonal change of basis in $\mathbb{R}^m$).

Copyright © 2021 Pearson Education, Inc.

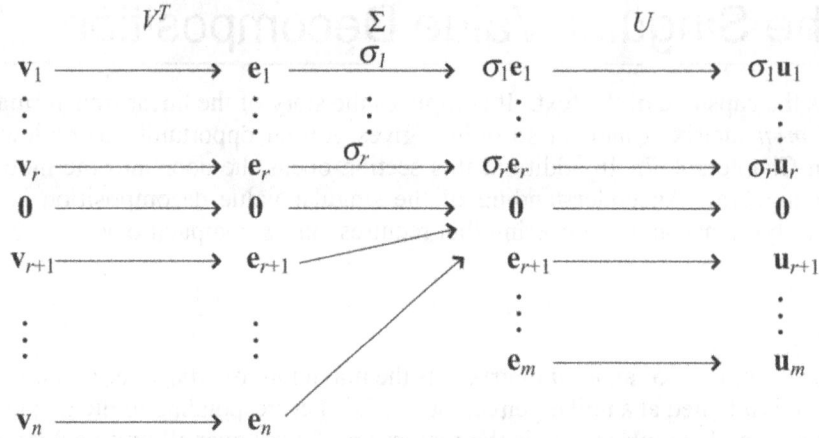

The Singular Value Decomposition: $A = U\Sigma V^T$

Note that the so-called *left* singular vectors of A are the columns of U (because U appears on the left in the factorization of A), even though $\mathbf{u}_1, \ldots, \mathbf{u}_n$ appear on the right side of the diagram above.

SOLUTIONS TO EXERCISES

1. $A = \begin{bmatrix} 1 & 0 \\ 0 & -3 \end{bmatrix}$, $A^TA = \begin{bmatrix} 1 & 0 \\ 0 & 9 \end{bmatrix}$. Eigenvalues and eigenvectors of A^TA are:

$$\lambda_1 = 9: \mathbf{v}_1 = \begin{bmatrix} 0 \\ 1 \end{bmatrix}; \quad \lambda_2 = 1: \mathbf{v}_2 = \begin{bmatrix} 1 \\ 0 \end{bmatrix}$$

(Remember to arrange the eigenvalues in decreasing order.) Thus

$$V = \begin{bmatrix} 0 & 1 \\ 1 & 0 \end{bmatrix}$$

The singular values are $\sigma_1 = \sqrt{9} = 3$ and $\sigma_2 = 1$. The matrix Σ is the same shape as A, and

$$\Sigma = \begin{bmatrix} \sigma_1 & 0 \\ 0 & \sigma_2 \end{bmatrix} = \begin{bmatrix} 3 & 0 \\ 0 & 1 \end{bmatrix}$$

Next, compute

$$A\mathbf{v}_1 = \begin{bmatrix} 1 & 0 \\ 0 & -3 \end{bmatrix}\begin{bmatrix} 0 \\ 1 \end{bmatrix} = \begin{bmatrix} 0 \\ -3 \end{bmatrix}, \quad A\mathbf{v}_2 = \begin{bmatrix} 1 & 0 \\ 0 & -3 \end{bmatrix}\begin{bmatrix} 1 \\ 0 \end{bmatrix} = \begin{bmatrix} 1 \\ 0 \end{bmatrix}$$

and normalize:

$$\mathbf{u}_1 = \frac{1}{\sigma_1} A\mathbf{v}_1 = \frac{1}{3}\begin{bmatrix} 0 \\ -3 \end{bmatrix} = \begin{bmatrix} 0 \\ -1 \end{bmatrix}, \quad \mathbf{u}_2 = \frac{1}{\sigma_2}\begin{bmatrix} 1 \\ 0 \end{bmatrix} = \begin{bmatrix} 1 \\ 0 \end{bmatrix}$$

Finally, $\{\mathbf{u}_1, \mathbf{u}_2\}$ is already a basis for $\mathbb{R}^2$, so the basis for $\mathbb{R}^2$ is complete, and

Copyright © 2021 Pearson Education, Inc.

This happens to
equal V.

$$U = \begin{bmatrix} 0 & 1 \\ -1 & 0 \end{bmatrix}, \text{ and } A = U\Sigma V^T = \begin{bmatrix} 0 & 1 \\ -1 & 0 \end{bmatrix}\begin{bmatrix} 3 & 0 \\ 0 & 1 \end{bmatrix}\begin{bmatrix} 0 & 1 \\ 1 & 0 \end{bmatrix}$$

7. $A = \begin{bmatrix} 2 & -1 \\ 2 & 2 \end{bmatrix}$. $A^TA = \begin{bmatrix} 8 & 2 \\ 2 & 5 \end{bmatrix}$. Find the eigenvalues of A^TA from the characteristic equation.

$$0 = \lambda^2 - 13\lambda + 36 = (\lambda - 9)(\lambda - 4); \quad \lambda_1 = 9, \quad \lambda_2 = 4$$

Corresponding unit eigenvectors for A^TA (calculations omitted) are:

$$\lambda_1 = 9: \mathbf{v}_1 = \begin{bmatrix} 2/\sqrt{5} \\ 1/\sqrt{5} \end{bmatrix}; \quad \lambda_2 = 4: \mathbf{v}_2 = \begin{bmatrix} -1/\sqrt{5} \\ 2/\sqrt{5} \end{bmatrix}$$

Take

$$V = \begin{bmatrix} 2/\sqrt{5} & -1/\sqrt{5} \\ 1/\sqrt{5} & 2/\sqrt{5} \end{bmatrix}$$

The singular values are $\sigma_1 = \sqrt{9} = 3$ and $\sigma_2 = \sqrt{4} = 2$. The matrix Σ is the same shape as A and $\Sigma = \begin{bmatrix} \sigma_1 & 0 \\ 0 & \sigma_2 \end{bmatrix} = \begin{bmatrix} 3 & 0 \\ 0 & 2 \end{bmatrix}$. Next, compute

$$A\mathbf{v}_1 = \begin{bmatrix} 2 & -1 \\ 2 & 2 \end{bmatrix}\begin{bmatrix} 2/\sqrt{5} \\ 1/\sqrt{5} \end{bmatrix} = \begin{bmatrix} 3/\sqrt{5} \\ 6/\sqrt{5} \end{bmatrix}, \quad A\mathbf{v}_2 = \begin{bmatrix} 2 & -1 \\ 2 & 2 \end{bmatrix}\begin{bmatrix} -1/\sqrt{5} \\ 2/\sqrt{5} \end{bmatrix} = \begin{bmatrix} -4/\sqrt{5} \\ 2/\sqrt{5} \end{bmatrix}$$

To check your work at this point, verify that $A\mathbf{v}_1$ and $A\mathbf{v}_2$ are orthogonal. (They are.) Then normalize:

$$\mathbf{u}_1 = \frac{1}{\sigma_1}A\mathbf{v}_1 = \frac{1}{3}\begin{bmatrix} 3/\sqrt{5} \\ 6/\sqrt{5} \end{bmatrix} = \begin{bmatrix} 1/\sqrt{5} \\ 2/\sqrt{5} \end{bmatrix}, \quad \mathbf{u}_2 = \frac{1}{\sigma_2} = \frac{1}{2}\begin{bmatrix} -4/\sqrt{5} \\ 2/\sqrt{5} \end{bmatrix} = \begin{bmatrix} -2/\sqrt{5} \\ 1/\sqrt{5} \end{bmatrix}$$

Since $\{\mathbf{u}_1, \mathbf{u}_2\}$ is a basis for $\mathbb{R}^2$, take $U = \begin{bmatrix} 1/\sqrt{5} & -2/\sqrt{5} \\ 2/\sqrt{5} & 1/\sqrt{5} \end{bmatrix}$. Thus

$$A = U\Sigma V^T = \begin{bmatrix} 1/\sqrt{5} & -2/\sqrt{5} \\ 2/\sqrt{5} & 1/\sqrt{5} \end{bmatrix}\begin{bmatrix} 3 & 0 \\ 0 & 2 \end{bmatrix}\begin{bmatrix} 2/\sqrt{5} & 1/\sqrt{5} \\ -1/\sqrt{5} & 2/\sqrt{5} \end{bmatrix} \text{ (use } V^T, \text{ not } V).$$

Study Tip: Your answer for a singular value decomposition may differ from that given in the text. To check your work, compute AV and $U\Sigma$. If $AV = U\Sigma$, then $A = U\Sigma V^T$ and your answer is correct (provided U and V truly are orthogonal matrices).

Copyright © 2021 Pearson Education, Inc.

13. The matrix $A^T A$ is 3×3. Because the text has not given you practice computing and solving a cubic characteristic equation, the *Hint* suggests that you consider A^T instead of A. (You are free to work on A itself, if you prefer.) Using A^T, compute

$$(A^T)^T A^T = AA^T = \begin{bmatrix} 3 & 2 & 2 \\ 2 & 3 & -2 \end{bmatrix} \begin{bmatrix} 3 & 2 \\ 2 & 3 \\ 2 & -2 \end{bmatrix} = \begin{bmatrix} 17 & 8 \\ 8 & 17 \end{bmatrix}$$

The characteristic equation is

$$0 = \lambda^2 - 34\lambda + 225 = (\lambda - 25)(\lambda - 9); \quad \lambda_1 = 25, \ \lambda_2 = 9$$

The corresponding unit eigenvectors and the matrix V are

$$\lambda_1 = 25: \mathbf{v}_1 = \begin{bmatrix} 1/\sqrt{2} \\ 1/\sqrt{2} \end{bmatrix}; \quad \lambda_2 = 9: \mathbf{v}_2 = \begin{bmatrix} -1/\sqrt{2} \\ 1/\sqrt{2} \end{bmatrix}; \quad V = \begin{bmatrix} 1/\sqrt{2} & -1/\sqrt{2} \\ 1/\sqrt{2} & 1/\sqrt{2} \end{bmatrix}$$

The singular values are $\sigma_1 = 5$ and $\sigma_2 = 3$. Thus Σ is $\begin{bmatrix} 5 & 0 \\ 0 & 3 \\ 0 & 0 \end{bmatrix}$, the same size as A^T. To get $\mathbf{u}_1$ and $\mathbf{u}_2$, compute $A^T\mathbf{v}_1$ and $A^T\mathbf{v}_2$,

$$A^T[\mathbf{v}_1 \quad \mathbf{v}_2] = \begin{bmatrix} 3 & 2 \\ 2 & 3 \\ 2 & -2 \end{bmatrix} \begin{bmatrix} 1/\sqrt{2} & -1/\sqrt{2} \\ 1/\sqrt{2} & 1/\sqrt{2} \end{bmatrix} = \begin{bmatrix} 5/\sqrt{2} & -1/\sqrt{2} \\ 5/\sqrt{2} & 1/\sqrt{2} \\ 0 & -4/\sqrt{2} \end{bmatrix}$$

and normalize:

$$\mathbf{u}_1 = \begin{bmatrix} 1/\sqrt{2} \\ 1/\sqrt{2} \\ 0 \end{bmatrix}, \quad \mathbf{u}_2 = \begin{bmatrix} -1/\sqrt{18} \\ 1/\sqrt{18} \\ -4/\sqrt{18} \end{bmatrix}$$

We need one more vector, orthogonal to $\mathbf{u}_1$ and $\mathbf{u}_2$. So write the equations $\mathbf{u}_1^T\mathbf{x} = 0$ and $\mathbf{u}_2^T\mathbf{x} = 0$ and solve for $\mathbf{x}$. Simpler equations are

$$\begin{array}{ll} \sqrt{2}\mathbf{u}_1^T\mathbf{x} = 0 & \\ \sqrt{18}\mathbf{u}_2^T\mathbf{x} = 0 & \end{array} \quad \text{or} \quad \begin{array}{rrrr} x_1 + x_2 & = 0 \\ -x_1 + x_2 - 4x_3 & = 0 \end{array}$$

The solution is $x_1 = -2x_3$, $x_2 = 2x_3$, x_3 free. A suitable unit vector is

$$\mathbf{u}_3 = \begin{bmatrix} -2/3 \\ 2/3 \\ 1/3 \end{bmatrix}$$

Thus an SVD of A^T is

Copyright © 2021 Pearson Education, Inc.

$$A^T = [\mathbf{u}_1 \quad \mathbf{u}_2 \quad \mathbf{u}_3] \begin{bmatrix} 5 & 0 \\ 0 & 3 \\ 0 & 0 \end{bmatrix} [\mathbf{v}_1 \quad \mathbf{v}_2]^T$$

So an SVD of A appears by taking transposes:

$$A = \begin{bmatrix} 1/\sqrt{2} & -1/\sqrt{2} \\ 1/\sqrt{2} & 1/\sqrt{2} \end{bmatrix} \begin{bmatrix} 5 & 0 & 0 \\ 0 & 3 & 0 \end{bmatrix} \begin{bmatrix} 1/\sqrt{2} & 1/\sqrt{2} & 0 \\ -1/\sqrt{18} & 1/\sqrt{18} & -4/\sqrt{18} \\ -2/3 \triangleleft & 2/3 & 1/3 \end{bmatrix}$$

This *is* an SVD because the outside matrices are orthogonal matrices, and the center matrix is a diagonal matrix of the proper type. Another way to find $\mathbf{u}_3$ is to realize that $\mathbf{u}_1$ and $\mathbf{u}_2$ form an orthonormal basis for Col A^T = Row A. The remaining $\mathbf{u}_3$ must be a basis for (Row A)$^{\perp}$ = Nul A.

Helpful Hint: The last remark in the solution of Exercise 13 applied to A^T because the main SVD construction was for A^T. In an SVD for A, the missing vectors $\mathbf{u}_{r+1}, \ldots, \mathbf{u}_n$ form an orthonormal basis for Nul A^T. (See Fig. 4) One way to obtain $\mathbf{u}_{r+1}, \ldots, \mathbf{u}_n$ is to construct a basis for the solution set of $A^T\mathbf{x} = \mathbf{0}$ and then perform the Gram-Schmidt process.

19. Let $A = U\Sigma V^T$. Then
$$\begin{aligned} A^T A &= (U\Sigma V^T)^T U\Sigma V^T = V\Sigma^T U^T U\Sigma V^T \\ &= V(\Sigma^T\Sigma)V^{-1} \qquad \text{Because } U \text{ and } V \text{ are orthogonal} \end{aligned}$$

If $\sigma_1, \ldots, \sigma_r$ are the nonzero diagonal entries in Σ, then $\Sigma^T\Sigma$ is diagonal, with diagonal entries $\sigma_1^2, \ldots, \sigma_r^2$ and possibly some zeros. Thus V diagonalizes $A^T A$. By the Diagonalization Theorem in Section 5.3, the columns of V are eigenvectors of $A^T A$, and $\sigma_1^2, \ldots, \sigma_r^2$ are the nonzero eigenvalues of $A^T A$. Hence $\sigma_1, \ldots, \sigma_r$ are the nonzero singular values of A. A similar calculation of AA^T shows that the columns of U are eigenvectors of AA^T.

23. From the proof of Theorem 10, $U\Sigma = [\sigma_1\mathbf{u}_1 \quad \cdots \quad \sigma_r\mathbf{u}_r \quad \mathbf{0} \quad \cdots \quad \mathbf{0}]$. The column-row expansion of a matrix product shows that

$$A = (U\Sigma)V^T = (U\Sigma) \begin{bmatrix} \mathbf{v}_1^T \\ \vdots \\ \mathbf{v}_n^T \end{bmatrix} = \sigma_1\mathbf{u}_1\mathbf{v}_1^T + \cdots + \sigma_r\mathbf{u}_r\mathbf{v}_r^T$$

This expansion generalizes the spectral decomposition in Section 7.1.

25. Consider the SVD for the standard matrix of T, say, $A = U\Sigma V^T$. Let $B = \{\mathbf{v}_1, \ldots, \mathbf{v}_n\}$ and $C = \{\mathbf{u}_1, \ldots, \mathbf{u}_m\}$ be bases constructed from the columns of V and U, respectively. Observe that, since the columns of V are orthonormal, $V^T\mathbf{v}_j = \mathbf{e}_j$, where $\mathbf{e}_j$ is the jth column of the $n \times n$ identity matrix. To find the matrix of T relative to B and C, compute

$$T(\mathbf{v}_j) = A\mathbf{v}_j = U\Sigma V^T\mathbf{v}_j = U\Sigma\mathbf{e}_j = U\sigma_j\mathbf{e}_j = \sigma_j U\mathbf{e}_j = \sigma_j\mathbf{u}_j$$

Copyright © 2021 Pearson Education, Inc.

So $[T(\mathbf{v}_j)]_C = \sigma_j \mathbf{e}_j$. Formula (4) in the discussion at the beginning of Section 5.4 shows that the "diagonal" matrix Σ is the matrix of T relative to $\mathcal{B}$ and C.

MATLAB   The Singular Value Decomposition

The command **[P D] = eig(A'*A)** produces an orthogonal matrix P of eigenvectors and a diagonal matrix D of eigenvalues of $A^T A$, but the eigenvalues in D may not be in decreasing order. In such a case, you will have to rearrange things to form V and Σ (denoted below by S). For instance, if P is 3×3, the command

  V = P(:,[1 3 2])

interchanges columns 2 and 3 of P to form V. The commands

    S = zeros(size(A));   S(2,2) = sqrt(D(3,3))

produce a zero matrix for "Σ" the same size as A and place the square root of the (3,3)-entry of D into the (2,2)-entry of S. Other diagonal entries for S can be entered similarly. To form U for the SVD, first normalize the nonzero columns of $A*V$ and place them in a matrix U. If U is square, you are finished. If U is not square, the missing columns must form an orthonormal basis for Nul A^T. (See Fig. 4.) The command **null(A')** produces this orthonormal basis. Thus, the square matrix U is given by

  U = [U   null(A')]

This construction of the SVD helps you to think about properties of the factorization. In practical work, however, you should use the much faster and more numerically reliable command **[U S V] = svd(A)**.

7.5 - Applications to Image Processing and Statistics

If you find remote sensing or image processing interesting, or if you plan to use multivariate statistics later in your career, then you will want to study this section thoroughly. You may have difficulty finding an elementary explanation of this material elsewhere. The idea for the application to image processing came from a student in David's linear algebra class—a geography major who was taking an undergraduate course in remote sensing. The book by Lillesand and Kiefer, referenced in the text, was one of the texts for her course.

Copyright © 2021 Pearson Education, Inc.

KEY IDEAS

The **first principal component** of the data in the matrix of observations is a unit eigenvector $\mathbf{u}_1$ corresponding to the largest eigenvalue of the covariance matrix S. If $\mathbf{u}_1 = (c_1, \ldots, c_p)$, then the entries in $\mathbf{u}_1$ are weights in a linear combination of the original variables, $x_1, \ldots, x_p$, that creates a new variable y_1 (sometimes called a composite score or *index*):

$$y_1 = \mathbf{u}_1^T \mathbf{X} = c_1 x_1 + \cdots + c_p x_p$$

The variance of the values of this index is the largest possible among all indices whose coefficients $c_1, \ldots, c_p$ form a unit vector. (The variance of y_1 is the largest eigenvalue of S.) The **second principal component** is the unit eigenvector corresponding to the second largest eigenvalue of S. The entries in the second principal component determine the index with greatest variance among all possible indices (determined by a unit vector) that are uncorrelated (in a statistical sense) with y_1. Additional principal components are defined similarly.

Checkpoints: (1) If the variables x_1 and x_3 are uncorrelated, what can you say about the covariance matrix S? (2) What is the covariance matrix of the new variables $y_1, \ldots, y_p$ formed from the principal components of S?

SOLUTIONS TO EXERCISES

1. The matrix of observations is $X = \begin{bmatrix} 19 & 22 & 6 & 3 & 2 & 20 \\ 12 & 6 & 9 & 15 & 13 & 5 \end{bmatrix}$, and the sample mean $\mathbf{M}$ is

$\begin{bmatrix} 12 \\ 10 \end{bmatrix}$. Subtract $\mathbf{M}$ from each column of X to obtain

$$B = \begin{bmatrix} 7 & 10 & -6 & -9 & -10 & 8 \\ 2 & -4 & -1 & 5 & 3 & -5 \end{bmatrix}$$

The sample covariance matrix is

$$S = \frac{1}{N-1} BB^T = \frac{1}{5}\begin{bmatrix} 7 & 10 & -6 & -9 & -10 & 8 \\ 2 & -4 & -1 & 5 & 3 & -5 \end{bmatrix}\begin{bmatrix} 7 & 2 \\ 10 & -4 \\ -6 & -1 \\ -9 & 5 \\ -10 & 3 \\ 8 & -5 \end{bmatrix}$$

$$= \frac{1}{5}\begin{bmatrix} 430 & -135 \\ -135 & 80 \end{bmatrix} = \begin{bmatrix} 86 & -27 \\ -27 & 16 \end{bmatrix} \quad \text{Usually, } S \text{ contains decimals.}$$

Study Tip: Note that the formula for the sample mean involves division by N, but for statistical reasons, the covariance matrix formula involves division by $N-1$.

Copyright © 2021 Pearson Education, Inc.

7. Let x_1, x_2 denote the variables for the two-dimensional data in Exercise 1. The characteristic equation of the covariance matrix S from Exercise 1 is $\lambda^2 - 102\lambda + 647 = 0$. By the quadratic formula, the roots of this equation are $\lambda_1 = 95.20$ and $\lambda_2 = 6.80$ (to two decimal places). The first principal component of the data is a unit eigenvector corresponding to λ_1, which turns out to be $(-.95, .32)$, or $(.95, -.32)$. The two possible choices for the new variable are $y_1 = -.95x_1 + .32x_2$ and $y_1 = .95x_1 - .32x_2$. The variance of y_1 is 95.20, while the total variance is $95.20 + 6.80 = 102$. Since $95.20/102 = .933$, the new variable y_1 explains about 93.3% of the variance in the data.

11. **a.** The solution in the text shows that the $\mathbf{Y}_k$ are in mean-deviation form, where $\mathbf{Y}_k = P\mathbf{X}_k$ for some $p \times p$ matrix P.

b. By part (a), the covariance matrix of $\mathbf{Y}_1, \ldots, \mathbf{Y}_N$ is

$$\frac{1}{N-1}[\mathbf{Y}_1 \cdots \mathbf{Y}_N][\mathbf{Y}_1 \cdots \mathbf{Y}_N]^T$$

$$= \frac{1}{N-1}P^T[\mathbf{X}_1 \cdots \mathbf{X}_N]\left(P^T[\mathbf{X}_1 \cdots \mathbf{X}_N]\right)^T$$

$$= P^T\left(\frac{1}{N-1}[\mathbf{X}_1 \cdots \mathbf{X}_N][\mathbf{X}_1 \cdots \mathbf{X}_N]^T\right)P$$

$$= P^T S P$$

because $\mathbf{X}_1, \ldots, \mathbf{X}_N$ are in mean-deviation form.

13. Let $\mathbf{M}$ be the sample mean of the data, and for $k = 1, \ldots, N$, write $\hat{\mathbf{X}}_k$ for $\mathbf{X}_k - \mathbf{M}$. Let $B = \left[\hat{\mathbf{X}}_1 \cdots \hat{\mathbf{X}}_N\right]$, the matrix of observations in mean deviation form. By the column-row expansion of BB^T, the sample covariance matrix is

$$S = \frac{1}{N-1}BB^T = \frac{1}{N-1}[\hat{\mathbf{X}}_1 \cdots \hat{\mathbf{X}}_N]\begin{bmatrix}\hat{\mathbf{X}}_1^T \\ \vdots \\ \hat{\mathbf{X}}_N^T\end{bmatrix}$$

$$= \frac{1}{N-1}\sum_1^N \hat{\mathbf{X}}_k\hat{\mathbf{X}}_k^T = \frac{1}{N-1}\sum_1^N (\mathbf{X}_k - \mathbf{M})(\mathbf{X}_k - \mathbf{M})^T$$

Answers to Checkpoints: (1) The (1, 3)-entry and (3, 1)-entry of S are zero. (2) The covariance matrix of $y_1, \ldots, y_p$ is the diagonal matrix formed from the eigenvalues of S. This matrix is diagonal because the new variables are pairwise uncorrelated. MATLAB Computing Principal Components.

Copyright © 2021 Pearson Education, Inc.

MATLAB   Computing Principal Components

The command **mean(X′)** produces a row vector whose jth entry lists the average of the jth row of X, and **diag(mean(X′))** creates a diagonal matrix whose diagonal entries are the row averages of X. (Be careful not to use **mean(X),** which lists the averages of the *columns* of X.) Finally, the command **diag(mean(X′))*ones(size(X))** creates a matrix the size of X, whose columns are all the same, each one listing the row averages of X. To convert the data in X into mean-deviation form, use

B = X - diag(mean(X′))*ones(size(X))

The sample covariance matrix is produced by

S = B*B′/(N-1)

The principal component data is produced by

   [U,D,V] = svd(B′/sqrt(N-l))

The columns of V are the principal components of the data, and the diagonal entries of **D^2** list the variances of the new variates.

Chapter 7 - Supplementary Exercises

In this chapter, Exercises 1-17 consist of true/false questions, whose level of difficulty varies. Some are similar to the ones that appear in many sections of the text, in which a word or phrase is sometimes missing or slightly misstated. Some follow fairly easily from a theorem: others may need careful reasoning. A few may require an argument that uses several ideas. In each case, think carefully about the statement and attempt to write a solution. The text provides the true/false answer, but you must supply the justification or counterexample.

19. If rank $A = r$, then $\dim \mathrm{Nul}\, A = n - r$ by the Rank Theorem. So 0 is an eigenvalue of A with multiplicity $n - r$, and of the n terms in the spectral decomposition of A exactly $n - r$ are zero. The remaining r terms (which correspond to nonzero eigenvalues) are all rank 1 matrices, as mentioned in the discussion of the spectral decomposition.

23. If $A = R^{T}R$, where R is invertible, then A is positive definite. Conversely, suppose that A is positive definite. Then $A = B^{T}B$ for some positive definite matrix B. Since the eigenvalues of B are positive, 0 is not an eigenvalue and so B is invertible. In particular, the columns of B are linearly independent. By Theorem 12 in Section 6.4, $B = QR$ for some $n \times n$ matrix Q

Copyright © 2021 Pearson Education, Inc.

with orthonormal columns and some upper triangular matrix R with positive elements on its diagonal. Since Q is square, $Q^TQ = I$. So

$$A = B^TB = (QR)^T(QR) = R^TQ^TQR = R^TR$$

and R has the required properties.

27. Start with an SVD decomposition, $A = U\Sigma V^T$. Since U is orthogonal, $U^TU = I$, and so $A = U\Sigma U^TUV^T = PQ$, where $P = U\Sigma U^T = U\Sigma U^{-1}$ and $Q = UV^T$. The matrix P is symmetric, because Σ is symmetric, and P has nonnegative eigenvalues because it is similar to Σ (which is diagonal with nonnegative entries). Thus P is positive semidefinite. The matrix Q is orthogonal because it is the product of orthogonal matrices.

Chapter 7 - Glossary Checklist

Check your knowledge by attempting to write definitions of the terms below. Then compare your work with the definitions given in the text's Glossary. Ask your instructor which definitions, if any, might appear on a test.

condition number (of A):   The quotient σ_1/σ_n, where . . . .

covariance (of variables x_i and x_j, for $i \neq j$):   The entry in the covariance matrix S for a matrix of observations, where x_i and x_j vary over the . . . coordinates, respectively of the observation vectors.

covariance matrix (or **sample covariance matrix**):   The $p \times p$ matrix S defined by $S = $ . . . , where B is a $p \times N$ matrix of observations . . . .

indefinite matrix: A symmetric matrix A such that . . . .

indefinite quadratic form: A quadratic form Q such that $Q(\mathbf{x})$ . . . .

left singular vectors (of A):   The columns of . . . in the singular value decomposition $A = $ . . . .

matrix of observations: A $p \times N$ matrix whose columns are . . . , each column listing p measurements made on . . . .

mean-deviation form (of a matrix of observations):   A matrix whose . . . vectors are . . . .

Moore-Penrose inverse: *See* pseudoinverse.

negative definite matrix: A symmetric matrix A such that . . . .

negative definite quadratic form: A quadratic form Q such that $Q(\mathbf{x})$ . . . .

negative semidefinite matrix: A symmetric matrix A such that . . . .

negative semidefinite quadratic form: A quadratic form Q such that . . . .

orthogonally diagonalizable: A matrix A that admits a factorization, $A = PDP^{-1}$, with P . . . and D . . . .

positive definite matrix: A symmetric matrix A such that . . . .

positive definite quadratic form: A quadratic form Q such that $Q(\mathbf{x})$ . . . .

Copyright © 2021 Pearson Education, Inc.

positive semidefinite matrix: A symmetric matrix A such that . . . .

positive semidefinite quadratic form: A quadratic form Q such that . . . .

principal axes (of a quadratic form $\mathbf{x}^T A \mathbf{x}$):   The orthonormal columns of an orthogonal matrix P such that . . . .

principal components (of the data in a matrix of observations B):   The . . . eigenvectors of a sample covariance matrix S for B, with the eigenvectors arranged so that the corresponding . . . .

projection matrix (or **orthogonal projection matrix**):   A symmetric matrix B such that . . . . A simple example is $B = $ . . . .

pseudoinverse (of A):   The matrix . . . , when UDV^T is a reduced singular value decomposition of A.

quadratic form: A function Q defined for $\mathbf{x}$ in R^n by $Q(\mathbf{x}) = $ . . . , where A is an $n{\times}n$ . . . matrix A (called the matrix of the quadratic form).

reduced singular value decomposition: A factorization $A = $ . . . , for an $m{\times}n$ matrix A of rank r, where U is __ $\times$ __ with orthonormal columns, D is __ $\times$ __ with . . . , and V is __ $\times$ __ with orthonormal columns.

right singular vectors (of A):   The columns of . . . in the singular value decomposition $A = $ . . . .

row sum: The sum of the entries . . . .

sample mean: The average M of a set of vectors, $\mathbf{X}_1, \ldots, \mathbf{X}_N$, given by $\mathbf{M} = $ . . . .

singular value decomposition (of an $m \times n$ matrix A): $A = $ . . . , where U is an __ $\times$ __ . . . matrix, V is an __ $\times$ __ . . . matrix, and Σ is an __ $\times$ __ . . . matrix with . . . .

singular values (of A):   The . . . of the eigenvalues of . . . , arranged . . . .

spectral decomposition (of A):   A representation $A = $ . . . , where . . . .

symmetric matrix: A matrix A such that . . . .

total variance: The . . . of the covariance matrix S of a matrix of . . . .

uncorrelated variables: Any two variables x_i and x_j (with $i \neq j$) that range over the ith and jth coordinates of the observation vectors in an observation matrix, such that . . . .

variance of a variable x_j: The diagonal entry . . . in the . . . matrix S for a matrix of observations, where x_j varies over the jth coordinates of the . . . .

Copyright © 2021 Pearson Education, Inc.

8 The Geometry of Vector Spaces

This chapter begins with geometric ideas that are already familiar to you. Affine combinations of vectors generalize the concepts of lines and planes to higher dimensions and describe the mathematical objects formed when subspaces are shifted away from the origin. Hyperplanes are a generalization of planes, and polytopes are a generalization of polygons.

The geometric objects studied in this chapter include line segments, solid triangles and polygons, specials types of curves and surfaces, and generalizations to sets in more than three dimensions. These objects form the building blocks for applications to computer graphics and to linear programming.

8.1 - Affine Combinations

This section introduces a special kind of linear combination used to describe useful geometric objects involving two or more variables.

KEY IDEAS

In this section, vectors are viewed as points in $\mathbb{R}^n$. Important definitions: *affine combinations*, *affine sets*, and *flats* (or, *translations* of subspaces). The affine combination of two linearly independent vectors (points) is just the line through these two points; the affine combination of three linearly independent vectors (points) is the plane through these three points. Theorems 1 to 4 give important properties of affine combinations and affine sets.

STUDY NOTES

As you work through this section, try to link the new concepts here to similar concepts from Chapter 1. To write $\mathbf{y}$ as an affine combination of a set of vectors $\{\mathbf{v}_1, \ldots, \mathbf{v}_p\}$, first write $\mathbf{y} - \mathbf{v}_1$ as a linear combination of $\mathbf{v}_2 - \mathbf{v}_1, \ldots, \mathbf{v}_p - \mathbf{v}_1$. To do this, row reduce the augmented matrix $[\mathbf{v}_2 - \mathbf{v}_1 \quad \mathbf{v}_3 - \mathbf{v}_1 \quad \cdots \quad \mathbf{v}_p - \mathbf{v}_1 \quad \mathbf{y} - \mathbf{v}_1]$ and read your solution from the result. Once you have written $\mathbf{y} - \mathbf{v}_1$ as a linear combination of $\mathbf{v}_2 - \mathbf{v}_1, \ldots, \mathbf{v}_p - \mathbf{v}_1$, you can solve for $\mathbf{y}$ as an affine combination of $\mathbf{v}_1, \ldots, \mathbf{v}_p$. Note that you can also translate theoretical questions about writing $\mathbf{y}$ as an affine combination of a set of vectors $\{\mathbf{v}_1, \ldots, \mathbf{v}_p\}$ into questions about writing $\mathbf{y} - \mathbf{v}_1$ as a linear combination of the vectors $\{\mathbf{v}_2 - \mathbf{v}_1, \ldots, \mathbf{v}_p - \mathbf{v}_1\}$.

Copyright © 2021 Pearson Education, Inc.

SOLUTIONS TO EXERCISES

1. $\mathbf{v}_1 = \begin{bmatrix} 1 \\ 2 \end{bmatrix}$, $\mathbf{v}_2 = \begin{bmatrix} -2 \\ 2 \end{bmatrix}$, $\mathbf{v}_3 = \begin{bmatrix} 0 \\ 4 \end{bmatrix}$, $\mathbf{v}_4 = \begin{bmatrix} 3 \\ 7 \end{bmatrix}$, $\mathbf{y} = \begin{bmatrix} 5 \\ 3 \end{bmatrix}$, First, translate the points by $\mathbf{v}_1$:

$$\mathbf{v}_2 - \mathbf{v}_1 = \begin{bmatrix} -3 \\ 0 \end{bmatrix}, \mathbf{v}_3 - \mathbf{v}_1 = \begin{bmatrix} -1 \\ 2 \end{bmatrix}, \mathbf{v}_4 - \mathbf{v}_1 = \begin{bmatrix} 2 \\ 5 \end{bmatrix}, \mathbf{y} - \mathbf{v}_1 = \begin{bmatrix} 4 \\ 1 \end{bmatrix}.$$

To find scalars c_2, c_3, and c_4 such that

$$c_2(\mathbf{v}_2 - \mathbf{v}_1) + c_3(\mathbf{v}_3 - \mathbf{v}_1) + c_4(\mathbf{v}_4 - \mathbf{v}_1) = \mathbf{y} - \mathbf{v}_1 \tag{*}$$

use the strategy of Example 1, and row reduce the augmented matrix having these translated points as columns:

$$\begin{bmatrix} -3 & -1 & 2 & 4 \\ 0 & 2 & 5 & 1 \end{bmatrix} \sim \bullet \bullet \bullet \sim \begin{bmatrix} 1 & 0 & -1.5 & -1.5 \\ 0 & 1 & 2.5 & .5 \end{bmatrix}.$$

From this conclude that the general solution of (*) is $c_2 = 1.5c_4 - 1.5$ and $c_3 = -2.5c_4 + .5$, with c_4 free. Set $c_4 = 0$ to obtain a simple solution: $c_2 = -1.5$ and $c_3 = .5$. Thus,

$$\mathbf{y} - \mathbf{v}_1 = -1.5(\mathbf{v}_2 - \mathbf{v}_1) + .5(\mathbf{v}_3 - \mathbf{v}_1), \text{ so that } \mathbf{y} = 2\mathbf{v}_1 - 1.5\mathbf{v}_2 + .5\mathbf{v}_3$$

Another solution is found by setting $c_4 = 1$. In this case, $c_2 = 0$, and

$$\mathbf{y} - \mathbf{v}_1 = -2(\mathbf{v}_3 - \mathbf{v}_1) + 1(\mathbf{v}_4 - \mathbf{v}_1), \text{ so that } \mathbf{y} = 2\mathbf{v}_1 - 2\mathbf{v}_3 + \mathbf{v}_4$$

If $c_4 = 3$, then

$$\mathbf{y} - \mathbf{v}_1 = 3(\mathbf{v}_2 - \mathbf{v}_1) - 7(\mathbf{v}_3 - \mathbf{v}_1) + 3(\mathbf{v}_4 - \mathbf{v}_1), \text{ and } \mathbf{y} = 2\mathbf{v}_1 + 3\mathbf{v}_2 - 7\mathbf{v}_3 + 3\mathbf{v}_4$$

Of course, many other answers are possible. Note that in all cases, the weights in the linear combination sum to 1.

7. The matrix $[\mathbf{v}_1 \ \mathbf{v}_2 \ \mathbf{v}_3 \ \mathbf{p}_1 \ \mathbf{p}_2 \ \mathbf{p}_3]$ reduces to

$$\begin{bmatrix} 1 & 0 & 0 & 2 & 2 & 2 \\ 0 & 1 & 0 & 1 & -4 & 2 \\ 0 & 0 & 1 & -1 & 3 & -3 \\ 0 & 0 & 0 & 0 & 0 & -5 \end{bmatrix}$$

Parts (a), (b), and (c) use columns 4, 5, and 6, respectively, as the "augmented" column.

   a. $\mathbf{p}_1 = 2\mathbf{v}_1 + \mathbf{v}_2 - \mathbf{v}_3$, so $\mathbf{p}_1$ is in Span S. The weights do not sum to 1, so $\mathbf{p}_1 \notin$ aff S.

   b. $\mathbf{p}_2 = 2\mathbf{v}_1 - 4\mathbf{v}_2 + 3\mathbf{v}_3$, so $\mathbf{p}_2$ is in Span S. The weights sum to 1, so $\mathbf{p}_2 \in$ aff S.

Copyright © 2021 Pearson Education, Inc.

   c. $p_3 \notin$ Span S because $0 \neq -5$ (the fourth equation), so p_3 cannot possibly be in aff S.

11. See the definition at the beginning of this section.

13. See the definition.

15. See equation (1).

17. See the definition prior to Theorem 3.

19. See the definition prior to Theorem 3.

21. Span $\{v_2 - v_1, v_3 - v_1\}$ is a plane if and only if $\{v_2 - v_1, v_3 - v_1\}$ is linearly independent. Suppose c_2 and c_3 satisfy $c_2(v_2 - v_1) + c_3(v_3 - v_1) = 0$. Then $c_2 v_2 + c_3 v_3 - (c_2 + c_3) v_1 = 0$. Then $c_2 = c_3 = 0$, because $\{v_1, v_2, v_3\}$ is a linearly independent set. This shows that $\{v_2 - v_1, v_3 - v_1\}$ is a linearly independent set. Thus, Span $\{v_2 - v_1, v_3 - v_1\}$ is a plane in $\mathbb{R}^3$.

27. If $p, q \in f(S)$, then there exist $r, s \in S$ such that $f(r) = p$ and $f(s) = q$. Given any $t \in \mathbb{R}$, we must show that $z = (1 - t)p + tq$ is in $f(S)$. Since f is linear,

$$z = (1 - t)p + tq = (1 - t)f(r) + t f(s) = f((1 - t)r + ts)$$

Since S is affine, $(1 - t)r + ts \in S$. Thus, z is in S and $f(S)$ is affine.

33. Since $(A \cap B) \subseteq A$, it follows from Exercise 30 that aff $(A \cap B) \subseteq$ aff A. Similarly, aff $(A \cap B) \subseteq$ aff B, so aff $(A \cap B) \subseteq ($aff $A \cap$ aff $B)$.

MATLAB   Affine Combinations

Given points v_1, v_2, v_3, v_4, and y, use **A=[v2-v1, v3-v1, v4-v1, y-v1]** to translate the points by v_1. To write y as an affine combination of v_1, v_2, v_3, and v_4, row reduce the matrix **A** to write $y - v_1$ as a linear combination of $v_2 - v_1$, $v_3 - v_1$, and $v_4 - v_1$. Then solve for y.

    If the chosen points b_1, b_2, ..., b_n form a basis for $\mathbb{R}^n$, then the calculations are less involved. For any y in $\mathbb{R}^n$, y is a unique linear combination of b_1, b_2, ..., b_n. This combination is an affine combination of the b's if and only if the weights sum to one. To determine if points p_1 or p_2 are affine combinations of a basis b_1, b_2, b_3 you can form the matrix **A=[b1, b2, b3, p1, p2]**. If **C=rref(A)**, the MATLAB command **sum(C(:,4))** will sum the weights in the fourth column of C whereas the command **sum(C)** will sum the entries in all of the columns of C.

Copyright © 2021 Pearson Education, Inc.

The standard homogeneous form of a point **v** is **[v;1]**. For calculation purposes it is easier to move the rows of 1's to the top in the first step. To use the homogeneous form of the points given by **A=[v1 v2 v3 p]**, the command **A=[ones(1,4); A]** will place a row of 1's in the top row. For an arbitrary number of columns, use **A=[ones(1,size(A,2)); A]**. The commands **C=ref(A)** and **sum(C)** will again be helpful here.

8.2 - Affine Independence

Affine dependence of a set of vectors is a restricted type of linear dependence, which means that affine independence is a weaker condition than linear independence—and this is just what is needed for applications to computer graphics.

KEY IDEAS

The important definitions here are *affine dependence*, *affine independence*, and *barycentric coordinates*. Theorem 5 concerns affine dependence. Theorem 6 provides an important property of affinely independent sets, which leads to barycentric coordinates of a point.

STUDY NOTES

Again it is important to link the new concepts you are learning to concepts you already know. It may be helpful to review Sections 1.7 and 4.4 before continuing with this section. How are barycentric coordinates related to the coordinates of a vector relative to a basis?

Notice that $\{\mathbf{v}_1, \ldots, \mathbf{v}_p\}$, is an affinely dependent set if and only if $\{\mathbf{v}_2 - \mathbf{v}_1, \ldots, \mathbf{v}_p - \mathbf{v}_1\}$ is a linearly dependent set. Use this idea both computationally and theoretically to help you solve problems in this section.

Finding barycentric coordinates is like finding the coordinates of a vector relative to a basis, except that you need an additional equation to ensure that the sum of the coefficients is 1. This adds a row of ones to the bottom of the augmented matrix you would use to find the coordinates of the vector relative to a basis.

SOLUTIONS TO EXERCISES

1. Let $\mathbf{v}_1 = \begin{bmatrix} 3 \\ -3 \end{bmatrix}$, $\mathbf{v}_2 = \begin{bmatrix} 0 \\ 6 \end{bmatrix}$, $\mathbf{v}_3 = \begin{bmatrix} 2 \\ 0 \end{bmatrix}$. Then $\mathbf{v}_2 - \mathbf{v}_1 = \begin{bmatrix} -3 \\ 9 \end{bmatrix}$, $\mathbf{v}_3 - \mathbf{v}_1 = \begin{bmatrix} -1 \\ 3 \end{bmatrix}$. Since $\mathbf{v}_3 - \mathbf{v}_1$ is a multiple of $\mathbf{v}_2 - \mathbf{v}_1$, these two points are linearly dependent. By Theorem 5, $\{\mathbf{v}_1, \mathbf{v}_2, \mathbf{v}_3\}$ is affinely dependent. Note that $(\mathbf{v}_2 - \mathbf{v}_1) - 3(\mathbf{v}_3 - \mathbf{v}_1) = \mathbf{0}$.

Copyright © 2021 Pearson Education, Inc.

A rearrangement produces the affine dependence relation $2\mathbf{v}_1 + \mathbf{v}_2 - 3\mathbf{v}_3 = \mathbf{0}$. (Note that the weights sum to zero.) Geometrically, $\mathbf{v}_1$, $\mathbf{v}_2$, and $\mathbf{v}_3$ are collinear.

7. Denote the given points as $\mathbf{v}_1$, $\mathbf{v}_2$, $\mathbf{v}_3$, and $\mathbf{p}$. Row reduce the augmented matrix for the equation $x_1\tilde{\mathbf{v}}_1 + x_2\tilde{\mathbf{v}}_2 + x_3\tilde{\mathbf{v}}_3 = \tilde{\mathbf{p}}$. Remember to move the bottom row of ones to the top as the first step to simplify the arithmetic by hand.

$$\begin{bmatrix} \tilde{\mathbf{v}}_1 & \tilde{\mathbf{v}}_2 & \tilde{\mathbf{v}}_3 & \tilde{\mathbf{p}} \end{bmatrix} \sim \begin{bmatrix} 1 & 1 & 1 & 1 \\ 1 & 2 & 1 & 5 \\ -1 & 1 & 2 & 4 \\ 2 & 0 & -2 & -2 \\ 1 & 1 & 0 & 2 \end{bmatrix} \sim \begin{bmatrix} 1 & 0 & 0 & -2 \\ 0 & 1 & 0 & 4 \\ 0 & 0 & 1 & -1 \\ 0 & 0 & 0 & 0 \\ 0 & 0 & 0 & 0 \end{bmatrix}$$

Thus, $x_1 = -2$, $x_2 = 4$, $x_3 = -1$, and $\tilde{\mathbf{p}} = -2\tilde{\mathbf{v}}_1 + 4\tilde{\mathbf{v}}_2 - \tilde{\mathbf{v}}_3$, so $\mathbf{p} = -2\mathbf{v}_1 + 4\mathbf{v}_2 - \mathbf{v}_3$, and the barycentric coordinates are $(-2, 4, -1)$.

ALTERNATE SOLUTION:

This problem can also be solved by "translating" it to the origin. That is, compute $\mathbf{v}_2 - \mathbf{v}_1$, $\mathbf{v}_3 - \mathbf{v}_1$, and $\mathbf{p} - \mathbf{v}_1$, find weights c_2 and c_3 such that

$$c_2(\mathbf{v}_2 - \mathbf{v}_1) + c_3(\mathbf{v}_3 - \mathbf{v}_1) = \mathbf{p} - \mathbf{v}_1$$

and then write $\mathbf{p} = (1 - c_2 - c_3)\mathbf{v}_1 + c_2\mathbf{v}_2 + c_3\mathbf{v}_3$. Here are the calculations for Exercise 7:

$$\mathbf{v}_2 - \mathbf{v}_1 = \begin{bmatrix} 2 \\ 1 \\ 0 \\ 1 \end{bmatrix} - \begin{bmatrix} 1 \\ -1 \\ 2 \\ 1 \end{bmatrix} = \begin{bmatrix} 1 \\ 2 \\ -2 \\ 0 \end{bmatrix}, \quad \mathbf{v}_3 - \mathbf{v}_1 = \begin{bmatrix} 1 \\ 2 \\ -2 \\ 0 \end{bmatrix} - \begin{bmatrix} 1 \\ -1 \\ 2 \\ 1 \end{bmatrix} = \begin{bmatrix} 0 \\ 3 \\ -4 \\ -1 \end{bmatrix},$$

$$\mathbf{p} - \mathbf{v}_1 = \begin{bmatrix} 5 \\ 4 \\ -2 \\ 2 \end{bmatrix} - \begin{bmatrix} 1 \\ -1 \\ 2 \\ 1 \end{bmatrix} = \begin{bmatrix} 4 \\ 5 \\ -4 \\ 1 \end{bmatrix}$$

$$\begin{bmatrix} \mathbf{v}_2 - \mathbf{v}_1 & \mathbf{v}_3 - \mathbf{v}_1 & \mathbf{p} - \mathbf{v}_1 \end{bmatrix} \sim \begin{bmatrix} 1 & 0 & 4 \\ 2 & 3 & 5 \\ -2 & -4 & -4 \\ 0 & -1 & 1 \end{bmatrix} \sim \begin{bmatrix} 1 & 0 & 4 \\ 0 & 1 & -1 \\ 0 & 0 & 0 \\ 0 & 0 & 0 \end{bmatrix}$$

Thus $\mathbf{p} - \mathbf{v}_1 = 4(\mathbf{v}_2 - \mathbf{v}_1) - 1(\mathbf{v}_3 - \mathbf{v}_1)$, and $\mathbf{p} = -2\mathbf{v}_1 + 4\mathbf{v}_2 - \mathbf{v}_3$.

9. See Theorem 5 and the discussion just before Example 1.

11. See Theorem 5.

Copyright © 2021 Pearson Education, Inc.

13. See the definition at the beginning of this section.

15. See the definition after Theorem 6, and see Figure 5.

17. See Example 6.

21. If $\{\mathbf{v}_1, \mathbf{v}_2\}$ is affinely dependent, then there exist c_1 and c_2, not both zero, such that $c_1 + c_2 = 0$, and $c_1\mathbf{v}_1 + c_2\mathbf{v}_2 = \mathbf{0}$. Then $c_1 = -c_2 \neq 0$ and $c_1\mathbf{v}_1 = -c_2\mathbf{v}_2 = c_1\mathbf{v}_2$, which implies that $\mathbf{v}_1 = \mathbf{v}_2$. Conversely, if $\mathbf{v}_1 = \mathbf{v}_2$, let $c_1 = 1$ and $c_2 = -1$. Then $c_1\mathbf{v}_1 + c_2\mathbf{v}_2 = \mathbf{v}_1 + (-1)\mathbf{v}_1 = \mathbf{0}$ and $c_1 + c_2 = 0$, which shows that $\{\mathbf{v}_1, \mathbf{v}_2\}$ is affinely dependent.

27. If $\{\mathbf{p}_1, \mathbf{p}_2, \mathbf{p}_3\}$ is an affinely dependent set, then there exist scalars $c_1, c_2,$ and c_3, not all zero, such that $c_1\mathbf{p}_1 + c_2\mathbf{p}_2 + c_3\mathbf{p}_3 = \mathbf{0}$ and $c_1 + c_2 + c_3 = 0$. Next, apply the linear transformation f to both sides of the vector equation, and obtain

$$c_1 f(\mathbf{p}_1) + c_2 f(\mathbf{p}_2) + c_3 f(\mathbf{p}_3) = f(c_1\mathbf{p}_1 + c_2\mathbf{p}_2 + c_3\mathbf{p}_3) = f(\mathbf{0}) = \mathbf{0},$$

since f is linear. This shows that the set $\{f(\mathbf{p}_1), f(\mathbf{p}_2), f(\mathbf{p}_3)\}$ is also affinely dependent.

MATLAB   Affine Independence

To determine if the points $\mathbf{v}_1, \mathbf{v}_2, \mathbf{v}_3, \mathbf{v}_4$ are affinely dependent, construct and row reduce the matrix **A=[v2-v1, v3-v1, v4-v1].** Row reducing the matrix **A** makes it possible to write $\mathbf{v}_4 - \mathbf{v}_1$ as a linear combination of $\mathbf{v}_2 - \mathbf{v}_1$ and $\mathbf{v}_3 - \mathbf{v}_1$. More algebra is needed to write $\mathbf{v}_4$ as a linear combination of $\mathbf{v}_1, \mathbf{v}_2,$ and $\mathbf{v}_3$. See the MATLAB note for section 8.1.

Given an affinely independent set $\{\mathbf{a}, \mathbf{b}, \mathbf{c}\}$, the following steps will find the barycentric coordinates of a point $\mathbf{p}$ determined by the set $\{\mathbf{a}, \mathbf{b}, \mathbf{c}\}$ by row reducing the augmented matrix of points in homogenous form:

format rat
A=[a b c p]
A=[ones(1,4); A]   %Place a row of ones in the top row.
C=ref(A)
sum(C)   %Sum the entries in each column of C.

8.3 - Convex Combinations

A convex combination of a set of vectors is a linear combination using *nonnegative* weights whose sum is 1. So a convex combination is a special type of affine combination.

Copyright © 2021 Pearson Education, Inc.

KEY IDEAS

The important terms here are *convex combination and convex hull*. Theorem 7 concerns convex combinations, Theorems 8 and 9 study properties of intersections of convex sets, and Theorem 10 gives an upper bound on the number of points required when writing a given point as a linear combination of points in a set.

SOLUTIONS TO EXERCISES

1. The set $V = \left\{ \begin{bmatrix} 0 \\ y \end{bmatrix} : 0 \le y < 1 \right\}$ is the vertical line segment from $(0,0)$ to $(0,1)$ that

includes $(0,0)$ but not $(0,1)$. The convex hull of S includes each line segment from a point in V to the point $(2,0)$, as shown in the figure. The dashed line segment along the top of the shaded region indicates that this segment is <u>not</u> in conv S, because $(0,1)$ is not in S.

7. Let $\mathbf{v}_1 = \begin{bmatrix} -1 \\ 0 \end{bmatrix}$, $\mathbf{v}_2 = \begin{bmatrix} 2 \\ 3 \end{bmatrix}$, $\mathbf{v}_3 = \begin{bmatrix} 4 \\ 1 \end{bmatrix}$, $\mathbf{p}_1 = \begin{bmatrix} 2 \\ 1 \end{bmatrix}$, $\mathbf{p}_2 = \begin{bmatrix} 3 \\ 2 \end{bmatrix}$, $\mathbf{p}_3 = \begin{bmatrix} 2 \\ 0 \end{bmatrix}$, $\mathbf{p}_4 = \begin{bmatrix} 0 \\ 2 \end{bmatrix}$, and $T = \{\mathbf{v}_1, \mathbf{v}_2, \mathbf{v}_3\}$.

   a. Use homogeneous forms for all of the vectors by adding a third row to each vector with 1 in the third entry. Next, create an augmented matrix (with four augmented columns) to write the homogeneous forms of $\mathbf{p}_1, \ldots, \mathbf{p}_4$ in terms of the homogeneous forms of $\mathbf{v}_1, \mathbf{v}_2$, and $\mathbf{v}_3$. Then, to simplify calculations, interchange rows 1 and 3, to put a row of 1's at the top. Finally, row reduction produces:

   $$\begin{bmatrix} \tilde{\mathbf{v}}_1 & \tilde{\mathbf{v}}_2 & \tilde{\mathbf{v}}_3 & \tilde{\mathbf{p}}_1 & \tilde{\mathbf{p}}_2 & \tilde{\mathbf{p}}_3 & \tilde{\mathbf{p}}_4 \end{bmatrix} \sim \begin{bmatrix} 1 & 1 & 1 & 1 & 1 & 1 & 1 \\ -1 & 2 & 4 & 2 & 3 & 2 & 0 \\ 0 & 3 & 1 & 1 & 2 & 0 & 2 \end{bmatrix} \sim \begin{bmatrix} 1 & 0 & 0 & \frac{1}{3} & 0 & \frac{1}{2} & \frac{1}{2} \\ 0 & 1 & 0 & \frac{1}{6} & \frac{1}{2} & -\frac{1}{4} & \frac{3}{4} \\ 0 & 0 & 1 & \frac{1}{2} & \frac{1}{2} & \frac{3}{4} & -\frac{1}{4} \end{bmatrix}$$

   The first four columns reveal that $\frac{1}{3}\tilde{\mathbf{v}}_1 + \frac{1}{6}\tilde{\mathbf{v}}_2 + \frac{1}{2}\tilde{\mathbf{v}}_3 = \tilde{\mathbf{p}}_1$ and $\frac{1}{3}\mathbf{v}_1 + \frac{1}{6}\mathbf{v}_2 + \frac{1}{2}\mathbf{v}_3 = \mathbf{p}_1$.

   Thus column 4 contains the barycentric coordinates of $\mathbf{p}_1$ relative to the triangle determined by T. Similarly, column 5 (as an augmented column) contains the barycentric coordinates of $\mathbf{p}_2$, column 6 contains the barycentric coordinates of $\mathbf{p}_3$, and column 7 contains the barycentric coordinates of $\mathbf{p}_4$.

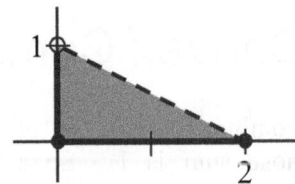

Copyright © 2021 Pearson Education, Inc.

b. $\mathbf{p}_3$ and $\mathbf{p}_4$ are outside conv T because in each case at least one of the barycentric coordinates is negative. $\mathbf{p}_1$ is inside conv T because all of its barycentric coordinates are positive. $\mathbf{p}_2$ is on the edge $\overline{\mathbf{v}_2\mathbf{v}_3}$ of conv T because its barycentric coordinates are nonnegative and its first coordinate is 0.

11. See the definition at the beginning of this section.

13. See Theorem 7.

15. See Theorem 10.

17. If $\mathbf{p}, \mathbf{q} \in f(S)$, then there exist $\mathbf{r}, \mathbf{s} \in S$ such that $f(\mathbf{r}) = \mathbf{p}$ and $f(\mathbf{s}) = \mathbf{q}$. The goal is to show that the line segment $\mathbf{y} = (1 - t)\mathbf{p} + t\mathbf{q}$, for $0 \leq t \leq 1$, is in $f(S)$. Since f is linear,
$$\mathbf{y} = (1 - t)\mathbf{p} + t\mathbf{q} = (1 - t)f(\mathbf{r}) + t f(\mathbf{s}) = f((1 - t)\mathbf{r} + t\mathbf{s})$$
Since S is convex, $(1 - t)\mathbf{r} + t\mathbf{s} \in S$ for $0 \leq t \leq 1$. Thus $\mathbf{y} \in f(S)$ and $f(S)$ is convex.

23. a. Since $A \subseteq (A \cup B)$, Exercise 22 shows that conv $A \subseteq$ conv $(A \cup B)$. Similarly, conv $B \subseteq$ conv $(A \cup B)$. Thus, $[(\text{conv } A) \cup (\text{conv } B)] \subseteq$ conv $(A \cup B)$.

b. One possibility is to let A be two adjacent corners of a square and B be the other two corners. Then $(\text{conv } A) \cup (\text{conv } B)$ consists of two opposite sides of the square, but conv $(A \cup B)$ is the whole square.

8.4 - Hyperplanes

Lines play an important but subtle role in ordinary plane geometry because a line divides the plane into two parts. In three dimensions, a plane divides $\mathbb{R}^3$ into two regions. In higher dimensions, a hyperplane divides $\mathbb{R}^n$ into two regions.

KEY IDEAS

Section 8.1 defined a hyperplane in $\mathbb{R}^n$ as a translate of a subspace of dimension $n - 1$. Computations with hyperplanes in Section 8.4 require an *implicit* description, involving a *linear functional*, defined before Example 1. Several important terms from topology appear in a box just before Example 7. Finally, the text explains what it means for a hyperplane to *separate* two sets. Theorems 12 and 13 provide sufficient conditions for when two sets can be separated.

Copyright © 2021 Pearson Education, Inc.

STUDY NOTES

It is convenient to represent a hyperplane in several different ways, and you want to be able to move comfortably between these different representations.

The implicit equation $a_1x_1 + a_2x_2 + \cdots + a_nx_n = d$ is one way to describe a hyperplane. Notice that $f(\mathbf{x}) = a_1x_1 + a_2x_2 + \cdots + a_nx_n$ is a linear transformation from $\mathbb{R}^n$ to $\mathbb{R}$, and such a transformation is referred to as a *linear functional*.

The set of solutions to $a_1x_1 + a_2x_2 + \cdots + a_nx_n = 0$ corresponds to the null space of the matrix $A = [a_1 \ a_2 \ \dots \ a_n]$, and hence can be written in parametric form (see Section 1.5). The solutions to $a_1x_1 + a_2x_2 + \cdots + a_nx_n = d$ can be described by adding a particular solution $\mathbf{p}$ to the null space of A.

Notice also that $\begin{bmatrix} a_1 \\ a_2 \\ \vdots \\ a_n \end{bmatrix} \cdot \begin{bmatrix} x_1 \\ x_2 \\ \vdots \\ x_n \end{bmatrix} = a_1x_1 + a_2x_2 + \cdots + a_nx_n$, so the equation

$a_1x_1 + a_2x_2 + \cdots + a_nx_n = d$ can be expressed as $\mathbf{n} \cdot \mathbf{x} = d$ where $\mathbf{n} = \begin{bmatrix} a_1 \\ a_2 \\ \vdots \\ a_n \end{bmatrix}$. Since the

solutions to $\mathbf{n} \cdot \mathbf{x} = 0$ describe the set of vectors orthogonal to $\mathbf{n}$, we refer to $\mathbf{n}$ as the *normal* vector to the hyperplane $\mathbf{n} \cdot \mathbf{x} = d$.

SOLUTIONS TO EXERCISES

1. $\mathbf{v}_2 - \mathbf{v}_1 = \begin{bmatrix} 3 \\ 1 \end{bmatrix} - \begin{bmatrix} -1 \\ 4 \end{bmatrix} = \begin{bmatrix} 4 \\ -3 \end{bmatrix}$, so let $\mathbf{n} = \begin{bmatrix} 3 \\ 4 \end{bmatrix}$. $f(x_1, x_2) = 3x_1 + 4x_2$ and

   $d = f(\mathbf{v}_1) = 3(-1) + 4(4) = 13$. This is easy to check by verifying that $f(\mathbf{v}_2)$ is also 13.

7. **a.** Let $\mathbf{v}_1 = \begin{bmatrix} 1 \\ 1 \\ 3 \end{bmatrix}$, $\mathbf{v}_2 = \begin{bmatrix} 2 \\ 4 \\ 1 \end{bmatrix}$, $\mathbf{v}_3 = \begin{bmatrix} -1 \\ -2 \\ 5 \end{bmatrix}$, $\mathbf{n} = \begin{bmatrix} a \\ b \\ c \end{bmatrix}$ and compute the translated points

$$\mathbf{v}_2 - \mathbf{v}_1 = \begin{bmatrix} 1 \\ 3 \\ -2 \end{bmatrix}, \quad \mathbf{v}_3 - \mathbf{v}_1 = \begin{bmatrix} -2 \\ -3 \\ 2 \end{bmatrix}.$$

Copyright © 2021 Pearson Education, Inc.

To solve the system of equations $(\mathbf{v}_2 - \mathbf{v}_1) \cdot \mathbf{n} = 0$ and $(\mathbf{v}_3 - \mathbf{v}_1) \cdot \mathbf{n} = 0$, reduce the augmented matrix for a system of two equations with three variables.

$$[1 \quad 3 \quad -2] \begin{bmatrix} a \\ b \\ c \end{bmatrix} = 0, \quad [-2 \quad -3 \quad 2] \begin{bmatrix} a \\ b \\ c \end{bmatrix} = 0.$$

Row operations show that $\begin{bmatrix} 1 & 3 & -2 & 0 \\ -2 & -3 & 2 & 0 \end{bmatrix} \sim \begin{bmatrix} 1 & 0 & 0 & 0 \\ 0 & 3 & -2 & 0 \end{bmatrix}.$

A suitable normal vector is $\mathbf{n} = \begin{bmatrix} 0 \\ 2 \\ 3 \end{bmatrix}.$

b. The linear functional is $f(x_1, x_2, x_3) = 2x_2 + 3x_3$, so $d = f(1, 1, 3) = 2 + 9 = 11$. As a check, evaluate f at the other two points on the hyperplane:

$$f(2, 4, 1) = 8 + 3 = 11 \text{ and } f(-1, -2, 5) = -4 + 15 = 11.$$

13. $H_1 = \{\mathbf{x} : \mathbf{n}_1 \cdot \mathbf{x} = d_1\}$ and $H_2 = \{\mathbf{x} : \mathbf{n}_2 \cdot \mathbf{x} = d_2\}$. Since $\mathbf{p}_1 \in H_1$, $d_1 = \mathbf{n}_1 \cdot \mathbf{p}_1 = 4$. Similarly, $d_2 = \mathbf{n}_2 \cdot \mathbf{p}_2 = 22$. Solve the simultaneous system $[1 \quad 2 \quad 4 \quad 2]\mathbf{x} = 4$ and $[2 \quad 3 \quad 1 \quad 5]\mathbf{x} = 22$:

$$\begin{bmatrix} 1 & 2 & 4 & 2 & 4 \\ 2 & 3 & 1 & 5 & 22 \end{bmatrix} \sim \begin{bmatrix} 1 & 0 & -10 & 4 & 32 \\ 0 & 1 & 7 & -1 & -14 \end{bmatrix}$$

The general solution provides one set of vectors, $\mathbf{p}$, $\mathbf{v}_1$, and $\mathbf{v}_2$. Other choices are possible.

$$\mathbf{x} = \begin{bmatrix} 32 \\ -14 \\ 0 \\ 0 \end{bmatrix} + x_3 \begin{bmatrix} 10 \\ -7 \\ 1 \\ 0 \end{bmatrix} + x_4 \begin{bmatrix} -4 \\ 1 \\ 0 \\ 1 \end{bmatrix} = \mathbf{p} + x_3\mathbf{v}_1 + x_4\mathbf{v}_2, \quad \text{where}$$

$$\mathbf{p} = \begin{bmatrix} 32 \\ -14 \\ 0 \\ 0 \end{bmatrix}, \mathbf{v}_1 = \begin{bmatrix} 10 \\ -7 \\ 1 \\ 0 \end{bmatrix}, \text{ and } \mathbf{v}_2 = \begin{bmatrix} -4 \\ 1 \\ 0 \\ 1 \end{bmatrix}.$$

Then $H_1 \cap H_2 = \{\mathbf{x} : \mathbf{x} = \mathbf{p} + x_3\mathbf{v}_1 + x_4\mathbf{v}_2\}$.

19. Theorem 3 in Section 6.1 says that $(\text{Col } B)^\perp = \text{Nul } B^T$. Since the two columns of B are clearly linear independent, the rank of B is 2, as is the rank of B^T. So $\dim \text{Nul } B^T = 1$, by the Rank Theorem, since there are three columns in B^T. This means that Nul B^T is one-dimensional, and any nonzero vector $\mathbf{n}$ in Nul B^T will be

Copyright © 2021 Pearson Education, Inc.

orthogonal to H and can be used as its normal vector. Solve the linear system $B^T\mathbf{x} = \mathbf{0}$ by row reduction to find a basis for Nul B^T:

$$\begin{bmatrix} 1 & 4 & -7 & 0 \\ 0 & 2 & -6 & 0 \end{bmatrix} \sim \begin{bmatrix} 1 & 0 & 5 & 0 \\ 0 & 1 & -3 & 0 \end{bmatrix} \Rightarrow \mathbf{n} = \begin{bmatrix} -5 \\ 3 \\ 1 \end{bmatrix}$$

Now, let $f(x_1, x_2, x_3) = -5x_1 + 3x_2 + x_3$. Since the hyperplane is a subspace, it goes through the origin and d must be 0.

The solution is easy to check by evaluating f at each of the columns of B.

21. See the definition at the beginning of this section.

23. See the discussion of (1) and (4).

25. See the comments after the definition of *strictly separate*.

27. See the sets in Figure 4.

8.5 - Polytopes

This section introduces a class of compact convex sets that are used to study optimization problems in a variety of fields, including engineering design, linear programming, and business management.

KEY IDEAS

The important definitions here are *polytope, supporting hyperplane, extreme point* (of a convex set), *profile* (of a convex set), *minimal representation* of a polytope, *simplex,* and *barycentric coordinates.* Theorem 14 describes the minimal representation of a polytope. Theorem 16 is a fundamental result for linear programming. (Linear programming is the primary subject of Chapter 9.) The second part of Section 8.5 discusses interesting geometric objects and shows how to visualize two special polytopes in four dimensions.

STUDY NOTES

In this section, it will help to think geometrically and to imagine or draw the shapes of the objects involved.

Copyright © 2021 Pearson Education, Inc.

SOLUTIONS TO EXERCISES

1. Evaluate each linear functional at each of the three extreme points of S. Then select the extreme point(s) that give the maximum value of the functional.
   a. $f(\mathbf{p}_1) = 1$, $f(\mathbf{p}_2) = -1$, and $f(\mathbf{p}_3) = -3$, so $m = 1$ at $\mathbf{p}_1$.
   b. $f(\mathbf{p}_1) = 1$, $f(\mathbf{p}_2) = 5$, and $f(\mathbf{p}_3) = 1$, so $m = 5$ at $\mathbf{p}_2$.
   c. $f(\mathbf{p}_1) = -3$, $f(\mathbf{p}_2) = -3$, and $f(\mathbf{p}_3) = 5$, so $m = 5$ at $\mathbf{p}_3$.

7. The three inequalities are (a) $x_1 + 3x_2 \leq 18$, (b) $x_1 + x_2 \leq 10$, and (c) $4x_1 + x_2 \leq 28$. Line (a) goes from (0,6) to (18,0). Line (b) goes from (0,10) to (10,0). And line (c) goes from (0,28) to (7,0). One vertex is (0,0). The x_1-intercepts (when $x_2 = 0$) are 18, 10, and 7, so (7,0) is a vertex. The x_2-intercepts (when $x_1 = 0$) are 6, 10, and 28, so (0,6) is a vertex. All three lines go through (6,4), so (6,4) is a vertex. The minimal representation is $\left\{ \begin{bmatrix} 0 \\ 0 \end{bmatrix}, \begin{bmatrix} 7 \\ 0 \end{bmatrix}, \begin{bmatrix} 6 \\ 4 \end{bmatrix}, \begin{bmatrix} 0 \\ 6 \end{bmatrix} \right\}$.

13. a. To determine the number of k-faces of the 5-dimensional hypercube C^5, look at the pattern that is followed in building C^4 from C^3. For example, the 2-faces in C^4 include the 2-faces of C^3 and the 2-faces in the translated image of C^3. In addition, there are the 1-faces of C^3 that are "stretched" into 2-faces. In general, the number of k-faces in C^n equals twice the number of k-faces in C^{n-1} plus the number of $(k-1)$-faces in C^{n-1}. The pattern is:
   $$f_k(C^n) = 2f_k(C^{n-1}) + f_{k-1}(C^{n-1}).$$
   For $k = 0, 1, \ldots, 4$, and $n = 5$, this gives $f_0(C^5) = 32$, $f_1(C^5) = 80$, $f_2(C^5) = 80$, $f_3(C^5) = 40$, and $f_4(C^5) = 10$. These numbers satisfy Euler's formula since,
   $$32 - 80 + 80 - 40 + 10 = 2.$$
   b. The general formula is $f_k(C^n) = 2^{n-k} \binom{n}{k}$, where $\binom{a}{b} - \dfrac{a!}{b!(a-b)!}$ is the binomial coefficient.

17. Check the number of facets (faces).

19. See Theorem 14.

21. See Theorem 16.

23. If follows from Euler's formula.

25. Let S be convex and choose $\mathbf{x}$ in $cS + dS$, where $c > 0$ and $d > 0$. Then there exist $\mathbf{s}_1$ and $\mathbf{s}_2$ in S such that $\mathbf{x} = c\mathbf{s}_1 + d\mathbf{s}_2$. But then

Copyright © 2021 Pearson Education, Inc.

$$\mathbf{x} = c\mathbf{s}_1 + d\mathbf{s}_2 = (c+d)\left(\frac{c}{c+d}\mathbf{s}_1 + \frac{d}{c+d}\mathbf{s}_2\right).$$

Now $\dfrac{c}{c+d}$ and $\dfrac{d}{c+d}$ are both nonnegative and sum to one. Since S is convex,

$$(c+d)\left(\frac{c}{c+d}\mathbf{s}_1 + \frac{d}{c+d}\mathbf{s}_2\right) \in S. \text{ Thus } \mathbf{x} \in (c+d)S.$$

Conversely, let $\mathbf{x} \in (c+d)S$. Then $\mathbf{x} = (c+d)\mathbf{s}$ for some $\mathbf{s} \in S$. But then $\mathbf{x} = c\mathbf{s} + d\mathbf{s} \in cS + dS$, as desired.

8.6 - Curves and Surfaces

This section moves beyond lines and planes to curves and surfaces used in engineering and CAD (computer-aided design).

KEY IDEAS

Control points for a curve are not necessarily points on the curve itself, but rather are points that appear in linear combinations whose weights are simple polynomials. Exercises 21 to 24 in Section 8.3 introduced quadratic and cubic Bézier curves. Section 8.6 studies matrix equations for both Bézier curves and Bézier surfaces. The practice problems in Section 8.6 introduce a related family of curves called B-splines.

STUDY NOTES

Think of a Bézier curve as a linear combination of the points $\mathbf{p}_0$, $\mathbf{p}_1,\ldots, \mathbf{p}_n$, with coefficients that contain the variable t.

SOLUTIONS TO EXERCISES

1. The original curve is $\mathbf{x}(t) = (1-t)^3\mathbf{p}_0 + 3t(1-t)^2\mathbf{p}_1 + 3t^2(1-t)\mathbf{p}_2 + t^3\mathbf{p}_3$ $(0 \le t \le 1)$. Since the curve is determined by its control points, it seems reasonable that to translate the curve, one should translate the control points. In this case, the new Bézier curve $\mathbf{y}(t)$ would have the equation

$$\mathbf{y}(t) = (1-t)^3(\mathbf{p}_0 + \mathbf{b}) + 3t(1-t)^2(\mathbf{p}_1 + \mathbf{b}) + 3t^2(1-t)(\mathbf{p}_2 + \mathbf{b}) + t^3(\mathbf{p}_3 + \mathbf{b})$$

$$= (1-t)^3\mathbf{p}_0 + 3t(1-t)^2\mathbf{p}_1 + 3t^2(1-t)\mathbf{p}_2 + t^3\mathbf{p}_3$$

$$+ (1-t)^3\mathbf{b} + 3t(1-t)^2\mathbf{b} + 3t^2(1-t)\mathbf{b} + t^3\mathbf{b}$$

A routine algebraic calculation verifies that
$$(1-t)^3 + 3t(1-t)^2 + 3t^2(1-t) + t^3 = 1$$

Copyright © 2021 Pearson Education, Inc.

for all t. Thus $\mathbf{y}(t) = \mathbf{x}(t) + \mathbf{b}$ for all t, and translation by $\mathbf{b}$ maps a Bézier curve into a Bézier curve.

7. From Exercise 3(b),

$$\mathbf{x}''(0) = 6(\mathbf{p}_0 - \mathbf{p}_1) + 6(\mathbf{p}_2 - \mathbf{p}_1) \quad \text{and} \quad \mathbf{x}''(1) = 6(\mathbf{p}_1 - \mathbf{p}_2) + 6(\mathbf{p}_3 - \mathbf{p}_2)$$

Use $\mathbf{x}''(0)$ with the control points for $\mathbf{y}(t)$, to get

$$\mathbf{y}''(0) = 6(\mathbf{p}_3 - \mathbf{p}_4) + 6(\mathbf{p}_5 - \mathbf{p}_4)$$

Set $\mathbf{x}''(1) = \mathbf{y}''(0)$ and divide by 6, to get

$$(\mathbf{p}_1 - \mathbf{p}_2) + (\mathbf{p}_3 - \mathbf{p}_2) = (\mathbf{p}_3 - \mathbf{p}_4) + (\mathbf{p}_5 - \mathbf{p}_4) \tag{*}$$

Since the curve is C^1 continuous at $\mathbf{p}_3$, the point $\mathbf{p}_3$ is the midpoint of the segment from $\mathbf{p}_2$ to $\mathbf{p}_4$, by Exercise 5(a). Thus $\mathbf{p}_3 = \frac{1}{2}(\mathbf{p}_2 + \mathbf{p}_4)$, which leads to $\mathbf{p}_4 - \mathbf{p}_3 = \mathbf{p}_3 - \mathbf{p}_2$. Substitution into (*) gives $\;(\mathbf{p}_1 - \mathbf{p}_2) + (\mathbf{p}_3 - \mathbf{p}_2) = -(\mathbf{p}_3 - \mathbf{p}_2) + \mathbf{p}_5 - \mathbf{p}_4$

$$(\mathbf{p}_1 - \mathbf{p}_2) + 2(\mathbf{p}_3 - \mathbf{p}_2) + \mathbf{p}_4 = \mathbf{p}_5$$

Finally, again from C^1 continuity, $\mathbf{p}_4 = \mathbf{p}_3 + \mathbf{p}_3 - \mathbf{p}_2$. Thus,

$$\mathbf{p}_5 = \mathbf{p}_3 + (\mathbf{p}_1 - \mathbf{p}_2) + 3(\mathbf{p}_3 - \mathbf{p}_2)$$

So $\mathbf{p}_4$ and $\mathbf{p}_5$ are uniquely determined by $\mathbf{p}_1$, $\mathbf{p}_2$, and $\mathbf{p}_3$. Only $\mathbf{p}_6$ can be chosen arbitrarily.

11. See equation (2).

13. See Example 1.

15. See Example 2.

17. **a.** From (12), $\mathbf{q}_1 - \mathbf{q}_0 = \frac{1}{2}(\mathbf{p}_1 - \mathbf{p}_0) = \frac{1}{2}\mathbf{p}_1 - \frac{1}{2}\mathbf{p}_0$. Since $\mathbf{q}_0 = \mathbf{p}_0$, $\mathbf{q}_1 = \frac{1}{2}(\mathbf{p}_1 + \mathbf{p}_0)$.

   b. From (13), $8(\mathbf{q}_3 - \mathbf{q}_2) = -\mathbf{p}_0 - \mathbf{p}_1 + \mathbf{p}_2 + \mathbf{p}_3$. So $8\mathbf{q}_3 + \mathbf{p}_0 + \mathbf{p}_1 - \mathbf{p}_2 - \mathbf{p}_3 = 8\mathbf{q}_2$.

   c. Use (8) to substitute for $8\mathbf{q}_3$, and obtain

   $$8\mathbf{q}_2 = (\mathbf{p}_0 + 3\mathbf{p}_1 + 3\mathbf{p}_2 + \mathbf{p}_3) + \mathbf{p}_0 + \mathbf{p}_1 - \mathbf{p}_2 - \mathbf{p}_3 = 2\mathbf{p}_0 + 4\mathbf{p}_1 + 2\mathbf{p}_2$$

   Then divide by 8, regroup the terms, and use part (a) to obtain

   $$\mathbf{q}_2 = \frac{1}{4}\mathbf{p}_0 + \frac{1}{2}\mathbf{p}_1 + \frac{1}{4}\mathbf{p}_2 = (\frac{1}{4}\mathbf{p}_0 + \frac{1}{4}\mathbf{p}_1) + (\frac{1}{4}\mathbf{p}_1 + \frac{1}{4}\mathbf{p}_2) = \frac{1}{2}\mathbf{q}_1 + \frac{1}{4}(\mathbf{p}_1 + \mathbf{p}_2)$$
   $$= \frac{1}{2}(\mathbf{q}_1 + \frac{1}{2}(\mathbf{p}_1 + \mathbf{p}_2))$$

21. **a.** $\mathbf{r}_0 = \mathbf{p}_0$, $\mathbf{r}_1 = \dfrac{\mathbf{p}_0 + 2\mathbf{p}_1}{3}$, $\mathbf{r}_2 = \dfrac{2\mathbf{p}_1 + \mathbf{p}_2}{3}$, $\mathbf{r}_3 = \mathbf{p}_2$

   b. *Hint:* Write the standard formula (7) in this section, with $\mathbf{r}_i$ in place of $\mathbf{p}_i$ for $i = 1, \ldots, 4$, and then replace $\mathbf{r}_0$ and $\mathbf{r}_3$ by $\mathbf{p}_0$ and $\mathbf{p}_2$, respectively:

Copyright © 2021 Pearson Education, Inc.

$$\mathbf{x}(t) = (1 - 3t + 3t^2 - t^3)\mathbf{p}_0 + (3t - 6t^2 + 3t^3)\mathbf{r}_1 + (3t^2 - 3t^3)\mathbf{r}_2 + t^3\mathbf{p}_2$$

Use the formulas for $\mathbf{r}_1$ and $\mathbf{r}_2$ from part (a) to examine the second and third terms in this expression for $\mathbf{x}(t)$.

MATLAB Bezier.m

The MATLAB program **bezier.m** draws graphs of Bézier curves of degrees 2, 3, and 4 in the plane. Type **bezier** at the command prompt. Enter the degree of the desired curve and the geometry matrix representing the control points. For a cubic curve, the geometry matrix will need to be a 2x4 matrix corresponding to the four control points. The program includes the command **hold on**, so rerunning the program will allow MATLAB to display any new graphs in the same window. If you wish to start a new figure, close the existing figure before running the program again.

Chapter 8 - Supplementary Exercises

In this chapter, Exercises 1-21 consist of true/false questions, whose level of difficulty varies. Some are similar to the ones that appear in many sections of the text, in which a word or phrase is sometimes missing or slightly misstated. Some follow fairly easily from a theorem; others may need careful reasoning. A few may require an argument that uses several ideas. In each case, think carefully about the statement and attempt to write a solution. The text provides the true/false answer, but you must supply the justification or counterexample.

25. Suppose $F_1 \cap F_2 \neq \varnothing$. Then there exist $\mathbf{v}_1$ and $\mathbf{v}_2$ in V such that $\mathbf{x}_1 + \mathbf{v}_1 = \mathbf{x}_2 + \mathbf{v}_2$. That is, $\mathbf{x}_1 = \mathbf{x}_2 + \mathbf{v}_2 - \mathbf{v}_1$ and $\mathbf{x}_2 = \mathbf{x}_1 + \mathbf{v}_1 - \mathbf{v}_2$. Then for all $\mathbf{v}$ in V we have $\mathbf{x}_1 + \mathbf{v} = \mathbf{x}_2 + (\mathbf{v}_2 - \mathbf{v}_1 + \mathbf{v}) \in \mathbf{x}_2 + V$ since V is a subspace. Thus $\mathbf{x}_1 + V \subseteq \mathbf{x}_2 + V$. Likewise, $\mathbf{x}_2 + \mathbf{v} = \mathbf{x}_1 + (\mathbf{v}_1 - \mathbf{v}_2 + \mathbf{v}) \in \mathbf{x}_1 + V$, so $\mathbf{x}_2 + V \subseteq \mathbf{x}_1 + V$. Hence, $F_1 = F_2$.

31. The positive hull of S is a cone with vertex $(0, 0)$ containing the positive y axis and with sides on the lines $y = \pm x$.

37. Algebraically, they all have one coefficient equal to zero. Geometrically, they all lie on one of the lines extending the sides of the triangle.

43. The sum of the "**abc**" coefficients in $(1, 1)$ is 3. So if we subtract 2 times the "**abc**" coefficients in $\mathbf{0}$, the new coefficients will sum to 1:

$$(1, 1) = \mathbf{a} + \mathbf{b} + \mathbf{c} - 2\left(\tfrac{2}{7}\mathbf{a} + \tfrac{2}{7}\mathbf{b} + \tfrac{3}{7}\mathbf{c}\right) = \tfrac{3}{7}\mathbf{a} + \tfrac{3}{7}\mathbf{b} + \tfrac{1}{7}\mathbf{c}.$$

Copyright © 2021 Pearson Education, Inc.

Once again the coefficients are all positive and the point $(1, 1)$ is inside triangle **abc**.

Chapter 8 - Glossary Checklist

Check your knowledge by attempting to write definitions of the terms below. Then compare your work with the definitions given in the text's Glossary. Ask your instructor which definitions, if any, might appear on a test.

affine combination A linear combination … .

affine dependence relation An equation of the form …where… .

affine hull of a set S The set of all … .

affinely dependent set A set $\{v_1,\ldots,v_p\}$ … .

affinely independent set A set $\{v_1,\ldots,v_p\}$ … .

affine set A set S of points such that … .

affine transformation A mapping … .

barycentric coordinates The set of weights … .

boundary point of a set S A point **p** such that … .

bounded set A set that is … .

closed ball A ball that contains all of its … .

closed set A set that contains all of its … .

convex combination A linear combination of vectors such that … .

convex hull The set of all … .

convex set A set with the property that … .

dimension of a flat The dimension of the corresponding … .

extreme point of a set S A point **p** in S such that … .

face A nonempty subset … .

facet A … dimensional face.

flat A translate of … .

hypercube Let I_j be the line segment from the origin to the j-th standard basis vector. Then … .

hyperplane A flat of … .

interior point A point **p** in S such that … .

linear functional A linear transformation … .

Copyright © 2021 Pearson Education, Inc.

open ball A set … .

open set A set … .

polygon A polytope in … .

polyhedron A polytope in … .

polytope The convex hull of … .

positive combination The set of linear combinations of a set of points with … .

positive hull The set of all … .

profile The set of … .

quadratic Bezier curve A curve whose description may be written in the form … .

simplex The convex hull of … .

supporting hyperplane A hyperplane … .

tetrahedron A three dimensional solid … .

Copyright © 2021 Pearson Education, Inc.

9 Optimization

There are many situations in business, politics, economics, military strategy, and other areas where one tries to optimize a certain benefit. This may involve maximizing a profit or the payoff in a contest or minimizing a cost or other loss. This chapter presents two mathematical models that deal with optimization problems. The fundamental results in both cases depend on properties of convex sets that were studied in Chapter 8. The first section in Chapter 9 introduces the theory of matrix games and develops strategies based on probability. Sections 9.2 to 9.4 explore techniques of linear programming and use them to solve a variety of problems.

9.1 - Matrix Games

A matrix game is a game between two players where each player has a choice of at least two possible moves or actions. The players each make a choice at the same time. The various outcomes of their moves are listed in a payoff matrix that displays what the row player R wins from the column player C for every possible combination of choices. The payoff matrix is known by both players, and the game is repeated several times with the same payoff matrix. Note that a negative entry in the payoff matrix would represent a loss by player R (and a win for player C) for that particular combination of plays. As the game proceeds, player R tries to choose a row each time that will maximize the amount won and player C tries to choose a column that will minimize the amount lost.

KEY IDEAS

In some matrix games the optimal strategy for player R is to choose the same row each time and likewise player C will always choose the same column. (See Example 1.) But a more interesting case arises when knowing what your opponent will play can be used to your advantage. In this case you do not want to make the same choice each time, or even make your choices in a predictable pattern. An optimal strategy is to chose rows and columns randomly with a certain probability assigned to each choice. (See Example 2.)

Copyright © 2021 Pearson Education, Inc.

Large matrix games with many rows and columns are difficult to solve. But if the payoff matrix has only two rows, then we show how an optimal solution can always be found. Furthermore, at the end of the section we show how some games with a large payoff matrix can be reduced to a matrix having only two rows.

STUDY NOTES

It is helpful to look carefully at Figure 2. The four straight lines come from taking the inner product of $\mathbf{x}(t)$ with each of the four columns of payoff matrix A. [See equation (2).] For each value of t, $v(\mathbf{x}(t))$ is the minimum value of the four linear graphs. This displays $v(\mathbf{x}(t))$ as the darkly colored bent line at the bottom. Row player R wants to choose the value of t that will maximize $v(\mathbf{x}(t))$. The highest point on the bent line is M. The first coordinate of M gives the value of t that determines the optimal strategy for R and the second coordinate gives the value of the game for R. Overall, the figure shows how the height of point M is the maximum of the minima of the four straight lines.

SECTION 9.1 GLOSSARY

dominate: Given vectors $\mathbf{a}$ and $\mathbf{b}$ in $\mathbb{R}^n$, with entries a_i and b_i, respectively, vector $\mathbf{a}$ is said to dominate vector $\mathbf{b}$ if $a_i \geq b_i$ for all $i = 1, \ldots, n$ and $a_i > b_i$ for at least one i.

expected payoff (of a game to player R for strategies $\mathbf{x}$ and $\mathbf{y}$): The sum of the expected payoffs to R over all possible pairs of choices that R and C can make. It is denoted by $E(\mathbf{x}, \mathbf{y})$ and computed as $E(\mathbf{x},\mathbf{y}) = \displaystyle\sum_{i=1}^{m}\sum_{j=1}^{n} x_i a_{ij} y_j = \mathbf{x}^T A \mathbf{y}$.

optimal strategy (for column player C): A strategy $\hat{\mathbf{y}}$ such that $v(\hat{\mathbf{y}}) = v_C$.

optimal strategy (for row player R): A strategy $\hat{\mathbf{x}}$ such that $v(\hat{\mathbf{x}}) = v_R$.

payoff matrix: A matrix that lists the amounts that the row player R wins from the column player C.

probability vector: A vector in $\mathbb{R}^n$ whose entries are nonnegative and sum to one.

pure strategy: A strategy in which one entry is 1 and the other entries are zeros.

recessive: Vector $\mathbf{b}$ is recessive to vector $\mathbf{a}$ if $\mathbf{a}$ dominates $\mathbf{b}$.

saddle point: An entry a_{ij} in a payoff matrix that is both the minimum of row i and the maximum of column j.

solution (of a matrix game): Any pair of optimal strategies $(\hat{\mathbf{x}}, \hat{\mathbf{y}})$.

strategy: A point in a strategy space.

Copyright © 2021 Pearson Education, Inc.

strategy space (for player R): The set of all probability vectors in $\mathbb{R}^m$.

(for player C): The set of all probability vectors in $\mathbb{R}^n$.

value (of C using strategy $\mathbf{y}$): The number $v(\mathbf{y})$ defined by $v(\mathbf{y}) = \max\limits_{\mathbf{x} \in X} E(\mathbf{x}, \mathbf{y})$.

value (of R using strategy $\mathbf{x}$): The number $v(\mathbf{x})$ defined by $v(\mathbf{x}) = \min\limits_{\mathbf{y} \in Y} E(\mathbf{x}, \mathbf{y})$.

value of the game: The common value $v = v_R = v_C$.

value of the game to column player C: The number v_C defined by $v_C = \min\limits_{\mathbf{y} \in Y} v(\mathbf{y})$.

value of the game to row player R: The number v_R defined by $v_R = \max\limits_{\mathbf{x} \in X} v(\mathbf{x})$.

SOLUTIONS TO EXERCISES

1.

$$
\begin{array}{cc}
 & \begin{array}{cc} d & q \end{array} \\
\begin{array}{c} d \\ q \end{array} & \begin{bmatrix} -10 & 10 \\ 25 & -25 \end{bmatrix}
\end{array}
$$

13. Given $A = \begin{bmatrix} 3 & 5 \\ 4 & 1 \end{bmatrix}$, graph $\begin{cases} z = 3(1-t) + (4)t = 3 + t \\ z = 5(1-t) + (1)t = 5 - 4t \end{cases}$. The lines intersect at $(t, z) = (\frac{2}{5}, \frac{17}{5})$.

The optimal row strategy is $\hat{\mathbf{x}} = \mathbf{x}(\frac{2}{5}) = \begin{bmatrix} 1 - \frac{2}{5} \\ \frac{2}{5} \end{bmatrix} = \begin{bmatrix} \frac{3}{5} \\ \frac{2}{5} \end{bmatrix}$, and the value of the game is $v = \frac{17}{5}$.

By Theorem 4, the optimal column strategy $\hat{\mathbf{y}}$ satisfies $E(\mathbf{e}_1, \hat{\mathbf{y}}) = \frac{17}{5}$ and $E(\mathbf{e}_2, \hat{\mathbf{y}}) = \frac{17}{5}$ because $\hat{\mathbf{x}}$ is a linear combination of both $\mathbf{e}_1$ and $\mathbf{e}_2$. From the first of these conditions,

$$
\tfrac{17}{5} = \begin{bmatrix} 1 & 0 \end{bmatrix} \begin{bmatrix} 3 & 5 \\ 4 & 1 \end{bmatrix} \begin{bmatrix} c_1 \\ 1 - c_1 \end{bmatrix} = \begin{bmatrix} 3 & 5 \end{bmatrix} \begin{bmatrix} c_1 \\ 1 - c_1 \end{bmatrix} = 5 - 2c_1
$$

From this, $c_1 = \frac{4}{5}$ and $\hat{\mathbf{y}} = \begin{bmatrix} \frac{4}{5} \\ \frac{1}{5} \end{bmatrix}$. As a check on this work, one can compute

$$
E(\mathbf{e}_2, \hat{\mathbf{y}}) = \begin{bmatrix} 0 & 1 \end{bmatrix} \begin{bmatrix} 3 & 5 \\ 4 & 1 \end{bmatrix} \begin{bmatrix} \frac{4}{5} \\ \frac{1}{5} \end{bmatrix} = \begin{bmatrix} 4 & 1 \end{bmatrix} \begin{bmatrix} \frac{4}{5} \\ \frac{1}{5} \end{bmatrix} = \frac{17}{5}
$$

19. a. Army: 1/3 river, 2/3 land; guerrillas: 1/3 river, 2/3 land; 2/3 of the supplies get through.

b. Army: 7/11 river, 4/11 land; guerrillas: 7/11 river, 4/11 land; 64/121 of the supplies get through.

Copyright © 2021 Pearson Education, Inc.

21. See the definition in the paragraph prior to Example 1.

23. Review the definition of a pure strategy.

25. See the paragraph before Example 4.

27. See the Minimax Theorem and the Fundamental Theorem for Matrix Games.

29. See Theorem 5

31. $\hat{\mathbf{x}} = \begin{bmatrix} \frac{1}{6} \\ \frac{5}{6} \\ 0 \end{bmatrix}, \ \hat{\mathbf{y}} = \begin{bmatrix} 0 \\ \frac{1}{2} \\ \frac{1}{2} \end{bmatrix}, \ v = 0$

9.2 - Linear Programming – Geometric Method

We begin our study of linear programming by considering the two-dimensional case. This will help us understand the essential nature of linear programming problems in a context that is easy to display (and solve) geometrically.

KEY IDEAS

Generally speaking, a linear programming problem involves a system of linear inequalities in variables $x_1, \ldots, x_n$ and a linear functional f from $\mathbb{R}^n$ into $\mathbb{R}$. The system of inequalities typically has many solutions, and the set of all these solutions is called the feasible set, denoted $\mathscr{F}$. The problem is to find a vector $\mathbf{x}$ in $\mathscr{F}$ that maximizes $f(\mathbf{x})$. In solving the problem we will use two important theorems from Chapter 8:

Theorem 8 tells us that the feasible set is convex since it is the intersection of a collection of closed half-spaces (which are convex). Furthermore, when $\mathscr{F}$ is bounded (which is typically the case), we may conclude that $\mathscr{F}$ is compact.

Theorem 16 says that when a linear functional f is defined on a nonempty compact convex set, the maximum value of the function on that set is attained at an extreme point of the set. So, Theorem 16 provides a simple way to solve the linear programming problem: Evaluate $f(\mathbf{x})$ at each of the extreme points of $\mathscr{F}$. Then select the $\mathbf{x}$ that yields the largest value.

STUDY NOTES

In displayed equation (1), the objective function f is defined as $f(\mathbf{x}) = \mathbf{c}^T\mathbf{x}$. It could also have been defined using the dot product notation as $f(\mathbf{x}) = \mathbf{c} \cdot \mathbf{x}$. (See Chapter 6.) Students who are experienced in working with the dot product may find that using the dot product notation makes the formulas easier to remember.

Copyright © 2021 Pearson Education, Inc.

SECTION 9.2 GLOSSARY

canonical linear programming problem: Maximize $f(\mathbf{x}) = \mathbf{c}^T\mathbf{x}$ subject to the constraints $A\mathbf{x} \le \mathbf{b}$ and $\mathbf{x} \ge \mathbf{0}$.

feasible set: The set of all the feasible solutions.

feasible solution (to a linear programming problem): A vector $\mathbf{x}$ that satisfies all the constraint inequalities.

infeasible (linear programming problem): The feasible set is empty.

objective function (in a linear programming problem): The linear function that is being maximized or minimized.

optimal solution: A feasible solution $\overline{\mathbf{x}}$ such that $f(\overline{\mathbf{x}}) = \max_{\mathbf{x} \in \mathscr{F}} f(\mathbf{x})$, where $\mathscr{F}$ is the feasible set.

unbounded (canonical linear programming problem): The objective function takes on arbitrarily large values in the feasible set.

SOLUTIONS TO EXERCISES

1. Let x_1 be the amount invested in mutual funds, x_2 the amount in CDs, and x_3 the amount in savings. Then $\mathbf{b} = \begin{bmatrix} 12{,}000 \\ 0 \\ 0 \end{bmatrix}$, $\mathbf{x} = \begin{bmatrix} x_1 \\ x_2 \\ x_3 \end{bmatrix}$, $\mathbf{c} = \begin{bmatrix} .11 \\ .08 \\ .06 \end{bmatrix}$, and $A = \begin{bmatrix} 1 & 1 & 1 \\ 1 & -1 & -1 \\ 0 & 1 & -2 \end{bmatrix}$.

7. First, find the intersection points for the bounding lines:

$$(1)\ 2x_1 + x_2 = 32, \qquad (2)\ x_1 + x_2 = 18, \qquad (3)\ x_1 + 3x_2 = 24$$

Even a rough sketch of the graphs of these lines will reveal that $(0, 0)$, $(16, 0)$, and $(0, 8)$ are vertices of the feasible set. What about the intersections of the lines corresponding to (1), (2), and (3)?

The graphical method will work, provided the graph is large enough and is drawn carefully. In many simple problems, even a small sketch will reveal which intersection points are vertices of the feasible set. In this problem, however, three intersection points happen to be quite close to each other, and a slight inaccuracy on a graph of size 3" × 3" or smaller may lead to an incorrect solution. In a case such as this, the following algebraic procedure will work well:

Copyright © 2021 Pearson Education, Inc.

When an intersection point is found that corresponds to two inequalities, test it in the other inequalities to see whether the point is in the feasible set.

The intersection of (1) and (2) is (14, 4). Test this in the third inequality: $(14) + 3(4) = 26 > 24$. The intersection point does not satisfy the inequality for (3), so (14, 4) is **not** in the feasible set.

The intersection of (1) and (3) is (14.4, 3.2). Test this in the second inequality: $14.4 + 3.2 = 17.6 \leq 18$, so (14.4, 3.2) **is** in the feasible set.

The intersection of (2) and (3) is (15, 3). Test this in the first inequality: $2(15) + (3) = 33 > 32$, so (15, 3) is **not** in the feasible set.

Next, list the vertices of the feasible set: (0, 0), (16, 0), (14.4, 3.2), and (0, 8). Then compute the values of the objective function $80x_1 + 65x_2$ at these points.

$$(0, 0): \qquad 80(0) + 65(0) = 0$$

$$(16, 0): \qquad 80(16) + 65(0) = 1280$$

$$(14.4, 3.2): \qquad 80(14.4) + 65(3.2) = 1360$$

$$(0, 9): \qquad 80(0) + 65(8) = 520$$

Finally, select the maximum of the objective function, which is 1360, and note that this maximum is attained at (14.4, 3.2).

11. See the definition.

13. See the definition.

19. Take any **p** and **q** in S, with $\mathbf{p} = \begin{bmatrix} x_1 \\ x_2 \end{bmatrix}$ and $\mathbf{q} = \begin{bmatrix} y_1 \\ y_2 \end{bmatrix}$. Then $\mathbf{v}^T\mathbf{p} \leq c$ and $\mathbf{v}^T\mathbf{q} \leq c$. Take any

scalar t such that $0 \leq t \leq 1$. Then, by the linearity of matrix multiplication (or the dot product if $\mathbf{v}^T\mathbf{p}$ is written as $\mathbf{v} \cdot \mathbf{p}$, and so on),

$$\mathbf{v}^T[(1 - t)\mathbf{p} + t\mathbf{q}] = (1 - t)\mathbf{v}^T\mathbf{p} + t\mathbf{v}^T\mathbf{q} \leq (1 - t)c + tc = c$$

because $(1 - t)$ and t are both positive and **p** and **q** are in S. So the line segment between **p** and **q** is in S. Since **p** and **q** were any points in S, the set S is convex.

9.3 - Linear Programming – Simplex Method

Having introduced linear programming in a simple setting where the solution can be found geometrically, we now consider more complicated problems that require an algebraic approach.

Copyright © 2021 Pearson Education, Inc.

KEY IDEAS

At the beginning of Section 9.3 there is an overview of the simplex method. It is repeated here because it provides an outline of where we are going conceptually. As you encounter the algebraic techniques that are used to perform each step, it will be helpful to keep in mind what step you are working on and how it fits into the broader picture.

1. Select an extreme point $\mathbf{x}$ of the feasible set $\mathscr{F}$.

2. Consider all the edges of $\mathscr{F}$ that join at $\mathbf{x}$. If the objective function f cannot be increased by moving along any of these edges, then $\mathbf{x}$ is an optimal solution.

3. If, on the other hand, f can be increased by moving along one or more of the edges, follow the path that gives the largest increase and move to the extreme point of $\mathscr{F}$ at the opposite end.

4. Repeat the process, beginning at step 2.

Since the value of f increases at each step, the path will not go through the same extreme point twice. And since there are only a finite number of extreme points, this process will end at an optimal solution (if there is one) in a finite number of steps.

Example 6 is particularly instructive because it solves a simple problem algebraically as well as geometrically and shows the correlation between these two techniques at each step.

STUDY NOTES

The Study Notes for Section 9.2 had a comment that related to using the dot product notation $f(\mathbf{x}) = \mathbf{c} \cdot \mathbf{x}$ instead of the matrix multiplication notation $f(\mathbf{x}) = \mathbf{c}^T\mathbf{x}$. That comment also applies in this section.

SECTION 9.3 GLOSSARY

basic feasible solution: Each variable is nonnegative and at most m of them are positive, where A has m rows. Geometrically, the basic feasible solutions correspond to the extreme points of the feasible set.

basic solution: A solution where matrix A has m rows and no more than m of the variables are nonzero.

cycling: The simplex method produces an infinite sequence of pivots that fails to lead to an optimal solution.

Copyright © 2021 Pearson Education, Inc.

pivot (on an entry): Transform the entry's coefficient into a 1 and then use it to eliminate corresponding terms in all the other equations.

slack variable: A nonnegative variable that is added to the smaller side of an inequality to convert it to an equality.

SOLUTIONS TO EXERCISES

1.

$$\begin{array}{cccccc} x_1 & x_2 & x_3 & x_4 & x_5 & M \\ \end{array}$$
$$\left[\begin{array}{cccccc|c} 2 & 7 & 10 & 1 & 0 & 0 & 20 \\ 3 & 4 & 18 & 0 & 1 & 0 & 25 \\ -21 & -25 & -15 & 0 & 0 & 1 & 0 \end{array}\right]$$

7. See the definition of a slack variable.

9. See the definition of feasible.

11. See the section on minimization.

13. First, bring x_2 into the solution; pivot with row 1. Then, bring x_1 into the solution; pivot with row 2. The maximum is 150, when $x_1 = 3$ and $x_2 = 10$.

$$\begin{array}{ccccc} x_1 & x_2 & x_3 & x_4 & M \\ \end{array}$$
$$\left[\begin{array}{ccccc|c} 2 & 3 & 1 & 0 & 0 & 36 \\ 5 & 4 & 0 & 1 & 0 & 55 \\ -10 & -12 & 0 & 0 & 1 & 0 \end{array}\right] \sim \left[\begin{array}{ccccc|c} \frac{2}{3} & 1 & \frac{1}{3} & 0 & 0 & 12 \\ \frac{7}{3} & 0 & -\frac{4}{3} & 1 & 0 & 7 \\ -2 & 0 & 4 & 0 & 1 & 144 \end{array}\right]$$

$$\begin{array}{ccccc} x_1 & x_2 & x_3 & x_4 & M \\ \end{array}$$
$$\sim \left[\begin{array}{ccccc|c} 0 & 1 & \frac{5}{7} & -\frac{2}{7} & 0 & 10 \\ 1 & 0 & -\frac{4}{7} & \frac{3}{7} & 0 & 3 \\ 0 & 0 & \frac{20}{7} & \frac{6}{7} & 1 & 150 \end{array}\right]$$

19. Begin with the same initial simplex tableau, bringing x_1 into the solution, with row 1 as the pivot row. This makes b_2 equal to -12, so the basic solution is still not feasible. To correct this, pivot on the negative entry -2 in the second column. This brings x_2 into the solution. The tableau now shows the optimal solution. The maximum of $-x_1 - 2x_2$ is -20, so the minimum of $x_1 + 2x_2$ is 20, when $x_1 = 8$ and $x_2 = 6$.

$$\begin{array}{ccccc} x_1 & x_2 & x_3 & x_4 & M \\ \end{array}$$
$$\left[\begin{array}{ccccc|c} -1 & -1 & 1 & 0 & 0 & -14 \\ 1 & -1 & 0 & 1 & 0 & 2 \\ 1 & 2 & 0 & 0 & 1 & 0 \end{array}\right] \sim \left[\begin{array}{ccccc|c} 1 & 1 & -1 & 0 & 0 & 14 \\ 0 & -2 & 1 & 1 & 0 & -12 \\ 0 & 1 & 1 & 0 & 1 & -14 \end{array}\right] \sim \left[\begin{array}{ccccc|c} 0 & 1 & -\frac{1}{2} & -\frac{1}{2} & 0 & 6 \\ 1 & 0 & -\frac{1}{2} & \frac{1}{2} & 0 & 8 \\ 0 & 0 & \frac{3}{2} & \frac{1}{2} & 1 & -20 \end{array}\right]$$

Copyright © 2021 Pearson Education, Inc.

9.4 - Duality

We conclude Chapter 9 by looking at a surprising connection between maximizing and minimizing problems. When a set of data is studied from one perspective you get a maximizing problem. When that same set of data is viewed from another perspective you get a minimizing problem.

KEY IDEAS

One of the joys of studying mathematics comes when you solve a problem and then discover that you have also solved a related problem that you did not realize was contained in the original problem. This is precisely what happens when you solve a maximizing linear programming problem (called the primal problem) using the simplex method. You discover that there is a related minimizing problem (called the dual problem) and the final tableau for the primal problem also contains the optimal solution to the dual problem. Furthermore, the optimal solution of the dual problem answers questions about the underlying data that you did not realize could even be asked.

Example 5 illustrates how a matrix game of any size can be solved using linear programming. In this case the solution of the primal maximization problem gives you the optimal strategy for the column player and (you guessed it) the solution of the dual minimization problem reveals the optimal strategy for the row player.

STUDY NOTES

CAUTION: When setting up a linear program to solve a matrix game, remember to use y's instead of x's for the variables in the maximization problem. We do this because the primal (maximization) problem produces the optimal strategy for the column player, not the row player.

SECTION 9.4 GLOSSARY

dual problem: If the primal problem P is to maximize $f(\mathbf{x}) = \mathbf{c}^T\mathbf{x}$ subject to $A\mathbf{x} \leq \mathbf{b}$ and $\mathbf{x} \geq \mathbf{0}$,
then the dual problem is to minimize $g(\mathbf{y}) = \mathbf{b}^T\mathbf{y}$ subject to $A^T\mathbf{y} \geq \mathbf{c}$ and $\mathbf{y} \geq \mathbf{0}$.

marginal value: The unit value of increasing or decreasing a variable in the objective function. These values are displayed in the optimal solution to the dual problem.

Copyright © 2021 Pearson Education, Inc.

SOLUTIONS TO EXERCISES

1. Minimize $36y_1 + 55y_2$

subject to $2y_1 + 5y_2 \geq 10$

$3y_1 + 4y_2 \geq 12$

and $y_1 \geq 0, y_2 \geq 0$.

7. The final tableau from Exercise 15 in Section 9.3 is

$$\begin{array}{cccccc|c} x_1 & x_2 & x_3 & x_4 & x_5 & M & \\ \hline 0 & 0 & 1 & -1 & 0 & 0 & 9 \\ 0 & 1 & 0 & 1 & -2 & 0 & 4 \\ 1 & 0 & 0 & -1 & 3 & 0 & 9 \\ \hline 0 & 0 & 0 & 1 & 2 & 1 & 56 \end{array}.$$

The solution of the dual problem is displayed by the entries in row 4 of columns 3, 4, 5, and 7. The minimum is $M = 56$, attained when $y_1 = 0$, $y_2 = 1$, and $y_3 = 2$.

9. It should be $A^T\mathbf{y} \geq \mathbf{c}$.

11. See Theorem 7.

13. See Theorem 7.

15 See Example 4.

19. The problem in Exercise 2 of Section 9.2 is to minimize $\mathbf{c}^T\mathbf{x}$ subject to $A\mathbf{x} \geq \mathbf{b}$ and $\mathbf{x} \geq \mathbf{0}$, where $\mathbf{x}$ lists the number of bags of Pixie Power and Misty Might, and

$$\mathbf{c} = \begin{bmatrix} 50 \\ 40 \end{bmatrix}, \; A = \begin{bmatrix} 3 & 2 \\ 2 & 4 \\ 1 & 3 \\ 2 & 1 \end{bmatrix}, \; \mathbf{b} = \begin{bmatrix} 28 \\ 30 \\ 20 \\ 25 \end{bmatrix}, \text{ and } \mathbf{x} = \begin{bmatrix} x_1 \\ x_2 \end{bmatrix}.$$

The dual of a minimization problem involving a matrix is a maximization problem involving the transpose of the matrix, with the vector data for the objective function and the constraint equation interchanged. Since the notation was established in Exercise 2 for a minimization problem, the notation here is "reversed" from the usual notation for a primal problem. Thus, the dual of the primal problem stated above is to maximize $\mathbf{b}^T\mathbf{y}$ subject to $A^T\mathbf{y} \leq \mathbf{c}$ and $\mathbf{y} \geq \mathbf{0}$. That is,

$$\text{Maximize } 28y_1 + 30y_2 + 20y_3 + 25y_4 \text{ subject to } \begin{bmatrix} 3 & 2 & 1 & 2 \\ 2 & 4 & 3 & 1 \end{bmatrix} \begin{bmatrix} y_1 \\ y_2 \\ y_3 \\ y_4 \end{bmatrix} \leq \begin{bmatrix} 50 \\ 40 \end{bmatrix}.$$

Copyright © 2021 Pearson Education, Inc.

Here are the simplex calculations for this dual problem:

$$
\begin{array}{ccccccc|c}
y_1 & y_2 & y_3 & y_4 & y_5 & y_6 & M & \\
3 & 2 & 1 & 2 & 1 & 0 & 0 & 50 \\
2 & 4 & 3 & 1 & 0 & 1 & 0 & 40 \\
\hline
-28 & -30 & -20 & -25 & 0 & 0 & 1 & 0
\end{array}
\sim
\begin{array}{ccccccc|c}
y_1 & y_2 & y_3 & y_4 & y_5 & y_6 & M & \\
2 & 0 & -\frac{1}{2} & \frac{3}{2} & 1 & -\frac{1}{2} & 0 & 30 \\
\frac{1}{2} & 1 & \frac{3}{4} & \frac{1}{4} & 0 & \frac{1}{4} & 0 & 10 \\
\hline
-13 & 0 & \frac{5}{2} & -\frac{35}{2} & 0 & \frac{15}{2} & 1 & 300
\end{array}
$$

$$
\sim
\begin{array}{ccccccc|c}
y_1 & y_2 & y_3 & y_4 & y_5 & y_6 & M & \\
\frac{4}{3} & 0 & -\frac{1}{3} & 1 & \frac{2}{3} & -\frac{1}{3} & 0 & 20 \\
\frac{1}{6} & 1 & \frac{5}{6} & 0 & -\frac{1}{6} & \frac{1}{3} & 0 & 5 \\
\hline
\frac{31}{3} & 0 & -\frac{10}{3} & 0 & \frac{35}{3} & \frac{5}{3} & 1 & 650
\end{array}
\sim
\begin{array}{ccccccc|c}
y_1 & y_2 & y_3 & y_4 & y_5 & y_6 & M & \\
\frac{7}{5} & \frac{2}{5} & 0 & 1 & \frac{3}{5} & -\frac{1}{5} & 0 & 22 \\
\frac{1}{5} & \frac{6}{5} & 1 & 0 & -\frac{1}{5} & \frac{2}{5} & 0 & 6 \\
\hline
11 & 4 & 0 & 0 & 11 & 3 & 1 & 670
\end{array}
$$

Since the solution of the original problem is the dual of the problem solved by the simplex method, the solution is given by the slack variables $y_5 = 11$ and $y_6 = 3$. The value of the objective function is the same for the primal and dual problems, so the minimum cost is \$670. This is achieved by blending 11 bags of PixiePower and 3 bags of MistyMight.

25. The game is $\begin{bmatrix} 1 & 2 & -2 \\ 0 & 1 & 4 \\ 3 & -1 & 1 \end{bmatrix}$. Add 3 to shift the game: $\begin{bmatrix} 4 & 5 & 1 \\ 3 & 4 & 7 \\ 6 & 2 & 4 \end{bmatrix}$.

The linear programming tableau for this game is

$$
\begin{array}{cccccc c|c}
x_1 & x_2 & x_3 & x_4 & x_5 & x_6 & M & \\
4 & 5 & 1 & 1 & 0 & 0 & 0 & 1 \\
3 & 4 & 7 & 0 & 1 & 0 & 0 & 1 \\
6 & 2 & 4 & 0 & 0 & 1 & 0 & 1 \\
\hline
-1 & -1 & -1 & 0 & 0 & 0 & 1 & 0
\end{array}
$$

Pivots:

$$
\begin{bmatrix}
0 & \frac{11}{3} & -\frac{5}{3} & 1 & 0 & -\frac{2}{3} & 0 & \frac{1}{3} \\
0 & 3 & 5 & 0 & 1 & -\frac{1}{2} & 0 & \frac{1}{2} \\
1 & \frac{1}{3} & \frac{2}{3} & 0 & 0 & \frac{1}{6} & 0 & \frac{1}{6} \\
0 & -\frac{2}{3} & -\frac{1}{3} & 0 & 0 & \frac{1}{6} & 1 & \frac{1}{6}
\end{bmatrix}
\sim
\begin{bmatrix}
0 & 1 & -\frac{5}{11} & \frac{3}{11} & 0 & -\frac{2}{11} & 0 & \frac{1}{11} \\
0 & 0 & \frac{70}{11} & -\frac{9}{11} & 1 & \frac{1}{22} & 0 & \frac{5}{22} \\
1 & 0 & \frac{9}{11} & -\frac{1}{11} & 0 & \frac{5}{22} & 0 & \frac{3}{22} \\
0 & 0 & -\frac{7}{11} & \frac{2}{11} & 0 & \frac{1}{22} & 1 & \frac{5}{22}
\end{bmatrix}
$$

Copyright © 2021 Pearson Education, Inc.

$$\sim \begin{bmatrix} 0 & 1 & 0 & \frac{3}{14} & \frac{1}{14} & -\frac{5}{28} & 0 & \frac{3}{28} \\ 0 & 0 & 1 & -\frac{9}{70} & \frac{11}{70} & \frac{1}{140} & 0 & \frac{1}{28} \\ 1 & 0 & 0 & \frac{1}{70} & -\frac{9}{70} & \frac{31}{140} & 0 & \frac{3}{28} \\ \hline 0 & 0 & 0 & \frac{1}{10} & \frac{1}{10} & \frac{1}{20} & 1 & \frac{1}{4} \end{bmatrix}$$

The optimal solution of the primal and dual problems, respectively, are

$$\overline{y}_1 = \frac{3}{28}, \ \overline{y}_2 = \frac{3}{28}, \ \overline{y}_3 = \frac{1}{28}, \text{ and } \ \overline{x}_1 = \frac{1}{10}, \overline{x}_2 = \frac{1}{10}, \overline{x}_3 = \frac{1}{20}, \text{ with } \lambda = \frac{1}{4}$$

The corresponding optimal mixed strategies for the column and row players, respectively, are:

$$\hat{\mathbf{y}} = \overline{\mathbf{y}} / \lambda = \overline{\mathbf{y}} \cdot 4 = \begin{bmatrix} \frac{3}{7} \\ \frac{3}{7} \\ \frac{1}{7} \end{bmatrix} \text{ and } \hat{\mathbf{x}} = \overline{\mathbf{x}} / \lambda = \overline{\mathbf{x}} \cdot 4 = \begin{bmatrix} \frac{2}{5} \\ \frac{2}{5} \\ \frac{1}{5} \end{bmatrix}$$

The value of the game with the shifted payoff matrix is $1/\lambda$, which is 4, so the value of original game is $4 - 3 = 1$.

Chapter 9 - Glossary Checklist

Check your knowledge by attempting to write definitions of the terms below. Then compare your work with the definitions given in the Glossary of each section of Chapter 9 in this Study Guide. Ask your instructor which definitions, if any, might appear on a test.

basic feasible solution   Each variable is … and … , where A has m rows. Geometrically, the basic feasible solutions correspond to … .

basic solution   A solution where matrix A has m rows and … .

canonical linear programming problem   Maximize … subject to the constraints … and … .

cycling   The simplex method … .

dominate   Given vectors $\mathbf{a}$ and $\mathbf{b}$ in $\mathbb{R}^n$, with entries a_i and b_i, respectively, vector $\mathbf{a}$ is said to dominate vector $\mathbf{b}$ if … .

dual problem   If the primal problem P is to maximize $f(\mathbf{x}) = \mathbf{c}^T \mathbf{x}$ subject to $A\mathbf{x} \le \mathbf{b}$ and $\mathbf{x} \ge \mathbf{0}$, then the dual problem is to … .

expected payoff (of a game to player R for strategies $\mathbf{x}$ and $\mathbf{y}$):   The sum of … . It is denoted by … and computed as … .

feasible set   The set of all … .

Copyright © 2021 Pearson Education, Inc.

feasible solution (to a linear programming problem)   A vector **x** that … .

infeasible (linear programming problem)   The feasible set is … .

marginal value   The unit value of … . These values are displayed in … .

objective function (in a linear programming problem)   The linear function that is … .

optimal solution   A feasible solution $\bar{\mathbf{x}}$ such that … .

optimal strategy (for column player C)   A strategy … .

optimal strategy (for row player R)   A strategy … .

payoff matrix   A matrix that lists … .

pivot (on an entry)   Transform the entry's coefficient into a 1 and … .

probability vector   A vector in $\mathbb{R}^n$ … .

pure strategy   A strategy in which … .

recessive   Vector **b** is recessive to vector **a** if … .

saddle point   An entry a_{ij} in a payoff matrix that is … .

slack variable   A … variable that is … .

solution (of a matrix game)   Any pair of … .

strategy   A point … .

strategy space (for player C)   The set of all … .

  (for player R)   The set of all … .

unbounded (canonical linear programming problem)   The objective function … .

value (of R using strategy **x**)   The number $v(\mathbf{x})$ defined by … .

value (of C using strategy **y**)   The number $v(\mathbf{y})$ defined by … .

value of the game: The common value … .

value of the game to column player C: The number v_C defined by … .

value of the game to row player R: The number v_R defined by … .

Copyright © 2021 Pearson Education, Inc.

www.ingramcontent.com/pod-product-compliance
Lightning Source LLC
Chambersburg PA
CBHW061404210326
41598CB00035B/6088

* 9 7 8 0 1 3 5 8 5 1 2 3 4 *